Elections and Exit Polling

Elections and Exit Polling

Fritz J. Scheuren
National Opinion Research Center (NORC)
University of Chicago
Chicago, IL

Wendy Alvey

WILEY

A JOHN WILEY & SONS, INC., PUBLICATION

Published by John Wiley & Sons, Inc., Hoboken, New Jersey.
Published simultaneously in Canada.

For general information on our other products and services or for technical support, please contact our Customer Care Department within the United States at (800) 762-2974, outside the United States at (317) 572-3993 or fax (317) 572-4002.

Wiley also publishes its books in a variety of electronic formats. Some content that appears in print may not be available in electronic format. For information about Wiley products, visit our web site at www.wiley.com.

Library of Congress Cataloging-in-Publication Data is available.

ISBN 978-0-470-29116-0

Printed in the United States of America.

10 9 8 7 6 5 4 3 2 1

My late husband Warren Mitofsky would have enjoyed the 2008 elections. He would have been filled with excitement, and he certainly would have had a lot to say about the new issues in surveying and exit polling during the primaries and in the lively fall campaign season yet to come. Warren was blessed because his work was his passion. He liked politics, surveys, statistics, learning, teaching, discussing, disputing, facing new challenges, and creating elegant solutions. Warren wasn't reverent, or modest or reticent, and he was at his best when he was sharing ideas and opinions with his friends and colleagues.

I don't want to try to name those of you who were so special to him for fear I should leave out someone. Many of you are in this book. Thank you for making his life so wonderful, and thank you for your kindnesses to me. I do want particularly to thank his last partner, Joe Lenski. With Joe's help, he did what he loved most, projecting election results.

Warren would have felt honored that his colleagues have put a volume together in his name. One of the things I like about the book, and hope you do too, is how much there is here in Warren's own words, quoted from a series of interviews he gave shortly before his death. I wish to personally acknowledge and thank the 50 or so contributors to this volume for their love and affection in this undertaking. I know that there was a real community effort made and I am indebted to all of you. I am particularly grateful to Fritz Scheuren and Wendy Alvey, who worked so hard to put the book together. It could not have happened without them.

Marianna (Mia) Mather
New York, March 2008

v

Foreword

Warren Mitofsky was my first boss twenty years ago when I joined the CBS News Election and Survey Unit in 1987 as a statistical associate straight out of college. Warren Mitofsky was also my partner in directing exit polling for the National Election Pool from 2003 to 2006. In those nineteen years I learned more from Warren Mitofsky than I have learned from anyone else with the possible exception of my own parents.

Warren was both a colleague and a friend, but always a teacher. He helped teach several generations of survey researchers and I was one of the many to be lucky enough to learn from the master. I, therefore, welcome this collection of papers on elections and exit polling honoring Warren and hope that this volume will help to inspire and teach current and future generations of survey researchers.

Warren was always willing and eager to answer questions about the important work that he had accomplished and the many methodological insights he had provided to the profession. His words should remind his any friends and colleagues of what we will miss and give an idea to those who were not lucky enough to meet him of what a unique and intelligent man he was.

I know Warren would not agree with every argument and theory presented in this book, but I also know that he would be the first to encourage continued research in order to deal with the many challenges facing both telephone survey research and exit polling.

Since his passing, Warren Mitofsky has been honored in many ways, including the naming of the AAPOR Innovators Award in his memory. I hope that this volume will serve to inspire the future innovators of the survey research industry.

Joe Lenski
March 16, 2008

Contents in Brief

CONTENTS

Preface

Elections and Exit Polling

Fritz Scheuren and Wendy Alvey

INTRODUCTION

On September 1, 2006, Warren J. Mitofsky died, at the age of 71. Mitofsky, who had had a long and influential career as a survey statistician, is best known for his work in telephone surveys and as the father of exit polling.

Who was Warren Mitofsky? A native New Yorker, Mitofsky spent much of his early career at the U.S. Census Bureau, where he lent his innovative energies to the design of sample surveys. It was while he was at the Census Bureau that he met Joe Waksberg. In 1967, Mitofsky left the federal government to return to New York, to do election projection and analysis for CBS News. While there, he and Waksberg developed the Random Digit Dialing techniques that have become standard practice today. Their approach permitted researchers to randomly select both listed and unlisted phone numbers more efficiently for telephone surveys. It was also during this period that Mitofsky and his friend, George Fine, first introduced exit polling, conducting interviews of voters as they left the polls in a gubernatorial contest in

Kentucky. By 1972, exit polling had become a staple of elections nationwide. While at CBS, Mitofsky worked with the Voter News Service consortium, directing exit polling during national elections in the 1990s and for the 2000 election. In 1993, he left CBS to establish Mitofsky International, and became a global consultant on exit polling. Mitofsky International subsequently joined forces with colleague Joe Lenski at Edison Media Research to analyze and report on the polling for the National Election Pool consortium in 2003. In recognition for his bold and innovative contributions to public opinion research, Mitofsky served as president of two major survey research associations—the American Association for Public Opinion Research (AAPOR) and the National Council on Public Polls—and was elected Fellow of the American Statistical Association (ASA).

This volume is dedicated to this colorful and remarkable man. It is a compilation of some of the latest information on studies of election surveys and exit polling. As such, this book is composed of material derived from a series of interviews with Mitofsky, held shortly before he died; connective text from the editors; and selected research papers written by current experts in public opinion polls on elections, voting, and exit polling covering the past decade.

Many of the papers included here are unpublished articles that were presented at meetings of the AAPOR and the ASA. Some have appeared in ASA-AAPOR proceedings volumes or are posted to Internet Web sites, but most have not appeared in scholarly journals or compendia such as this one. For the most part, they document research efforts since the 2000 U.S. presidential election, including some of the controversies and debates that have arisen due to discrepancies between polling estimates and projections and subsequent official vote tallies.

As it would be impossible to provide a comprehensive collection of the work that has appeared in this rich field, the editors have taken the liberty of selecting readings that describe some of the current issues and tried to point out other areas that the readers might find useful. The intent is mainly to provide readers with an introduction to some of the concerns and solutions that have been addressed—and continue to draw our attention. It is our hope that this book will prove useful to a broad range of readers, from public opinion pollsters to statisticians and survey researchers, political scientists, election reform advocates, the media, students, and the general public.

FORMAT

This book consists of seven chapters. The first chapter lays the groundwork for the volume and provides some general background information. Chapters 2 to 4 cover work that documents activities around the U.S. national elections of 2000 and 2004, and the mid-term elections in 2006. Chapter 5 highlights some articles that focus on election and exit polling elsewhere in the world, in an attempt to point out some of the challenges that differ from those faced by pollsters in the United States. Chapter 6 looks ahead to this year's election (2008), providing some past recommendations, work in progress, and suggestions for improving future election polls. The book concludes with Chapter 7, which provides a few examples of more technical articles that describe statistical methods—some are techniques that are used in papers that appear in this volume and others describe new methodologies that may enhance polling estimates.

The chapters, except for Chapter 1, all begin with some introductory text provided by the editors. This is set in gray shading, so that it is clear who wrote the information provided there. Next, most of the chapters present excerpts from the Mitofsky interviews. These are set in text boxes, to distinguish them from the editors' text and the contributed articles. Then, each chapter presents a number of selected readings, written by current researchers in the field. Some of these are full papers by the contributing authors and others are excerpts from longer papers. In some cases, part of the paper may appear in one place and another excerpt may appear elsewhere in the chapter or book. Notes from the editors provide cross-references when this occurs.

Finally, each chapter concludes with a section on Summary Observations and some Editors' Additional References. The observations attempt to tie the chapter readings to other chapters and articles in the book. Since this material is written by the editors, it appears in gray, as well. The additional references sometimes cite material that is very current—related to the 2008 primary elections—in an attempt to capture the latest issues and discussions that have arisen.

For the most part, the contributed papers appear essentially as written by the authors and provided to the editors. Formatting changes, however, were made to ensure that the book has a cohesive look and feel. Any errors discovered were brought to the attention of the authors and corrected per their suggestions. In addition, minor editorial changes were made, on occasion, primarily of a cosmetic nature. Tables, charts, equations, and footnotes are each numbered sequentially in each chapter. On occasion, some clarifying text is provided in square brackets [such as these] by the editors. Any additional footnotes that appear in square brackets were also added by the editors and references cited there will appear in the Editors' Additional References at the end of each chapter.

COPY PREPARATION

The selection of the material for this volume was the sole responsibility of the editors. Responsibility for the content of the individual readings, though, belongs to the authors. While there was no formal peer review process, to provide systematic approval of the statistical and mathematical content of the papers, an informal perusal was provided by several individuals who are well-known and regarded in the fields of survey research and public opinion polling. The editors are especially indebted to Mike Brick, Kathy Frankovic, Thomas B. Jabine, Joe Lenski, and Clyde Tucker for their time and attention to this matter in the midst of the busy primary election season.

MITOFSKY AWARD

All editors' royalties from *Elections and Exit Polling* will be donated to the Warren J. Mitofsky Award for Excellence in Survey Research. This award is being co-sponsored by the AAPOR and the ASA and will be managed by The Roper Center for Public Opinion Research, where Mitofsky served on the Board of Directors for 27 years, the last year as its Chair. The award distinguishes important work on public opinion or survey methodology that has been published in a book, journal, magazine, or newspaper, or presented at a professional conference. Special consideration is given to work that is based on data obtained by the researcher or author directly from The Roper Center's vast public opinion archive, as well as to work that utilizes multiple data sources or compares survey results over time. The 2007 winner of the Mitofsky Award was Professor John Mueller of Ohio State University, for his book *War, Presidents and Public Opinion*. For more information about the Mitofsky Award, see the following Internet site: *http://www.ropercenter.uconn.edu/center/mitofsky_overview.html*.

Acknowledgments

The editors of this volume, Fritz Scheuren and Wendy Alvey, would like to begin with a special acknowledgment of Warren Mitofsky for sharing his limited time and inimitable recollections of his innovative career in election polling. Next, we would like to thank all of the contributing authors for their dedicated efforts to explain and enhance this field of research and their creative and sound statistical attempts to address the problems that elections and exit polling present. In particular, we appreciate their cooperation and timely responses to our requests. We also wish to express our gratitude to Joe Lenski for his thoughtful comments in the Foreword, despite the fact that this effort came right in the midst of the 2008 election primaries. Further, we extend our sympathies and thanks to Mia Mather, for sending her personal Dedication to this book during such a difficult time for her. As noted in the Preface, Thomas B. Jabine, Joe Lenski, and Clyde Tucker also provided valuable feedback regarding this volume.

A very special appreciation goes to Eric Sampson, who provided the layout for this book, taking on what, no doubt, was a much larger job than he (and we) anticipated. Thanks also go to Megan Murphy, at the American Statistical Association (ASA) Office, for her assistance in accessing articles from ASA journals; Jim Dickey, Ron Wasserstein, and Lars Lyberg, for roles in permitting inclusion of copyrighted material; Marilyn Ford, at the National Opinion Research Center (NORC) at the University of Chicago, for her help in transcribing the Mitofsky interviews and her many other efforts on behalf of this book; and our publication editors, Steve Quigley and Jacqueline Palmieri, at John Wiley & Sons, Inc., for their guidance and patience with us throughout this effort. Lastly, we would like to express our appreciation to Elizabeth Lam Scheuren and Paul R. Alvey, Jr. for their continued support and understanding during this project.

Contributors

Francisco Abundis, University of Connecticut

Vittorio Addona, Macalester College

Wendy Alvey, Independent Consultant

Arlene S. Ash, Boston University

Joseph Bafumi, Dartmouth College

Mark Baldassare, Public Policy Institute of California

Christopher Barnes, Pulsar Research and Consulting

Matt Barreto, University of Washington

Mary Batcher, Ernst and Young, LLP

René Bautista, University of Nebraska, Lincoln

Joel David Bloom, University of Albany, State University of New York

Richard Brady, University of California, San Diego

Fares Braizat, University of Jordan

Mario Callegaro, Knowledge Networks

Jason Cawley, Coady Diemar Partners

LinChiat Chang, Opinion Research Corp.

Howard Christensen, Brigham Young University (retired)

Murray Edelman, Rutgers University

Robert S. Erikson, Columbia University

David P. Fan, University of Minnesota

Gösta Forsman, Swedish National Road Administration and Linköping University, Sweden

Edward Gracely, Drexel University

Chase Harrison, Harvard University

Keating Holland, CNN

Annica Isaksson, Linköping University, Sweden

Diana Jergovic, National Opinion Research Center, University of Chicago

Olena Kaminska, University of Nebraska, Lincoln

Cheryl Katz, Independent Consultant and Journalist

Scott Keeter, Pew Research Center

Susan Kyle, Department of Health and Human Services

Joseph Lenski, Edison Media Research

Elizabeth Liddle, University of Nottingham

Mark Lindeman, Bard College

Yan Liu, Internal Revenue Service

Eiji Matsuda, *Asahi Shimbun*, Japan

John McCarthy, Berkeley National Laboratory

Warren J. Mitofsky, Mitofsky Research International

Whitney Moore, National Opinion Research Center, University of Chicago

Marco A. Morales, New York University

Edward Mulrow, National Opinion Research Center, University of Chicago

Joe Murphy, Research Triangle Institute

Gregg R. Murray, Texas Tech University

Colm O'Muircheartaigh, University of Chicago

Jennie Elizabeth Pearson, University of Nebraska, Lincoln

Francisco Pedraza, University of Washington

Jeri Piehl, Opinion Research Corp.

Chris Riley, State University of New York, Brockport

Wendy Rotz, Ernst and Young, LLP

Douglas A. Samuelson, InfoLogix

Fritz Scheuren, National Opinion Research Center, University of Chicago

Anthony Scime, State University of New York, Brockport

Melaney Slater, HSBC Bank

Paul Sommers, Middlebury College

Howard Stanislevic, E-Voter Education Project

Nicolaos Synodinos, University of Hawaii, Manoa

Jose Alberto Vera, Instituto Nacional de Estadística, Geografía e Informática, Mexico

Nicole Vicinanza, Aguirre International

Christopher Wlezien, Temple University

Kirk Wolter, National Opinion Research Center, University of Chicago

CHAPTER 1

INTRODUCTION

Fritz Scheuren and Wendy Alvey

1.1 INTRODUCTION

We compiled this book on *Elections and Exit Polling* to honor a great and gifted individual, Warren Mitofsky. Sadly, Mitofsky is not here to see the results contributed by so many of his colleagues and friends.

Those who knew Mitofsky undoubtedly miss him deeply.[1] His blunt, principled fierceness and persistence made him a force of nature. He could—and did—inspire fear. You had to love him, however, especially when he was on a crusade against what he (usually rightly) considered bad polling practice. There, perhaps, may never again be anyone else as authoritative as Warren Mitofsky on exit polling or on still other practical arts in our business, such as the conduct of telephone surveys.

It is hoped that those who knew Mitofsky may find material here that contributes to improvements in exit polling methodology, thus keeping his efforts and memory alive. Those who missed the opportunity to know him personally may find in these pages at least a taste of his pioneering genius—maybe even a "starter set" to move the technology he created to a higher level, as the authors here have already done.

---

[1] For those who do not know Warren Mitofsky, a brief biographical sketch is provided in the Preface. For more on the man and his contributions, see Morin (2006), Brick and Tucker (2007), and Fienberg (2007).

In a very real sense, the collection of insights provided in this volume is a community effort. We were fortunate to interview Warren Mitofsky shortly before he died and selected recollections from his career are presented in "boxed in" text in each chapter. Our role, as editors, was basically to shape the work, mostly of others. We added connective material and context (presented in gray shaded text), in an effort to create a united whole. After this general introduction (Chapter 1), we look at exit polling during three of the last four U.S. election cycles: 2000, 2004, and 2006 (Chapters 2 to 4, respectively). To round out our effort, we added a chapter on election polling elsewhere in the world (Chapter 5). We follow that with some predictions and recommendations for the upcoming 2008 U.S. elections (Chapter 6) and conclude with some more methodological selections (Chapter 7).

While we recognize that we are unable to be comprehensive, this compilation attempts to highlight some of the key issues facing pollsters, their media clients, and the rest of us. We hope you will find here some of the newer approaches to the puzzles we all face. This book is a distillation of the wisdom of experts, both in the U.S. and elsewhere; also there is a prediction of things to come.

In this chapter we start by defining a few terms (Section 1.2), say something about the history of election polling (Section 1.3), and present a description of an exit polling experience (Section 1.4). We then sketch the sampling techniques that support such surveys (Section 1.5).

Election polling, in general, and exit polling, in particular, have many limitations. We describe some of them in Section 1.6. While their impact may be obvious to experienced practitioners, they may be inadequately dealt with if the methods are not applied by someone like a Warren Mitofsky. After all, in polling we are talking about something that is more of an art than a science. Applications of election polls (Section 1.7) are well known, but we still summarize them briefly, if only as lessons learned for those who might otherwise misuse them. (See, also, Chapter 4.)

By design—and with considerable effort—this volume appears before the 2008 U.S. presidential election. The co-evolution of polling and the election process, itself, may well be beyond any specific predictions that might have been made here. Even so, we could not resist repeating general recommendations and predictions (Section 1.8) that we thought worthy of consideration. (See, especially, Chapter 6.)

New surprises will continue to make elections in the U.S.—and other countries—fascinating methodologically. We hope for everyone's sake, however, that whatever surprises occur, they are ones that can be taken in stride and survived, leading us back, ultimately, to the trust—deserved, we hope, this time—which we had in the election system before the 2000 Florida debacle opened our eyes.

Some readers may want to skip what comes immediately below and go right to the research papers that follow (starting on page 16)—most of which are being published for the first time.

1.2   DEFINITIONS

Polling is a seemingly indispensable tool in the running of modern democracies. While public opinion polls can cover a broad range of topics, *election polling* focuses in on data collected by independent groups—such as the media—to obtain information on voting intent from respondents. Election polls can be administered in person or by mail. Initially the polls done by George Gallup were conducted in person.[2] One notorious mail survey that was conducted was *The Literary Digest* poll, conducted during the 1936 election, which predicted Alf Landon would win over Franklin Roosevelt.[3] Today, pre-election polling is done mostly by telephone, although some Internet polling has started. This book touches on all of these approaches to some extent, but focuses mainly on exit polling—a technique made popular by its best known practitioner, Warren Mitofsky.

Exit polling is a form of intercept polling that has come into its own beginning with the 1967 primaries. Some practitioners actually conduct short interviews, but now usually every *n*th individual leaving a polling location is handed a clip-board and asked to fill out a one-page questionnaire. In addition to asking some demographic questions—and even, occasionally, some queries about the election process—the main focus of the questionnaire is to learn for whom the respondent has just cast a vote.

These questionnaires are then sent immediately to a central location where they are added to a statistical database for ongoing analysis. Technologies vary, but the transmission of the questionnaire results can be by conventional telephone or fax; by cell phone, usually as a text message; or via the Internet. Because settings vary outside polling locations, sometimes more than one method can be used in a given election.

For the data transmission to be successful it has to be accurate and fast. Projections based on the results are traditionally developed throughout the day, with full summaries available about the time the polls close. Typically, analysts evaluate exit poll data using previously developed statistical models. They also consider available information obtained on Election Day, such as actual outcomes and reports from the field team about irregularities, as well as information obtained before Election Day, such as absentee ballots, early voter surveys, and data from pre-election polls or previous electoral results, to eventually project the winner on election night. When information is not enough to make a projection or the race is very tight, forecasters may not "call" a winner on election night.

Figure 1.1 contains other key terms that are commonly used in exit polling. For definitions of response and nonresponse concepts specific to exit polling, see Slater and Christensen (2002) in this volume (2.3.1).

Some other sources of information about election and exit polling include Edison Media Research (2008), Traugott and Lavrakas (2004), and Zukin (2004). In addi-

---

[2] See Lee (1999) for information on George Gallup.
[3] See Grier (2002).

Important Exit Polling Terms

Exit Poll   A sample survey of voters exiting a collection of polling places.

Precinct   The smallest political division in the state consisting of registered voters residing in a geographic region defined by the county clerk.

Polling Place   A location specified by the county clerk where voters from one or more precincts cast votes.

Post-Election Survey   Sample survey of registered voters after Election Day (typically done by telephone) to confirm information about the election and voting decisions.

Pre-Election Survey   Sample survey of likely registered voters prior to Election Day (typically done by telephone) to measure voting intentions and opinions on election issues.

Stratification   The division of a population into non-overlapping subgroups called strata. It is done either for administrative purposes and/or to improve efficiency of estimation.

Ultimate Sampling Unit (USU)   The basis unit in a sample survey from which information is to be gathered or on which measurements are to be made. Examples: A person, a voter, a household head, a classroom in a school.

Primary Sampling Unit (PSU)   A set of ultimate units, often belonging to a geographic area, that represents a unit to be sampled as the first step in selecting a sample. Examples: A county consisting of voters, a school district consisting of students, a housing unit consisting of households.

Secondary Sampling Unit (SSU)   A collection of ultimate units, created by dividing a primary sampling unit, which is to be sampled once a PSU is chosen. Examples: A polling place chosen within a sampled county, a classroom of students chosen within a sampled school.

Two-Stage Sampling   A sample design where the secondary sampling unit is the ultimate sampling unit, so that there are only two stages of sampling: the selection of PSUs and the selection of SSUs sampled.

Systematic Sampling   A sample design where sample units are selected sequentially, following the order of the sample frame. Examples: including every 14th voter exiting a polling place or every 5th name in a telephone directory in the sample. Systematic random sampling is choosing the first sample element at random.

Certainty Units   A PSU that is included in the set of sampled units with a probability of inclusion of one. Certainty units are usually determined by administrative decision. Very large or important PSUs may be chosen as certainty units. Examples: The counties containing the largest population centers and those in which a participating college or university are located are certainty units in an exit poll, to guarantee their inclusion in the sample.

PPS Sampling   Assigning the probability of inclusion for PSUs equal to the size of the PSU. PSUs with a large number of USUs are more likely to be included in the sample than PSUs with a small number of USUs. Examples: The probability of selecting a county in an exit poll is proportionate to the number of voters, or the probability of selecting a school for a school district survey is proportionate to the number of students in the school.

Sample Weights   The weight attached to each ultimate unit in the sample that represents the inverse of the probability of selection. The sample weights provide a simple expression for producing unbiased estimates in complex designs.

Complex Sample Design   A sample design involving more than one of the sampling methods of stratification, cluster sampling, systematic sampling, and so forth.

Figure 1.1   Important exit polling terms.
Source: Twenty Years of the Utah Colleges Exit Poll: Learning by Doing, by Scott Grimshaw, Howard Christensen, David Magleby, and Kelly Patterson. Reprinted with permission from *Chance*. Copyright ©2004: American Statistical Association. All rights reserved.

tion, among the blogs and listserves that encourage discussion regarding election and exit polling, see especially *http://www.pollster.com* or join *AAPORnet*. AAPOR also provides a free, online course for journalists and their readers—*Understanding and Interpreting Polls*.[4] For more general information on surveys, see also Frankel and Frankel (1987) and the American Statistical Association's (ASA) pamphlet series *What Is a Survey?*[5]

1.3 BRIEF HISTORY

As early as the 1940s, a modest exit poll took place in Denver, Colorado, where voters were interviewed outside polling stations (Frankovic, 1992); however, such a methodological exercise was overlooked and the benefits were not evident in those days, mainly because neither the polling nor the media industry were developed as we know them today.[6] The early polls were not very reliable. In fact, some pre-election polls were clearly unsuccessful, as was the case of the 1936 *Literary Digest* poll, which erroneously "called" electoral outcomes despite its use of a large mail "sample" of almost 2.3 million people (Squire, 1988). Problems occurred in the 1948 election, as well, even though much better methods were used.

It was not until the 1960s that a major television network devoted significant resources to the collection of information from actual voters in order to "call" electoral results (Lindeman, 2006). Before such exit polling attempts, journalistic predictions were based mainly on known patterns in "key" or "tag" precincts from previous elections (Mitofsky, 1991; Sudman, 1986). In the context of the 1967 Kentucky gubernatorial election, an exit poll was conducted and sponsored by the CBS television network; the survey design was proposed by Warren Mitofsky (Mitofsky, 1991, 2004). An interesting Mitofsky anecdote, published by Richard Morin of *The Washington Post*, describes the inspiration of the exit polling technique:

> "In 1967, CBS was preparing its coverage of the Kentucky governor's race. Mitofsky had hired a market researcher, George Fine, to help him collect voting data on Election Day. During a conversation, Fine happened to mention some work he was doing for the movie industry. 'The movie people wanted to test a film before they released it for distribution,' Mitofsky recalled, 'so they would show it in test theaters, show it to a test audience. And George decided to interview the people as they left the theaters.... I can't swear whether he suggested it or we put two and two together. And we said, 'Why don't we interview [Kentucky residents] leaving the polling places?'" (Morin, 2006).

Towards the end of the 1960s—certainly by the beginning of the 1970s—exit polling had become a widespread way of collecting data from actual voters. In 1968 CBS expanded exit polls to twenty states for the presidential election. In 1973, the

---

[4] See *http://www.newsu.org/Angel/section/default.asp?format=course&id=aapor_polling07*.

[5] See *http://www.whatisasurvey.info/* on the ASA's Web site, *www.amstat.org*.

[6] Fienberg (2007) dates early election night forecasting efforts back to the 1951 use of the UNIVAC I computer. In addition to Mitofsky's contributions, Fienberg also describes the role of John Tukey in 1960, when he used "swing-o-metric" precincts to predict the election outcome.

NBC television network conducted its first exit poll and started incorporating exit poll data into its projections—before that, projections were exclusively based on a sample of actual tallies, also known as "quick counts." As exit polling estimates became more expedited and precise, they came to be seen as primarily being collected to call electoral results;[7] as a consequence, their use became highly controversial in U.S elections.

For instance, in 1980 NBC made its call for the presidential winner (about 90 minutes before the ABC television network, and two hours before CBS) almost three hours before the West Coast polls closed. Such anticipation in calling results raised questions about whether announcing early predictions of the presidential winner depressed voter turnout in the West, thereby affecting the outcomes of several close congressional races. This hypothesis was, according to Sudman (1986), "not a statistical, but a political, issue."

In the early 1990s the Voter News Service (VNS) was created by the major TV networks in the United States: ABC, CBS, CNN, Fox, and NBC; the Associated Press (AP); and 19 newspapers. They pooled funds to get a larger and theoretically more accurate sample from voters on Election Day. However, there was no central decision-maker; each partner wanted to decide on how things should be done, resulting in a complicated arrangement among participants. On Election Day 2000, the networks and the VNS miscalled the winner of the Florida presidential contest. "What happened on election night 2000 was that the television networks and wire services first mistakenly called Al Gore the winner of Florida's 25 electoral votes; six hours later, the networks declared them the property of George W. Bush—only to retract that call [later] as well." (Frankovic, 2003). David Moore asserts such miscalls were partly triggered by fierce competition among the networks (Moore, 2006). Then, on Election Day 2002, VNS was unable to deliver the exit poll data; the VNS was subsequently disbanded.

For Election Day 2004, the networks' relied on a newly created entity, the National Election Pool (NEP), which hired Joe Lenski of Edison Media Research and Warren Mitofsky of Mitofsky International to provide the exit polling results to the consortium and its subscribers. After the experience on Election Day 2000, and despite the careful and conservative criteria adopted by NEP for calling results, in 2004 early and incomplete exit polling data were improperly leaked to part of the journalistic-political community and quickly spread over the Internet. Before polling stations closed, the 2004 election had been characterized positive to one candidate and negative to the other, but towards the end of the day electoral outcomes were at odds with those early expectations. Confusion around early results leaked to some media were mainly due to a misinterpretation of exit polling data. Furthermore, the difference between exit poll results and official tallies in 2000 and 2004 were seen by some analysts and journalists as evidence of electoral fraud (e.g., Freeman and Bleifuss, 2006).

---

[7]Lost from this understanding was the, perhaps, even more desirable benefit that exit polls could provide the demographic make-up of each candidates' voters and the issues they rated as most important in making their selection. This use of the exit polls in the 2008 primaries, arguably, has been central.

1.4   A DAY IN THE LIFE OF AN EXIT POLL

In the excerpt below, Mitofsky and Edelman (2002) describe a typical day in the life of a network pollster on Election Day. While they do not detail the trials of the interviewer trying to capture respondents on a blustery November day in Minnesota, they do provide a sense of what it is like to peek behind the scene on Election Day.

1.4.1   Excerpt from: Election Night Estimation[8]
Warren J. Mitofsky and Murray Edelman

What It's Like on the Front Line.   Our talk is about election night. For the most part we will tell you about what we did at CBS News.... We will give you some understanding of what it takes to simultaneously conduct 51 surveys of over 100,000 respondents and present the results to millions of viewers—and almost never get the outcome wrong...

We thought the place to start was to get you in the mood. We want you to know what it feels like on a typical Election Day to be one of us. It starts the night before, when we hope all the last-minute details have been dealt with. Everyone has his or her assignment. Everyone has been trained and rehearsed: the vote collectors at the precincts, the exit poll interviewers, the analysts, those entering and reviewing vote returns and data, those who use the results in the television and radio studios. There are literally tens of thousands of people in a myriad of jobs. All our reference materials are where they should be. All the phone numbers we need are handy. Every computer system works. We are ready for Election Day, the culmination of the last two years' work.

A good night's sleep and we can be ready for a day that will last from the time we get up on Election Day until after the next night's network news broadcasts—some 36 hours later. Unfortunately, the night's sleep is not very restful. Usually the phone rings too early in the morning about yet another problem and a few more missed details cross our thinking. The area where we worked was in the studio during our CBS days. Now Murray is in a neutral site and is working for all the networks, the Associated Press, and most of the large newspapers in the country, as well as an assortment of television stations in cities around the country. Warren is in a similar place, but only has to worry about CNN and CBS.

Shortly after noon, we are anxious to get our first look at the first wave of exit poll results. This is the time when we confirm that the computer system really works; that the interviewers are doing their jobs; and that no official has kept our interviewer away from the exit to the polling place. We also want to see if the contests we were told were landslides really are landslides and the close races really are close. We do it when the first round of exit poll results comes in. We try not to forget that sometimes the results change over the day.

---

[8]Excerpt from Election Night Estimation, by Warren J. Mitofsky and Murray Edelman, reprinted with permission from the *Journal of Official Statistics*. Copyright ©2002: Statistics Sweden. All rights reserved.

During the hour before the polls close in the first states, we have to be ready for the third and last wave of exit poll results. Kentucky and Indiana close first. The best way to handle the buildup yet to come is to triage. Someone else on the decision desk can confirm the winner of an expected landslide as we get closer to poll closing. And the very close races can wait, too. They cannot be called from an exit poll. Exit polls are not that precise. It is those races that are in between, the ones with about eight point margins, that we must concentrate on in the half hour before the polls close.

We review detailed results for each race on our list. If we can make a call, we do; and if not, we wait for more sample precinct vote returns. The biggest problem in this time period is not to lose track of races that have still to be decided. Even though the computer screen lists all races in priority order, a race can slip between the cracks for a period of time. We try not to let this happen.

What we are trying to do is give the results and not make a mistake doing it. Forget those cynics who tell you that this is a reckless race to be first on the air with a winner. That is not the goal. We want to get projections on the air as soon as we can, once we feel certain we have correctly identified the winner. Fear of being wrong is the overriding emotion, not racing through projections.

Once those first two states [Kentucky and Indiana] close, we hurry to look at the nine states that close at 7:00 p.m. and 7:30 p.m. That number is manageable. Any states not called at poll closing are assigned to other members of the team to watch, unless they are races of national interest. Then we watch them ourselves. These are races where an incumbent senator might be defeated, like Ashcroft in Missouri or Abraham in Michigan in the [2000] election, or states that are key to an electoral victory, like Florida.

But for the most part, we want to get started on the 8 o'clock states. That is the big rush. Eighteen states close their polls in the 8 o'clock hour and another twelve the next hour. Those two hours will be the big crush. It will test our capacity to collect and process all that [sic] data. It also will test the organizing we have done so we, or the analysts working with us, can review everything carefully enough to be confident about what we project. If things go as usual, there will be lots of senate, governor, and presidential states where the winner is known with enough precision from the exit poll. If we cannot make a projection at poll closing, we will have to keep monitoring the race until there are enough real vote returns in our sample precincts to enable a projection.

We have help doing this. First, there are fine statisticians working with us on election night. Next, there is a computer screen that lists all the races in states where the polls have closed that have not been called. As more sample precinct returns come in, the color for each state indicates the current status of the results. Yellow means all criteria for a call have been met, and light blue means the results are getting close to call status. With any luck we will not see a state colored red. Red states go to the top of the list automatically. That means a race that has been called may no longer meet all the criteria for a projection. For the states in red, we look to see if we have to make a retraction. Usually this early warning sign turns out to be of no consequence. One of many criteria for making a call may have been marginal. One precinct more or less may turn this indicator red temporarily.

Occasionally, we have a mistake. The sooner we recognize it and announce on air that we are retracting a projection, the less trouble the errant projection will cause the news reporting. It is bad enough to have made a mistake, but it must be corrected as publicly as it was made. If a state is crucial to control of the Senate or to an electoral vote victory, the projection can seriously mislead the election reporting. If the state is not key to some trend, the wrong projection still has misled viewers. On the other hand, we do not want to retract a call too quickly. We do not want to issue a retraction just because the leading candidate is not ahead by a big enough margin to satisfy our call criteria. We want to be reasonably sure we were wrong.

As the night progresses, it becomes clear that some races are just too close to call from sample precinct returns. Our next source is the county vote tallies. In the New England states, they report from towns or cities. These county and town returns trickle in after the polls close and eventually reach 100 percent of the precincts reporting in the early hours of the morning. We model these returns also. They can be used in combination with the precinct samples or they can be used separately. For the closest races, this is what we rely upon for projections. It can be in the middle of the night or sometimes it takes until the next day. And sometimes the best call we can make when all the votes are counted is 'Too Close to Call.'"

1.5 THEORY AND METHODOLOGY

Grimshaw et al. (2004:32) explain that

> "An exit poll is a sample of voters exiting the polls on Election Day. The purpose is to find out whom the electorate voted for and why they voted as they did. This allows pollsters to predict the winners for various political races of interest without having to wait for a complete count. For the television news media, concerned about ratings and market share, that is sufficient justification for their extensive involvement in exit polls. But early projections are of less interest than the academic study of the election process and investigating the reasons people vote as they do. Traugott and Lavrakas (2000) point out that an exit poll's respondents are interviewed immediately after voting and before any results are announced, so they are not prone to the bandwagon effect that tends to inflate winners' percentages in post-election surveys. Also, because the sample takes place as voters exit the voting place, the problem of misreporting actual voter response is reduced, compared to the responses of other post-election surveys."

In Fienberg (2007:9), Philip Meyer of *USA Today* points out that "It is the combination of fast relays of official returns and exit poll data that make the broadcasters' election night projections so spookily fast and accurate." (Figure 1.2 summarizes the steps for modern exit polling.)

The methodological approach for election projections made using the UNIVAC in the 1950s was based on regression analysis using historical data and early returns. By 1960, Tukey was using key precincts in each state that represented how the state would vote. "What mattered was how closely the swings from year to year in that precinct reflected the swing in the state total." (Fienberg, 2007:11) (Alternate precincts were

Interviewer-respondent phase	After voters have cast their ballots in a given polling booth, interviewers select respondents based on a systematic interval (every nth voter), which is previously calculated on probabilistic principles, to hand them a questionnaire that may include questions on vote choice, public opinion and demographic characteristics. Depending on the design, the questionnaire can be self- or interviewer-administered, or a combination of both modes.
Interviewer-data entry personal phase	Interviewers transmit collected exit polling data to a Data Center facility on the telephone at pre-scheduled times. Usually three or four hours are set in advanced. Toll free numbers are the typical way fieldwork personnel call to the Data Center facility. Data-entry personnel receive and capture exit polling data by means of ad hoc software computer. After transmitting exit polling data, interviewers resume exit polling interviews until the polling booth closes. The interviewer may report some observed irregularities in these calls, may call to a different telephone number, or make such reports after the fact.
Data analysis phase	As information comes, data are formatted to be analyzed using standard or ad hoc statistical software packages. Procedures for statistical analyses vary across exit polling agencies. Once analysts have run statistical tests and other less formal analysis, such as comparison with previous election results and pre- election polls, exit poll estimates are prepared to be presented at the end of the Election Day. Sometimes the election is *too close to call* or there is not enough information, so analysts await information from actual tallies, also known as *quick counts*.

Figure 1.2   Typical exit poll process on Election Day.
Source: Bautista et al. (forthcoming).

selected, though, if the swing precincts didn't use machine ballots.) These data were used to predict the national electoral vote totals, adding in the impact of historical data, pre-election polls, political scientists' predictions, and early voting returns. (Fienberg, 2007:12) The forecasting models that resulted were later recognized as "hierarchical Bayesian methods with the use of empirical Bayesian techniques at the top level."[9] In the meantime, from 1964 to 2000, Jack Moshman and his team of statisticians and political scientists had good results at ABC, using regression models to predict the winning candidates.

Beginning around 1967, Mitofsky, Murray Edelman, and Joe Waksberg introduced survey research methods, by initiating exit polling. Samples were based on a:

> "stratified probability sample with precincts selected in a single stage with probability proportional to the vote total in a recent election. They also used a weighting technique to control the variation in a party's vote, a form of regression estimation to allow for reporting from early return precincts, and a form of ratio estimation to utilize information from a past election in the denominator." (Fienberg, 2007:14)

In exit polling, large samples are generally possible at the precinct level. As done by Edison-Mitofsky, the precincts are selected at random with probability proportional to size (PPS).[10] The number of sample precincts, however, is usually on the small side—perhaps 20 or so in the smaller states or states predicted to have one-sided outcomes. In larger states and states with close contests, more precincts may be employed. For example, for the 2004 presidential election, Ohio had 50 sample precincts.

Exit polls, like surveys in general, are affected by both sampling and nonsampling errors. Sampling errors can be estimated using conventional cluster sampling formulas. A common mistake made by many users of exit polls is to focus on the large number of completed interviews and ignore the clustering which typically can lower the effective sample size by a third to a half.[11]

Nonresponse is a big problem in exit polls, with less than half the sampled voters responding. The results of the survey are thus very sensitive to how the nonresponse is handled. A number of the contributing authors in the following chapters address nonresponse issues.

1.6   RECENT ELECTORAL APPLICATIONS

Public opinion polls have a long history. As noted in section 1.3, election polling was conducted as early as the 1940s, but didn't come into its own until the mid- to late-1960s. Such polling was first used to predict election results for the media in presidential and statewide elections. Congressional projections were added later, when election polling was able to focus in on precinct-specific data. Today election and exit polling are used to project winners in all types of election races. In addition to

---

[9]Fienberg (2007:13)

[10]See Zukin (2004) for a general discussion on sample selection in election polls.

[11]For data by ethnicity, unlike by gender, this can be particularly important.

winning candidates, the outcomes of ballot initiatives can also be forecast using such techniques. Demographic information and multi-mode designs permit collection of supplemental data that provide insights into who is voting for whom and what some of the potential motivations are for their votes. This helps candidates target voters more effectively and serves as input for development and support for public policy.

The collection of demographic data and information on the polling process also provides the ability for data users to apply the results to assess the veracity and accuracy of the election. In some other countries, the impetus for exit polling was actually to reassure the public of a fair election—see, for example, Bautista et al. (2005) and Bautista et al. (forthcoming).[12] More recently, in the 2008 primary in New Hampshire, the fact that polling results failed to predict Hillary Clinton as the winner led to extensive review of the data and a host of explanations for the discrepancies which occurred, ranging from machine fraud to failure to account for likely voters.[13]

1.7   SOME MAJOR LIMITATIONS

Like all survey research, pre-election polls and exit polling are subject to a number of major limitations. In section 1.5 we mentioned two of the principal limitations of exit polling—sampling and nonsampling error. Two important ways of reducing polling error are to ensure that the sample selected is representative of the voting population and to ensure that the random sampling scheme that the design depends upon is, indeed, followed. Since not all persons vote—especially in primary elections and mid-term elections—developing stratified samples for pre-eleciton polls that reflect those who will vote is quite challenging. In the case of exit polls, once the selection scheme is developed, it is necessary to be sure that those intercepted to report their vote are, in fact, the ones who should be in the sample. One of the biggest concerns that can occur arises when the selection discipline of taking every *n*th person breaks down and interviewers approach individuals whom they think will respond, rather than keeping to the prescribed selection order. A breakdown here can introduce selection bias that has not been accounted for in the sample design. Another problem is that of co-location of precincts. *Co-location* occurs when two or more precincts are in the same physical facility—a school, for example. When someone is approached upon leaving such a facility, it is usually not clear which precinct the voter voted in, which makes it difficult to maintain the desired selection probabilities.

Limitations due to nonsampling error can include missed voters who are not asked to respond for various reasons—either due to the respondent or the interviewer; respondents who refuse to cooperate or are unable to respond for a host of reasons; failure to interpret the questions accurately; transmission errors, and data entry problems. Site-specific problems that interfere with the voters' intent to cast an accurate ballot are also concerns that can impact exit polling results. For example, the design of the butterfly ballot in Palm Beach County, Florida in the 2000 presidential election

---

[12]For Bautista et al. (2005), see 5.2.1 in this volume.
[13]See Dopp (2008); Thompson. (2008); and Erikson and Wlezien (2008).

led many voters to believe they had cast a ballot for Al Gore, when, in fact, they had voted for Patrick Buchanan.[14]

Many of the contributing authors in the chapters that follow discuss limitations they have dealt with and describe their approaches to improving the quality of the election and exit polling data. Chapter 6, in particular, addresses problem areas that still persist today or that have arisen as polling efforts have become more widespread. In so doing, that chapter looks at the quality of the voting data, as well as that of the polling information.[15]

1.8 RECOMMENDATIONS AND A FEW PREDICTIONS

Growth and changes in polling technology and methods have led to improvements in current techniques. Recent challenges posed by the elections of 2000 and 2004, in particular, have also raised questions about current methods and the need for further refinements. Chapter 6 looks at some of the recommendations that followed those events and likely issues that remain to be addressed. This book went to the printer near the end of the primary season for the 2008 presidential election in the United States. Even though the 2008 election process is still in its early stages, we have already seen areas that have raised concerns. Fortunately, there is much data available to examine these problems and there is an active and creative audience that is eager to help. Warren Mitofsky would surely be pleased.

1.9 REFERENCES

In addition to works cited by the contributing authors, each chapter provides some additional references provided by the editors. For more complete coverage of the topic, also see:

- Walden, Graham R. (1996). *Polling and Survey Research Methods, 1935-1979: An Annotated Bibliography*, Westport, CT: Greenwood Press.

- Walden, Graham R. (1990). *Public Opinion Polls and Survey Research: A Selected Annotated Bibliography of U.S. Guides and Studies from the 1980s,* New York, NY: Garland.

- Walden, Graham R. (2002). *Survey Research Methodology, 1990-1999: An Annotated Bibliography*, Westport, CT: Greenwood Press.

Editors' Additional References

Adams, Greg (2001). Voting Irregularities in Palm Beach, Florida, *Chance*, 14 (1), 22-24.

---

[14]For example, see Pomper (2001), Adams (2001), and Dillman (2000).
[15]See Rotz and Gracely (2008) in this volume (6.5.1).

Bautista, R.; Callegaro, M.; Vera, J.A.; and Abundis, F. (2005). Nonresponse in Exit Poll Methodology: A Case Study in Mexico, paper presented at the meeting of the American Association for Public Opinion Research, *Proceedings of the Section on Survey Research Methods*, American Statistical Association, 3802-3809.

Bautista, R.; Callegaro, M.; Kharchenko, N.; Paniotto, V., and Scheuren, F. (forthcoming). Exit Polling Methodologies Across Nations: Some Constraints and Measurement Problems, paper to be presented at the 3MC Conference in Berlin, Germany, May 2008.

Brick, Michael J. and Tucker, Clyde (2007). Mitofsky-Waksberg: Learning from the Past, *Public Opinion Quarterly*, 71 (5), 703-716.

Dillman, Donald (2000). Statement on Palm Beach County Florida November 2000 Ballot, testimony, November 9, 2000, available on the Internet at: *http://www.sesrc.wsu.edu/dillman/palmbeach_statement. htm*.

Dopp, K. (2008). New Hampshire Democratic Primary Election Results Are Suspicious, available on the Internet at: *http://electionarchive.org/ucvData/NH/ReleaseReNHPrimary2008.pdf*, January 14, 2008.

Edison Media Research (2008). Exit Polls, responses to frequently asked questions, available on the Internet at: *http://www.exit-poll.net/exit_polling.html*.

Erikson, R.S. and Wlezien, C. (2008). Likely Voter Screens and the Clinton Surprise in New Hampshire, available on the Internet at *http://www.pollster.com/blogs/likely_voter_screens_and_the_c.php*, January 13, 2008.

Fienberg, S. (2007). Memories of Election Night Predictions Past: Psephologists and Statisticians at Work, *Chance*, 20 (4), 8-17.

Frankel, Martin R. and Frankel, Lester R., (1987). Fifty Years of Survey Sampling in the United States, *Public Opinion Quarterly*, 51, part 2: Supplement, S127-S138.

Frankovic, K. A. (1992). Technology and the Changing Landscape of Media Polls, *Media Polls in American Politics,* Mann, T.E. and Orren, G.R. (eds.), Washington, DC: The Brookings Institution, 32-54.

Frankovic, K. A. (2003). News Organizations' Responses to the Mistakes of Election 2000: Why They Will Continue to Project Elections, *Public Opinion Quarterly,* 67(1), 19-31.

Freeman, S. and Bleifuss, J. (2006). *Was the 2004 Presidential Election Stolen? Exit Polls, Election Fraud, and the Official Count,* New York, NY: Seven Stories Press.

Grier, David Alan (2002). Predicting the Unpredictable Election, *Chance*, 15 (2), 30-35.

Grimshaw, S. D.; Christensen, H. B.; Magleby, D. B.; and Patterson, K. D. (2004). Twenty Years of the Utah Colleges Exit Poll: Learning by Doing, *Chance*, 17 (2), 32-38.

Lee, Byung-Kwan (1999). George Gallup: Gallup Poll, Austin, TX: University of Texas, available on the Internet at: *http://www.ciadvertising.org/studies/student/99_spring/interactive/bklee/theory2/poll.html*.

Lindeman, M. (2006). Beyond Exit Poll Fundamentalism: Surveying the 2004 Election Debate, paper presented at the 61st annual conference of the American Association for Public Opinion Research, Montreal.

Mitofsky, W.J. (1991). A Short History of Exit Polls, *Polling and Presidential Election Coverage*, Lavrakas, P.J. and Holley, J.K. (eds.), Newbury Park, CA: Sage Publications, 83-99.

Mitofsky, W. J. (2004). Exit Polls and Election Projections, *Public Opinion and Polling Around the World: A Historical Encyclopedia*, Geer, J.G. (ed.), 1, Santa Barbara, CA: ABC-Clio, 396-401.

Mitofsky, Warren J. and Edelman, Murray (2002). Election Night Estimation, *Journal of Official Statistics*, 18 (2), 165-179.

Moore, D. W. (2006). *How to Steal an Election*. New York, NY: Nation Books.

Morin, R. (2006). The Pioneer Pollster Whose Credibility You Could Count On. *The Washington Post*, September 6, 2006, C01, available on the Internet at: *http://www.mitofskyinternational.com/WP-Morin. html*.

Pomper, G.M. (2001). The Presidential Election, *The Election of 2000*, Pomper, G. M. et al. (eds.), New York, NY: Chatham House Publishers, Seven Bridges Press, 126-154.

Rotz, W. and Gracely, E. (2008). General Principles for Statistical Election Auditing, unpublished paper.

Slater, M. and Christensen, H. (2002). Applying AAPOR's Final Disposition Codes and Outcome Rates to the 2000 Utah College's Exit Poll, *American Statistical Association Proceedings*, AAPOR ASA Section on Survey Research Methods, 540-545.

Squire, P. (1988). Why the 1936 *Literacy Digest* Poll Failed. *Public Opinion Quarterly*, 52, 125-133.

Sudman, S. (1986). Do Exit Polls Influence Voting Behavior? *Public Opinion Quarterly*, 50 (3), 331-339.

Thompson, C. (2008). Can You Count on Voting Machines? available on the Internet at: *http://www.nytimes.com/2008/01/06/magazine/06Vote.html?_r=2&ref=magazine&oref=slogin&oref=slogin*, January 6, 2008.

Traugott, Michael W. and Lavrakas, Paul J. (2004). *The Voter's Guide to Election Polls*, 3rd ed. Lanham, MD: Rowman and Littlefield.

Zukin, Cliff (2004). Sources of Variation in Publisher Election Polling: A Primer, American Association for Public Opinion Research article, available on the Internet at: *http://www.aapor.org/uploads/zukin_election_primer.pdf*.

Frank and Ernest ©Thaves/Dist. by Newspaper Enterprise Association, Inc.

CHAPTER 2

THE INFAMOUS 2000 ELECTION

2.1 INTRODUCTION

Although not the first election to take the media by surprise,[16] the 2000 U.S. presidential election was, indeed, an anomaly. That is the year that added new terms such as "butterfly ballots"[17] and "pregnant chads" to the American public's vocabulary. Votes, overvotes, conflicting announcements of exit poll results, and recounts ultimately led to George W. Bush being declared the winner over Albert E. Gore in the State of Florida—and, hence, the United States—an issue that was not decided until December 12, 2004, by the U.S. Supreme Court. Kathleen Frankovic and Monika

---

[16] *The Literary Digest* projected Alf Landon to beat Franklin D. Roosevelt in 1936 and the *Chicago Tribune* miscalled the 1948 election for Thomas Dewey instead of winning candidate, Harry Truman. (See Grier, 2002.)
[17] Adams (2001).

Editors' Note: Text in gray shading is provided by the editors of this volume. Boxed-in text represents excerpts from interviews with Warren Mitofsky. The balance of the chapter is made up of papers prepared by contributing authors. [Text in square brackets within the selected readings was added by the editors.]

Elections and Exit Polling. Edited by Fritz Scheuren and Wendy Alvey

McDermott describe the political climate in the United States that led to this turn of events.[18] Gerald Pomper's account of the Florida election, itself, is a lively recap, which lays the foundation for the papers in this chapter.[19]

The television networks added to the confusion. First, most of the networks announced that Gore had won the election in Florida—a projection retracted later that day. Then, John Ellis (a Bush cousin), working for Fox News, announced that Bush had won the Florida vote and, according to David W. Moore, several other television stations promptly followed suit.[20] The closeness of the official tallies, complaints about the unusual Palm Beach County ballot design, and anecdotal information about voting process irregularities led to further questions about the Florida results. Both mechanical and manual recounts ensued. In the end, the Florida Supreme Court and the U.S. Supreme Court were brought into the act, with the U.S. Supreme Court ultimately reversing the Florida court's decision and ending the recounts. This effectively awarded Florida's 25 electoral votes to George Bush, giving him the edge to win a second term in office.

The need to avoid a repeat of the confusion of 2000 was, indeed, an underlying motivation for much statistical research that followed. Here, we begin with some brief recollections by Warren Mitofsky about the exit poll predictions. Mitofsky emphasizes the need for better coordination and data quality. The contributed papers for this chapter then discuss quality control issues from two standpoints—a look back at the quality of the data collected and a look forward at ways to improve data quality. First, we begin with a paper on the evaluation of the Florida voting systems that compares different voting methods in use for the 2000 election. The next reading examines press reports of the candidates to assess the issues that impacted voter preference. The final section presents two papers that hone in on statistical methods to enhance quality—the first focuses on the need for common terminology and standardized coding of outcome rates and the second looks at weighting techniques that take party identification into account.

Below Mitofsky remembers Florida exit poll predictions from election night 2000.

> "The election night mistake in Florida naming Bush as the winner, was made by five networks. I was consulting for two of them, and I made one of those mistaken calls. The two that didn't make the mistake were VNS [Voter News Service]—they never said Bush won—and the Associated Press (AP). I complimented them for not making the call, but I criticized them because they were both the sources for misleading county data. Joe Lenski and I had called VNS an hour earlier, when it was not clear there was a mistake in saying Bush was the winner in Florida and, therefore, the presidency, and we said to VNS, "Get everybody in the place to verify every last vote in every county in Florida." We said this race is going to decide the presidency and we think you've got mistakes in there. And lo and

---

[18] See Frankovic and McDermott (2001).
[19] See Pomper (2001).
[20] See Moore (2006).

behold we had already called a winner with a mistake in one of the counties, but we did not know that at the time. At the time VNS entered the mistaken county data, the AP's independent count had just corrected the mistake. The AP had entered the mistaken county vote an hour and a half earlier than VNS, and they had taken it out of their system at about the same time VNS put it in. Neither of them told anyone about the correction of the bad data. Isn't this worth a bulletin from the AP? You've just found a major mistake in a state that is going to decide the presidency, and they didn't find a way to tell anyone! And the other criticism was that neither organization had any statistical quality controls in the processing of the county data."

2.2 ANALYSIS: LOOKING BACK

In the aftermath of the 2000 presidential election, there were many researchers and statisticians who sought to explain the problems that had led to such an unexpected result.[21] One such effort was an evaluation of the different voting systems used in the state of Florida for the 2000 presidential election. In this assessment, Wolter et al. (2003) examines the comparative quality of the different voting methods.

2.2.1 Excerpt from: Reliability of the Uncertified Ballots in the 2000 Presidential Elections in Florida[22]
Kirk Wolter, Diana Jergovic, Whitney Moore, Joe Murphy, and Colm O'Muircheartaigh

Introduction. The presidential election of 2000 was among the closest and most interesting elections in American history. In the state of Florida, 6,138,120 ballots were cast. At the time the U.S. Supreme Court stopped the vote counting, 175,010 ballots (or approximately 2.9%) remained uncertified for the presidential race, including 61,190 undervotes and 113,820 overvotes. The uncertified rate is comparable to rates in the presidential elections of 1996 and 1992, with 2.5% and 2.3% of the ballots uncertified, respectively. However, the 2000 election was so close that every ballot could potentially have made the difference.[23]

Following the 36-day, presidential-election crisis, a group of the nation's most respected media organizations (*The New York Times*; *The Washington Post*; it The Wall Street Journal; CNN; The Associated Press; The Tribune Publishing Company, represented by *The Los Angeles Times*, *The Chicago Tribune*, and *The Orlando Sentinel*;

---

[21] Don Dillman, among others, provided testimony regarding the election process. See, for example, Dillman (2000).

[22] Reliability of the Uncertified Ballots in the 2000 Presidential Election in Florida, by Kirk Wolter, Diana Jergovic, Whitney Moore, Joe Murphy, and Colm O'Muircheartaigh, reprinted with permission from *The American Statistician*. Copyright ©2004: The American Statistical Association. All rights reserved.

[23] [Finklestein and Levin (2003) discuss some statistical assumptions which related to the Florida and U.S. Supreme Court decisions on recounting the Florida ballots.]

Cox News Service, representing *The Palm Beach Post*; and *The St. Petersburg Times*) hired the National Opinion Research Center (NORC) at the University of Chicago to conduct a thorough review of Florida's uncertified ballots. This team recognized the historical significance of the uncertified ballots, and worked together to create a definitive historical archive of these ballots. This formidable task was enabled by Florida's sunshine law (P.L. 101.572), which authorizes anyone to view ballots from a state election.

The Media Group and NORC adopted two main analytic goals. The Media Group took as its principal goal the determination of which uncertified ballots could be assigned as votes to specific presidential candidates, and, consequently, of which candidate might have won the state of Florida, given alternative voting counting scenarios or standards. NORC's analytic goal was to measure and compare the reliability of the various voting systems in use in Florida, with the aim of providing elections officials with a base of information to guide the improvement of future elections. On November 12, 2001, NORC released the archive to the public via its Web site (*NORC.ORG*) and the Media Group published or aired the results of [its] analysis of vote totals. This [excerpted] article mainly contains [highlights of] the results of NORC's analysis of reliability.[24]

The state of Florida is divided into 67 counties, which used five different voting systems. The majority of counties (41) used a system by which voters used a specified pen or pencil to fill in arrows or ovals to select candidates. An optical scanner would read the ballots and register votes for the candidates. Fifteen counties used a Votomatic punch card voting system. This system required the voter to insert a computer punch card, containing many prescored chads (a small area of approximately 1/16 inch in diameter with a perforated border), into a device and then to use a stylus to punch out the chad for the selected candidate. A machine—which recorded the number of chads punched out for each candidate—counted the votes. Nine counties used a Datavote system, which was similar to Votomatic technology in that voters were required to insert a computer punch card into a device. For the Datavote system, however, the voter aligned a mechanical punch tool with the candidate of choice and punched a hole into the ballot. There were no prescored chads. A machine then read the punch cards and counted the votes. One county used a lever voting system—a system by which voters did not use ballots. A large voting apparatus simply tallied votes. Finally, one county used paper ballots on which voters used any pen or pencil to indicate selected candidates by marking an *X* in a box next to the candidate's name. Paper ballots were counted by hand.

None of the voting systems are perfect and uncertified ballots can result from mistakes by the voters, errors by the counting system (machine or canvassing board), or intentional actions of the voters. An undervoted ballot means the machine count (or, in some cases, the subsequent hand count) records no vote for President, such as when a voter intends not to vote in the presidential race. Undervotes from Votomatic technology may also occur when a chad is not completely punched out or when a pen

---

[24][For full details on the methodology used, see Chapter 7 (7.2.2 in this volume); the complete paper by Wolter et al. (2003) is also available upon request through *http://www.votingsystems.us*.]

is used to mark a chad (instead of a stylus used to punch out a chad). Undervotes from optical scan systems result when something other than the specified pen or pencil is used to vote. Because scanners can read only certain markings, using the designated writing implement is crucial to casting a successful vote. Optical scan undervotes can also result when marks are made outside of the designated oval or arrow or when the marks made within the oval or arrow are not dark or complete enough. Datavote and lever systems can produce only intentional undervotes. However, absentee ballots in these counties can be undervoted if a pen is used to indicate selected candidates. Finally, paper voting systems produce only intentional undervotes. The critical issue with an undervoted ballot is that the machine (or in some cases the county canvassing board) did not detect a vote for the presidential race.

Ballots are overvoted when more than one candidate is selected for the presidential race. Intentional and unintentional overvoting is possible with all voting systems except the lever system, which deactivates the first candidate if a second candidate is selected in a race. Votomatic and Datavote ballots are subject to overvoting if a correction is attempted by simply punching for another candidate after the first candidate has been indicated. Optical scan and paper ballots are subject to overvoting when a correction is attempted by crossing out or otherwise negating one selection and indicating another selection on the same ballot. The optical scanning machines cannot determine which mark is the intended vote. Paper ballots are all reviewed by humans and, thus, present fewer ballot errors of this type. Finally, the misuse of the write-in section of the ballot may create an overvote. A voter who voted for a candidate and then wrote in that candidate's name or the name of another candidate created an overvote that could not be certified for any candidate. The critical issue with an overvoted ballot is that the machine (or county canvassing board) detected more than one vote in the presidential race.

Rates of ballot error differ by type of voting system. While only 22% of Florida's counties used Votomatic technology, approximately 79% of the uncertified ballots were from Votomatic counties and 3.8% of Votomatic ballots were uncertified. Optical-scan counties (61% of all Florida counties) and one paper county produced approximately 18% of the uncertified ballots, and 1.3% of their total ballots were uncertified. Datavote counties (13% of all Florida counties) and one lever county accounted for approximately 1% of the uncertified ballots, and 3.7% of their total ballots were uncertified. Votomatic had an undervote error rate of 1.5% versus 0.6% and 0.3% for Datavote and optical scan, respectively. Details appear in Table 2.1. These simple statistics suggest that optical-scan systems may present fewer problems to voters than punch card systems. (Optical scanning and other high-tech solutions, however, are still no panacea, as was demonstrated recently in the 2002 Democratic primary election in Florida.)

Methodology. To pursue our analytical goals, we devised teams of three workers (called coders) who reviewed and recorded the nature of the markings on every one of the uncertified ballots. This type of information allowed the Media Group to examine vote totals, given a variety of suggested vote counting standards, and NORC to study the variability in the coding or the reliability of the ballot systems. Three coders per

Table 2.1   Florida voting systems by number of counties, number of uncertified ballots examined by NORC, and total number of ballots

Voting system	Number of counties	Undervotes	Undervotes as a percent of total certified and uncertified ballots	Overvotes	Overvotes as a percent of total certified and uncertified ballots	Total uncertified ballots	Uncertified ballots as a percent of total certified and uncertified ballots	Total certified and uncertified ballots
Votomatic (punch card with chad)	15	53,215	1.5	84,822	2.3	138,037	3.8	3,642,160
Datavote (punch card without chad)	9	771	0.6	4,427	3.2	5,198	3.7	138,869
Optical scan	41	7,204	0.3	24,571	1.0	31,775	1.3	2,357,091
Lever	1	†						
Paper	1	‡						
Total	67	61,190	1.0	113,820	1.9	175,010	2.9	6,138,120

† Ballot totals for the lever county are included in totals for the Datavote counties.

‡ Ballot totals for the paper county are included in the totals for the optical scan counties.

ballot also provided a certain redundancy that protected against operational problems. For example, if one coder fell ill and was unable to complete the assignment, there would still be two records of the markings to measure both vote totals and reliability.

We trained the coders to work independently of one another and to record what they saw in the form of a numerical code (e.g., code 1 signifies a chad with a single corner detached) for each chad or candidate position on each ballot. For example, with three coders, C chad or candidate positions, and B ballots, there were $3CB$ codes recorded. Experts from NORC and the Media Group created the numeric coding scheme to describe the types of marks on ballots.

The coders were Florida residents who worked in the counties in which they lived and possibly in adjoining counties. They were drawn from past and present NORC field staff and new employees hired specifically for this assignment. Each candidate for employment was asked to respond to a political activities screener (to ensure that no individuals with undue political bias were hired as coders) and to pass a near-point vision test (to ensure that his or her near-point vision was sufficient for ballot examination). Recruits were also informed that the work would require many hours of sitting and focusing on a very small object held at arm's length. Those who did not pass the political screener or the near-point vision test were not hired. Those who did not wish to work under the anticipated physical conditions passed on this assignment. Those selected were assigned to teams of three.

We assigned senior-level staff to act as team leaders and to provide on-site supervision for the coding operation. Thus, each team consisted of four people: one team leader who was responsible for supervision and management issues and three coders who were responsible for coding the marks on the uncertified ballots.

We developed and implemented an integrated data capture system that involved paper coding forms and a computerized data-entry system. We designed three coding forms, one for each of the three main voting systems used in Florida (Votomatic, Datavote, and optical scan), to capture chad-level or candidate-level information from the presidential and U.S. Senate sections of the ballots. The forms also allowed us to capture some information from remaining sections of the ballot (e.g., whether there were dimples in other races on the ballot, whether the voter used colored ink, whether the ballot was torn, and verbatim transcription of any notes written on the ballot by the voter).

We trained the team leaders on the technical and administrative aspects of the project over a two-day period. Subsequently, team leaders were responsible for training and evaluating their teams of coders. Team leaders and coders participated in mock coding sessions prior to the start of work. Workers who did not demonstrate acceptable coding in these sessions were replaced.

The intention of the project was not to mimic or replicate the work of county canvassing boards. Our coders examined each chad or candidate position independently of all other chads and candidates and made a judgment about the chad being observed. Conversely, canvassing board members examine a ballot as a committee and discuss the markings. They examine all markings on a ballot in relation to each other with the goal of assigning a vote to a candidate. NORC coders did not attribute votes to candidates.

The ballot examination period extended from February 5, 2001, to May 30, 2001. Between one and four teams worked in parallel in each of the counties, with four used in the largest counties and one in the smallest. Prior to the arrival of the teams, representatives from the Media Group and local election officials worked to prepare the ballots for examination. In general, the uncertified ballots were organized by ballot type and precinct. Thus, coding teams were able to view all undervotes (by precinct) and then all overvotes (by precinct) in an efficient manner.

Florida law stipulates that only county elections officials may handle ballots. Thus, coders sat on one side of a table and examined ballots that were displayed by officials sitting on the other side of the table. The coders used light boxes to examine dimples and other nuances on the ballots and they instructed officials to display the ballots over the light at advantageous viewing angles.

Team leaders monitored the work of the coders to assure that the forms were being completed as required and that the coders demonstrated careful recording of identifying information. Yet team leaders did not check the accuracy of the coding— rather, the aim was to record accurately the *independent* judgments of the coders.

During planning stages for the project, experts from the Media Group theorized that overvote ballots would be easier to code with less variability than undervote ballots. They reasoned that full punches and fully filled ovals are typical of overvote ballots and should be very easy to identify. We tested their theory in Polk, Nassau, and Pasco counties—which employed optical scan, Datavote, and Votomatic systems— and found that the three coders agreed on the codes assigned almost always. Because the reliability of the overvote ballots was apparently so high, the experts decided there was little benefit to using three codings in all counties. Subsequently, as a cost-saving measure, we used only one coder for overvote ballots in the remaining 64 counties.

At the end of each day, team leaders conducted a second review of the day's work to ensure that all forms were properly completed and organized for transmittal to NORC's data entry facility in Chicago. Team leaders shipped the forms to Chicago via Federal Express, including a special transmittal form detailing the county worked, the number of forms included, and the identification numbers associated with the shipment.

In Chicago, we entered identifying information about the coding forms into an electronic tracking system and prepared the forms for data entry. Trained clerks reviewed every cell on every form for completeness and legibility. They assigned special codes, if necessary, to missing data or illegible fields.

Trained operators key-entered the coding forms, followed by 100% verification (i.e., all forms were key-entered independently a second time). Managers conducted an adjudication/reconciliation process to resolve any discrepancies.

We developed two primary databases for the ballot-level information. One database contains the coded information for every chad or candidate space on every ballot across 67 counties. This file does not attempt to reconcile candidate information across ballots, it simply reflects the reality of the disparate ballot designs used throughout the state of Florida. (By this, we mean that our raw data reflect the fact that the candidates names are not in the same field on each ballot for each county.) The second or aligned database reconciles the coded information for every ballot for each presidential and

U.S. Senate candidate. (This means each coder/candidate is in the same field in each ballot record.)

Characteristics of Florida Coders. Overall, 153 coders worked on the project. Once hired, each coder completed a short demographic questionnaire. For comparison, we can use the Year 2000 General Social Survey, a national sampling of householders 18 years of age and over. Table 2.2 shows the results.

When compared with the general population, the coders are more often female and older and have more years of formal education. When one turns to variables with potential political or ideological implications, there are small differences in family income, ethnicity, and party identification. In terms of presidential vote in 2000, coders split right down the middle.

It would be far fetched to claim the coders are a representative sample. To begin with, a truly representative sample of workers would not be available for such temporary employment. But Table 2.2 supports two propositions:

1. The coders, as a group, seem to have no political tilt which might influence their collective judgments and

2. The results here seem roughly applicable to the people who become election officials at the precinct level, provided they have training and supervision similar to that of the project and are not political partisans.

Description of Coder Reliability. The objective of the coding of the uncertified ballots was to obtain an accurate (reliable and valid) record of the marks on the ballots. Knowing that no coding operation can ever be flawless, our objective was to assess the quality of the coding operation. This quality has two principal dimensions: reliability and validity.

Reliability is a measure of the consistency in the data; it can be described using a variety of measures, some of which we discuss in what follows. Bailar and Tepping (1972) and Kalton and Stowell (1979) discussed coder reliability studies. The first dimension, *validity* (average correctness), is much more difficult to assess, and there is little opportunity to measure validity from internal evidence within our data. We attempted to avoid invalid data by not hiring biased coders and by coder training and supervision.

Table 2.3 shows the number of undervote ballots by mark status and by type (Votomatic, Datavote, and Optical Scan). We use the term *mark status* to signify whether a ballot was completely blank on the presidential race (as determined by all coders who reviewed the ballot) or not. Almost 56% of total ballots were completely blank. The table also shows the number of undervote ballots that were marked (by at least one coder) for each presidential candidate. For example, 13,113 ballots were marked for Bush and 13,283 were marked for Gore.

Blank ballots are uninteresting and uninformative regarding the reliability of

Table 2.2   Distribution of Florida coders and of U.S. adults in the 2000 General Social Survey (GSS)

Coder characteristics	Coders (in percent)	GSS 2000 (in percent)
Sex	$N = 152^*$	$N = 2,817$
Male	29.6	44
Female	70.4	56
Total	100.0	100
Age (recoded from date of birth)	$N = 148$	$N = 2,832$
63+	23.6	16
53–62	25.0	12
41–52	26.4	24
20–40	25.0	45
18–19	0.0	2
Total	100.0	99
Education	$N = 152$	$N = 2,799$
Graduate degree	9.9	8
Bachelor's degree	23.7	16
High school graduate	65.1	61
Less than high school	1.3	16
Total	100.0	101
Family Income	$N = 150$	$N = 2,456$
$100,000+	6.0	7
$75,000–100,000	9.3	9
$50,000–75,000	17.3	18
$25,000–50,000	34.7	29
Less than $25,000	32.7	36
Total	100.0	99
Race/Ethnicity	$N = 150$	
Hispanic (of any race)	7.0	16.8**
Non-Hispanic white alone	80.1	65.4**
Black Alone	10.6	14.6**
Some Other Race Alone	2.0	5.1**
Two or More Races	0.7	2.4**
Total	100.4	104.3**

*Deviations from $N = 153$ are due to nonresponse.
**Data from the 2000 Census for the total population of Florida. Total adds to more than 100% because the categories are nonoverlapping.

continued next page

Table 2.2   Distribution of Florida Coders and of U.S. adults in the 2000 General Social Survey (GSS) (*continued from previous page*)

Coder characteristics	Coders (in percent)	GSS 2000 (in percent)
Party Identification	$N = 152$	$N = 2,805$
Democratic	36.2	33
Independent	30.3	41
Republican	29.6	24
Other	3.9	2
Total	100.0	100
2000 Vote	$N = 151$	Survey was conducted
Democratic	34.4	before the election.
Republican	35.1	
Other	6.0	
Didn't vote	24.5	
Total	100.0	

*Deviations from $N = 153$ are due to nonresponse.
**Data from the 2000 Census for the total population of Florida. Total adds to more than 100% because the categories are nonoverlapping.

Florida's ballot systems Thus, in the balance of this section, we present statistics only for the universe of 27,125 marked ballots...[25]

...In what follows, we only present reliability statistics for the Republican, Democratic, and Libertarian candidates: Bush, Gore, and Browne. Results for other candidates mirror those for Browne. To simplify the presentation, we present reliability statistics only for the dimple or greater and two corners or greater variables.[26]

While Bush and Gore reliabilities are similar for Votomatic and optical scan ballots, Bush reliability is lower than Gore reliability for Datavote ballots. Votomatic ballots tend to follow the pattern by absentee status observed overall. Datavote ballots, however, reveal larger differences, but still regular ballots are the more reliable. For example, given the dimple or greater standard, the Bush pairwise agreement statistics are 0.77 for absentee ballots and 0.98 for regular ballots, respectively. The differences even continue for the two corners or greater standard. Datavote absentee ballots also reveal a relatively sizable difference between Bush and Gore reliability. This finding is probably due to the fact that more of these ballots were blank for Gore...

[25] [More detailed discussion of the reliability testing is provided in 7.2.2 in Chapter 7 of this volume.]
[26] These are conditions that were coded by the coders who examined the ballots.

Table 2.3   Number of undervote ballots by mark status by type of ballot

Ballots by mark status	Votomatic ballots	Datavote ballots	Optical ballots	Total ballots
Total Ballots	53,193	765	7,202	61,150
Completely Blank Ballots *	28,970	466	4,599	34,025
Marked Ballots **	24,223	299	2,603	27,125
Marked Ballots by Candidate #				
Marked for Bush	11,735	168	1,210	13,113
Marked for Gore	11,726	135	1,422	13,283
Marked for Browne	599	8	141	748
Marked for Nader	617	4	164	785
Marked for Harris	347	2	122	471
Marked for Hagel	479	1	107	587
Marked for Buchannan	572	2	125	699
Marked for McReynolds	221	2	107	330
Marked for Phillips	359	4	102	465
Marked for Moorehead	285	2	102	389

* All coders saw no marks for any presidential candidate.
** At least one coder saw a mark for at least one presidential candidate.
\# At least one coder saw a mark for the named candidate. Because some ballots were marked for more than one candidate and some were judged marked for different candidates by different coders, marked ballots by candidate do not add to total marked ballots.

Results of Recount Analysis.   On November 12, 2001, the Media Group published vote totals for Bush and Gore based upon nine possible standards (or scenarios) for what constitutes a vote, with two variations for two of the standards. Their vote totals were defined as the certified votes from the presidential election plus presumed "new votes" discovered among the uncertified ballots, given the various standards. For each scenario, they examined vote totals under two levels of agreement:

1. A unanimous level that required agreement by all three coders, and

2. A majority-rules level that required agreement by at least two of the coders.

Table 2.4 details the nine scenarios analyzed by the Media Group and corresponding margins of victory. For example, given the Gore Request, Bush's margin is +225 votes. It is important to note that no one scenario was a likely outcome of the actual election. Indeed, a time- and labor-intensive examination and recount of the 175,010 uncertified ballots (such as the one conducted for this project) was not an option in the hectic days following the presidential election. With that caveat in mind, the results of the recount analysis are fascinating and three points merit mention. First, with margins ranging from 42 to 493 votes, no recount scenario produces a margin of victory greater than that of the official certified results. Second, it appears that each candidate's preferred recount standard would have yielded a defeat for that candidate; for example, Gore wins by 105 votes, given the Bush Standard. Third, it is clear that the election was an exceedingly close contest, regardless of which standard would have been used in recounting votes. Readers interested in a more comprehensive or interpretive analysis of the data should see any Media Group member's November 12, 2001 publication and Keating (2002). The online databases from this project allow future researchers to both replicate these findings and investigate alternate scenarios.

Leaving aside the substance of the various standards, what is striking to us is that there was apparently such a large volume of new votes among the uncertified ballots. Even under the unanimous level of agreement (which finds fewer new votes than the majority-rules level), there are between 1,369 and 14,281 new votes, which translate into between 0.8% and 8.2% of all uncertified ballots, respectively.

Summary. In this article, we obtained the following findings regarding coder reliability. For ballots classified as undervotes:

- Reliability is similar for the two major candidate, Bush and Gore, both statewide and by county, except for some of the smaller counties where small sample size can lead to divergent results.

- Reliability is similar and actually higher for all remaining candidates (than for the major candidates), reflecting the fact that most ballots are blank for these candidates and most coders readily discern and record a blank.

- Reliability is lower for the all-codes variable than for the dimple or greater variable and is highest for the two-corners or greater variable.

- Reliability is lower for the two punch card systems, Votomatic and Datavote, than for the optical scan systems.

- Overall, regular ballots are more reliable than absentee ballots; Datavote absentee ballots are considerably worse than the corresponding regular ballots; absentee and regular ballots are essentially equally reliable for optical scan systems; and Votomatic absentee ballots are slightly less reliable than the corresponding regular ballots.

- Among optical scan ballot designs, the oval is slightly preferred to the arrow.

- Among oval designs, there is little to choose between the single- and split-column formats.

Coder reliability is extremely high for ballots classified as overvotes. Also, undervote and overvote error rates are higher for punch card systems than for optical scan systems. We recommend states consider these findings as they look for ways to reform ballot systems for future elections.

We also found that extra-role characteristics of coders affect coding outcomes in Votomatic, but not optical scan, counties. In Votomatic counties, our analysis shows

- Men see more marks than women.

Table 2.4   Vote margins for official results and nine recount scenarios

Scenario	Description	Margin, majority agreement	Margin, unanimous agreement
Official results	Vote totals certified by Florida election officials	Bush: +537	
3. Florida Supreme Court Complex	Accepts completed recounts in eight counties, certified results from four counties that said they would not have recounted, and applies 55 county-reported standards; includes overvotes where counties report intent to count them. If the U.S. Supreme Court had upheld the Florida Supreme Court ruling, this scenario would have resulted.	Bush: +493	Bush: +323
2. Florida Supreme Court Simple	Accepts completed recounts in four counties; elsewhere punch card undervotes with at least one corner detached; any affirmative mark on optical scan ballots; no overvotes	Bush: +430	Bush: +369
8. Gore Request	Accepts certified results in 65 counties, including hand-counts in Broward and Volusia; uncertified hand-counts in Palm Beach and parts of Miami-Dade; and one-corner detachment in remaining Miami-Dade; no overvotes	Bush: +225	Bush: +212
4. 65 County Custom	Accepts whatever each individual county considered a vote, includes overvotes where appropriate; certified results accepted for Broward and Volusia	Gore: +171	Gore: +81

continued on next page

Source: *Chicago Tribune*, November 12, 2001.

Table 2.4 Vote margins for official results and nine recount scenarios *continued from previous page*

Scenario	Description	Margin, majority agreement	Margin, unanimous agreement
7. Most restrictive	Requires unequivocal punches and complete fills on optical scan ballots; no overvotes; accepts certified results Volusia	Gore: +115	Gore: +127
6. Most Inclusive	Accepts all dimpled chads; any affirmative mark on optical scan ballots; includes overvotes in optical counties if prevailing standard dictates; accepts certified results for Volusia	Gore: +107	Bush: +110
5. Bush Standard	Accepts chads with two corners detached; any affirmative mark on optical scan ballots, includes overvotes; accepts certified results for Volusia	Gore: +105	Gore: +146
1. Prevailing Statewide	Requires one corner detached on punch cards; any affirmative mark on optical scan ballots; includes overvotes; accepts certified results for Volusia	Gore: +60	Gore: +145
9. Palm Beach Rules	Accepts dimpled chads only when other dimples present on ballot; includes overvotes in optical scan counties if prevailing standard dictates; accepts certified results for Volusia	Gore: +42	Gore: +114

Source: *Chicago Tribune*, November 12, 2001.

- Self-identified Republicans are more likely to code a mark in favor of Bush than are self-identified Democrats, while Republicans are less likely to code a mark in favor of Gore than are Democrats.

- Both of the aforementioned findings are stronger in the Bush direction.

This analysis does not tell us which group of coders is correct; it merely identifies systematic differences between groups (or categories) of coders. Neither does it tell us that the characteristics of the coders that we have identified are causally connected to the outcome. It does, however, suggest that the coding of Votomatic ballots is subject to variation that is associated with the characteristics of the coders. This renders adjudication of disputed ballots much more problematic for ballots using the Votomatic system.

Given a number of alternative standards as to what constitutes a vote, analysts working for the Media Group found thousands of potential new votes among the uncertified ballots in NORC's database. These results suggest that state elections officials should consider wider use of new technology that edits the completed ballot in real time as the voter submits it. This technology provides the voter a chance to correct the ballot if it is deemed to be spoiled. Officials should also consider a statewide, consistently applied procedure for human review and counting, where appropriate, of ballots that remain uncertified after the initial machine count.

Arising out of the material here, an important question is the degree to which systems of lower reliability will be tolerated in our democratic system. In this regard, discrepancies between regular versus absentee ballot reliability might provide a benchmark or reference point for future decisions about whether varying ballot methods conform with the constitutional "equal-protection" principal that figured in the ultimate decision by the U.S. Supreme Court.

Finally, in the course of doing this project, we observed two other possible limitations in American elections. First, ballots are not usually pre-tested on real voters prior to their use in the general election. County election officials, some with considerable experience, design ballots under state law. But even experts may create confusing designs or unclear instructions. In the field of survey research, experts draft the questionnaire, pre-test it on a small sample, and make appropriate adjustments prior to finalizing it and using it in the main survey. This is parallel to the proof-of-concept stage in technological fields. We recommend the practice of pre-testing be extended to the development of ballots prior to their use in a general election. Second, voters are not usually trained prior to an election. Precincts often display voting instructions at the polling place, and precinct workers often offer assistance to voters who make a specific request. But many people do not take time to read instructions or are embarrassed to ask for help. At the same time, some voters are voting for the first time ever, for the first time in years, or for the first time in a new location under a different ballot system. Some voters' eyesight may be failing. And there can be many other problems, too. To address this range of concerns, we recommend states look for cost-effective means of training voters on the ballot system prior to the election.

Acknowledgment. All authors were with NORC at the time of the Florida Ballots Project. The authors thank Jim Davis for collaboration in the development the discussion of coder characteristics and the general discussion and analysis of coder reliability; Barbara Bailar for collaboration in the development of the section on and general discussion and analysis of coder reliability; and Anirban Basu, Anthony Bryk, and Pamela Campanelli for helpful discussions on the application of hierarchical models to this problem. We also thank the referees for helpful comments.

References

Bailar, B. A., and Tepping, B. J. (1972). *Effects of Coders*, Series ER 60 (9), Washington, DC: U.S. Bureau of the Census.

Chicago Tribune (2001). Still Too Close to Call; Conclusion Not Clear Even if Recount Allowed, November 12, 14–15.

Kalton, G., and Stowell, R. (1979). A Study of Coder Variability, *Applied Statistics*, 28, 276.

Keating, D. (2002). Democracy Counts: The Media Consortium's Florida Ballot Project, paper presented at the annual meeting of the American Political Science Association.

Basson (2008) examines the impact of depth of political knowledge and partisanship on the accuracy of predicting voter intent. Another factor that could have impacted the vote results is the news reports prior to Election Day. In the following article, David Fan (2001) uses time series analysis to examine the support for the top candidates in the period leading up to the election. The excerpt below presents highlights of his results. We include this reading as illustrative of the kinds of analysis that are conducted after an election.

2.2.2   Excerpt from: Predicting the Bush/Gore Election from the Press: The Cost to George W. Bush of His November Surprise of an Old Arrest
David P. Fan

Introduction. This paper uses the ideodynamic model to predict the time trend of voter preference for George W. Bush over Al Gore in the 2000 United States presidential election. This time trend was predicted from 32,000 newspaper and news wire stories scored as pro and con for the two candidates. The quality of the fit was demonstrated by the predicted time series staying within the 95 percent confidence intervals of the poll points from April 4 through November 7, Election Day. These results show that press discussion of the candidates in the pro and con frames successfully included information from conditions such as the state of the economy and from events such as the November 2 revelation of Bush's arrest for drunken driving in 1976. Furthermore, the economy was not an important factor in the 2000 election, in contrast to those of 1984 to 1996.

Dependent variable. The dependent variable time series consisted of 383 polls obtained from the Roper database from April 4, 2000 to Election Day, asking this question or a close variant: "Still thinking about the November (2000) election for

President—if the election for President were held today and you had to make a choice, for whom would you vote ... Republican George W. Bush for President with Dick Cheney for Vice President or Democrat Al Gore for President and Joe Lieberman for Vice President?" If the poll permitted leaners, then the leaner and committed responses were summed. Early in the poll series, the vice-presidential candidates were not part of the question. [27]

Explanatory variables. The major persuasive information relevant to the Bush and Gore race was obtained, as was done in prior ideodynamic analyses, from stories from the Associated Press news wire and 136 newspapers from around the country. These stories were obtained from the NEXIS electronic database, which was searched every two weeks from March 29 through Election Day, November 7, 2000. The search was for stories with at least two mentions of Al Gore, Joe Lieberman, Ralph Nader, Pat Buchanan, George W. Bush, Dick Cheney, or any pair of these candidates. All stories were scored using the InfoTrend computer method (Watts et al., 1999). The retrieved text was further scored for the appearance of a word referring to Bush in the same paragraph as a word referring to arrest or drunken driving. The reason was that the press revealed his 1976 (DWI) drunken driving conviction on Thursday night, November 2, just four days and a few hours before the election.

The Ideodynamic Model. The time trend of candidate preference was computed using the ideodynamic model based on this theoretical statement: Changes in the population of Bush supporters result from an increase due to Gore partisans being recruited by information able to convert Gore supporters to favor Bush and from a decrease due to Bush adherents being converted to prefer Gore by information able to recruit Bush supporters.

Results. Having argued that candidate preference could be driven mainly by valence variables, the first step was to score paragraphs as pro and con for Bush and Gore (Table 2.5, right hand column). The numbers of paragraphs both pro and con for Gore were very slightly higher than those which were pro-Bush, with less than 10 percent separating the three numbers. The odd score out was con-Bush, which was only 2/3 of that of pro-Bush, so Bush benefited from a relatively higher ratio of favorable coverage, even though his total scores were lower.

There was a peak for Bush in early May, due to John McCain's endorsement, followed by a peak at the end of July and the beginning of August, corresponding to the Republican convention. For Gore, there was an equivalent increase around the time of the Democratic convention, with an earlier spike due to favorable publicity surrounding his choice of Joe Lieberman for vice-president. After both conventions there was a relatively quiescent period with the Olympic games holding the media's attention. Afterwards, campaign coverage increased, with a focus being the televised debates in October. Although coverage of both candidates increased after the onset of the debates, there was a more noticeable increase in pro-Bush information than for the

---

[27] [For a more complete discussion of the methodology for this paper, see the complete paper by Fan (2001), available upon request through *http://www.votingsystems.us*.]

Table 2.5 Constants and fit for predictions of the time trend of pre-election candidate preference polls using valence explanatory variables
Persistence constants k are presented as values relative to the value for Pro-Bush for the 5 constant model and to OtherPro-Bush for the 9 constant model. The pairs of numbers underneath *95% interval* give the 95 percent confidence interval computed at the calibrated values for the other constants. The fit to the pre-election poll series is given by the R^2 and the Root Mean Squared Deviation (RMSD) in percent.

Item	Value	Confidence	Significance	Paragraphs
Valence variables alone; 5 constant model using equations (3)				
(Estimated Constant)		*(95% interval)*		*(Total)*
Persistence p	8.1 minutes	(0,0.6 days)	> 0.1	
Persuasibility k				
Pro-Bush= k_{PB}	1.00	(0.98,1.03)	< 0.005	57616
Con-Bush= k_{CB}	0.91	(0.88,0.95)	< 0.005	39191
Pro-Gore= k_{PG}	0.63	(0.61,0.66)	< 0.005	62076
Con-Gore= k_{CG}	0.37	(0.34,0.39)	< 0.005	63693
R^2	0.46			
RMSD	2.50			
Valence variables in economy/other frames; 9 constant model using equations (4)				
(Estimated Constant)		*(95% interval)*		*(Total)*
Persistence p	145 minutes	(0,0.7 days)	> 0.1	
Persuasibility k				
OtherPro-Bush= k_{NPB}	1.00	(0.96,1.04)	< 0.005	55049
OtherCon-Bush= k_{NCB}	0.99	(0.94,1.04)	< 0.005	37171
OtherPro-Gore= k_{NPG}	0.64	(0.60,0.66)	< 0.005	57918
OtherCon-Gore= k_{NCG}	0.38	(0.35,0.41)	< 0.005	60433
EconPro-Bush= k_{EPB}	0.91	(0.87,0.92)	< 0.005	2567
EconCon-Bush= k_{ECB}	0.002	(0,0.80)	> 0.1	2020
EconPro-Gore= k_{EPG}	0.29	(0,0.73)	> 0.1	4158
EconCon-Gore= k_{ECG}	0.14	(0,0.73)	> 0.1	3260
R^2	0.47			
RMSD	2.48			

other three valence scores. In early November, there was a sharp, but very transient, drop in both pro- and con-Bush coverage, just after the disclosure of his DWI arrest.

The bad news came in two separate disclosures about a day apart. First there was the news of his arrest, and then it became known that he had withheld the information to protect his daughters. As a result, the DWI information rose over both Friday, November 3 and Saturday, November 4, before finally starting to fall on Sunday, November 5. The time trend in Bush preference was the result of interactions of the persuasive force functions as revealed using the ideodynamic model. Bush's predicted gain on Gore occurred after the second debate, when the press consistently gave Bush relatively more positive coverage while keeping it balanced for Gore. As a result, opinion moved from less than 50 percent for Bush up to 54.0 percent. Then, valence information dropped his preference to 50.1 percent over the two days of the revelations, before a recovery to 52.7 percent on Election Day. That left the Bush preference about 1.3 percent below his peak value. Interestingly, the 2000 election showed much smaller economic effects than before. First of all, economic valence information only accounted for about one twentieth of all valence paragraphs, meaning that the economy really was not an important issue (Table 2.5).

Discussion. As with earlier studies, the analyses in this paper indicate that the press has retained its ability to represent the principal election information available to the public with, at most, minor contributions from such other sources as campaign advertising, televised debates, and Internet information, which was much less of a factor for earlier elections. That is because press-based pro/con persuasive force functions for the major presidential candidates could be entered alone into the ideodynamic model to give good predictions of the time trend of pre-election polls.

Since the dependent variable time series was an indiscriminant pool of all Roper database polls from different survey organizations asking approximately the same question, a substantial part of the variance was due to different polls giving somewhat different results. Nevertheless, the R^2 was 0.46, and, more importantly, the predicted time trend did not exceed the limits of the 95 percent confidence intervals of the surveys throughout the entire seven months of the election. Therefore, it will be difficult to improve on the fit.

If the economy-valence information had higher persuadability constants, then the populace would have been primed to rate pro/con news in the economic context as more influential. That was exactly what had been found for the 1984 to 1996 elections,[28] but was untrue for 2000, where the only significant economy-valence information was pro-Bush. Although Gore had more economy/valence coverage than Bush, none of that coverage was significant.

The final prediction for Election Day was 52.7 percent for Bush, meaning that he should have won the popular vote, but the prediction was of the poll time trend and not the vote. The last surveys used for the prediction were from November 5, with the normalized values for Bush being 52.9, 52.5, and 49.4 percent after removal of the

---

[28] See Watts et al. (1999).

minor candidates and the "don't knows." These average to 51.5 percent, compared to 52.0 predicted from press coverage for the same day.

Therefore, valence coverage predicted the polls with high accuracy just before Election Day. However, neither the pre-election polls nor the media time trends designed to match them account for turnout, which could have accounted for some of the difference between Bush's predicted preference of 52.7 percent on Election Day and his actual vote share of 49.7 percent after removal of the minor candidates. Besides turnout, the DWI information, itself, could have been able to shift Bush opinion to Gore. Unfortunately, it was not possible to assess the effect of the DWI news alone, because there were not enough poll points after the disclosure for a serious test. Nevertheless, the time trend predictions in this paper did show a rapid and very noticeable drop followed by a quick recovery, indicating that the effects of the DWI news would likely have dissipated completely even if the disclosure had been just a week earlier.

Reference

Watts, M. D.; Shah, D. V.; Domke, D., Fan, D. P.; and Fibison, M (1999). News Coverage, Economic Cues and the Public's Presidential Preferences, 1984–1996, *Journal of Politics*, 61, 914–942.

2.3 DATA QUALITY

Even before the challenges of 2000, data quality was an issue of major concern with regard to election and exit polling. One area of rising concern in all surveys—and especially exit polls—is nonresponse. In the paper below, Slater and Christensen (2002) discuss the need for consistent terminology and recommend standardized coding adopted by the American Association for Public Opinion Research (AAPOR), adapted specifically for exit polling.

2.3.1 Applying AAPOR's Final Disposition Codes and Outcome Rates to the 2000 Utah Colleges' Exit Poll
Melaney Slater and Howard Christensen

Introduction. A major concern in exit polling, as well as other types of polling, is properly informing the public about the accuracy of survey results. At least three organizations—the American Association for Public Opinion Research (AAPOR), the National Council on Public Polls (NCPP), and the Council of American Survey Research Organizations (CASRO)—have published disclosure standards. AAPOR (2001) declared that disclosure standards provide "essential information about how the research was conducted to ensure that consumers of survey results have an adequate basis for judging the reliability and validity of the results reported."

AAPOR's, NCPP's, and CASRO's disclosure standards expect the inclusion of response and nonresponse rates. Under article five in AAPOR's Standards for Minimal Disclosure, AAPOR declares that researchers should include "completion rates" in their surveys (AAPOR, 1986). Moreover, the NCPP discusses in the last point of their Principles of Disclosure that survey findings should contain the "percentages upon which conclusions are based" (NCPP, 1979). Finally, CASRO states under section 6 of article B in their Responsibilities in Reporting to Clients and the Public that "the number not reached" and "the number of refusals" should be presented in the research organization's report to the public (CASRO, 2001). Complete statements of the disclosure standards for AAPOR, NCPP, and CASRO are available on their Web sites[29]. Unfortunately, election pollsters have been slow to adopt the nonresponse disclosure standards. Miller, Merkle and Wang (1991: 207) say that nonresponse is "rarely or never treated" in election poll reports.

Even when response and nonresponse rates have been reported, many researchers have been using their own definitions to report these figures. For example, in Lohr's research (1999: 281), she has found survey researchers using five different formulas that have all been defined as response rate. The disclosure of nonresponse should be consistent across surveys and, therefore, comparable.

Filling the Need for Standards.   To address the confusion of the various nonresponse methods, AAPOR has recently published "Standard Definitions: Final Dispositions of Case Codes and Outcome Rates for Surveys" (2000). We will refer to this recent publication as Standard Definitions throughout the remainder of this paper. Standard Definitions has been built on prior research and recent technology to establish a range of general cases. AAPOR found many discrepancies regarding the final disposition codes between surveys. AAPOR discovered that researchers are using more than 24 classifications of final disposition codes and [no two] were exactly alike. They also learned that many of the codes were "unique to a particular study and categories often were neither clearly defined nor comparable across surveys" (AAPOR, 2000). Standard Definitions discusses the inconsistency of calculating outcome rates across different studies. AAPOR refers to outcome rates as including response, cooperation, refusal, and contact rates. They state that in some cases the outcome rates "commonly cited in survey reports have the same names, but are used to describe fundamentally different rates" (AAPOR, 2000). In other cases, "different names are sometimes applied to the same rates" (AAPOR, 2000). As a result, a confusion is created because survey researchers are "rarely doing things in a comparable manner and frequently are not even speaking the same technical language" (AAPOR, 2000).

AAPOR hopes to establish a "common language and definitions that the research industry can share" (2000). This will allow researchers to "more precisely calculate response rates and use those calculations to directly compare the response rates of different surveys" (AAPOR, 2000). Standard Definitions brings a universal interpretation of survey nonresponse methodology that can be compared over a variety of sampling modes.

Extension of AAPOR Standards to Exit Polls.   The most common methods for gathering survey data are in-person interviews, telephone interviews, and mail-administered surveys (Glynn, Herbst, O'Keefe, and Sharpiro, 1999:72). AAPOR's Standard Definitions account for all three methods, but no combination thereof. Glynn

---

[29][See *www.aapor.org/ethics/stddef.html/*, *www.ncpp.org/disclosure.htm*, and *www.casro.org/codeofstandards.cfm.*]

et al. (1999:74) discuss an increase in blended data collection methods by mentioning, "multiple data collection techniques are now more in use." An extension of AAPOR's codes to account for a broader spectrum of data collection methods would be very useful to survey researchers. In this study, we adapted AAPOR's Standard Definitions to general exit polls and, specifically, to the Utah Colleges' Exit Poll (UCEP 2000).

Review of Exit Poll Methodology.   Exit polls are "conducted on Election Days at polling places while voting is in progress" (Mitofsky, 1991:93). A typical exit poll begins with a probability sample of polling places or precincts inside a designated area. The probability sample ensures that every precinct or polling place has a known probability of being chosen that is greater than zero (Forman, 1991:19). Exit pollsters typically use a stratified sample of precincts within some geographical area, which are selected proportionally to the votes cast in the previous election (Mitofsky and Edelman, 1995:82; Merkle and Edelman, 2000:69). Within each precinct or polling place, an interviewer intercepts a systematic sample of every *n*th voter as they leave their polling place and invites them to fill out a survey (Mitofsky, 1991:93; Mitofsky and Edelman, 1995:82; Traugott and Lavrakas, 2000:21). Mitofsky (1991: 93) affirms that when conducted correctly, an "exit poll is very much like any other scientific sample survey."

In exit polls, nonrespondents can be separated into refusals and misses (Merkle and Edelman, 2000: 73). Merkle and Edelman state that a refusal "occurs when a sampled voter is asked to fill out a questionnaire and declines," and a miss takes place when the "interviewer is too busy to approach the voter or when the voter does not pass the interviewer" (Merkle and Edelman, 2000:73).

Nonresponse is usually very high in exit polls. As stated by Mitofsky and Edelman (1995:95), Voter Research and Surveys (VRS) reported an average of 42 percent nonresponse in the 1992 presidential exit poll, which compares to 40 percent that CBS reported in 1988. According to Mitofsky and Edelman (1995: 95), one-fourth of the nonresponse was due to interviewers missing voters who should have been included in the sample and the rest of the nonresponse was due to respondents refusing to participate in the survey. Merkle and Edelman (2000:73) report that, in the 1996 presidential exit poll, Voter News Service (VNS) averaged a statewide refusal rate of 33 percent and miss rate of 10 percent.

Modification of Standard Definitions.   This paper provides a general case scenario of final disposition codes and outcome rates from which any individual exit poll may commence. Some disposition codes and outcome rates can be modified due to a particular methodology of an exit poll.

By examining the household, mail, and telephone options of Standard Definitions, we determined that a combination of the household and mail final disposition codes would best represent the final disposition codes for an exit poll. However, many of the household interview and mail codes do not apply to an exit poll. For instance, such mail and household codes as: refused to pay postage, no mail receptacle, not a housing unit, and vacant housing unit are obviously not appropriate for an exit poll.

In addition, there are unique elements that pertain to an exit poll. In contrast to other data collection methods, an exit poll consists of an interviewer contact outside of a polling place, and a self-administered survey (Mitofsky, 1991:93; Traugott and Lavrakas, 2000:21). Instead of an actual list of names of selected respondents, an exit poll has a sampling interval that systematically selects respondents. The field period for an exit poll is extremely short, consisting of only one day. Consequently, there is no possibility of performing callbacks, a technique that typically lowers the nonresponse rate.

Beyond data collection differences, exit polls have to take into account unique classifications of respondents. These include absentee voters and missed voters. Traugott and Lavrakas (2000:14) report that one-fourth of all of the votes in the 1996 presidential election were cast using absentee ballots. However, absentee ballots are not included in the sampling frame of exit polls because they do not vote at their precincts on Election Day. Currently, absentee voters are assumed to have a negligible effect on the voting results.

On the other hand, missed voters are included in most exit polls. VRS and VNS reported from the 1992 and 1996 presidential exit poll data, respectively, that an average of about 10 percent of the nonresponse was categorized as missed voters (Mitofsky and Edelman, 1995:95; Merkle and Edelman, 2000:73).

Final Disposition Codes. In this section, the definitions of the final disposition codes for household interview and mail surveys of AAPOR are adapted to apply to general exit polls. An entire list of the final disposition codes is provided in Table 2.6.

The following terms are typical of exit polls. An *interviewer* is a pollster that contacts selected voters and invites them to fill-out a questionnaire. A *contact* is when an interviewer approaches a respondent and asks them to fill-out a questionnaire. A *returned questionnaire* (1.0) refers to a respondent accepting a questionnaire from an interviewer, at least partially completing the questionnaire, and returning it to the interviewer. An *eligible case* (2.0) is when there is a possible contact, but the respondent does not fill-out the questionnaire enough to be considered a partial questionnaire. Cases of *unknown eligibility* (3.0) occur where there is no contact and no returned questionnaire. *Not eligible* cases (4.0) differ from unknown eligible cases in that some contact has been made with the respondent to determine that they are not eligible, or it is completely obvious that they are not eligible (e.g., children 0–10 years old).

The following sections are explanations of the final disposition codes. These sections are parallel to the "RDD Telephone Surveys of Households" and "In-Person Household Surveys" sections of Standard Definitions. Therefore, in many situations we use the same language; however, we do not constrain our paper to put these similarities into quotes.

Returned questionnaires. Returned questionnaires are separated into two categories:

1. Complete (1.1) and

2. Partial (1.2) questionnaires.

Table 2.6   Final disposition codes for a general exit poll

1. Returned questionnaire	(1.0)
Complete	(1.1)
Partial	(1.2)
2. Eligible, contact or no contact	(2.0)
Refusal and break-off.	(2.10)
Refusal.	(2.11)
Other person refusal.	(2.111)
Known respondent-level refusal	(2.112)
Break-off.	(2.12)
Non-contact/missed voter	(2.20)
Did not pass by interviewer	(2.21)
Interviewer too busy	(2.22)
Other	(2.30)
Precinct not attempted or worked	(2.31)
Unable to reach/unsafe area	(2.311)
Unable to locate an address	(2.312)
Physically or mentally unable/incompetent . . .	(2.32)
Language	(2.33)
Respondent language problem	(2.332)
Wrong language questionnaire	(2.333)
No interviewer available for needed language	(2.334)
Literacy.	(2.34)
Micellaneous	(2.35)
3. Unknown eligibility, non-contact	(3.0)
Unknown if eligible respondent.	(3.20)
No screener completed	(3.21)
Missed voter	(3.22)
Did not pass by interviewer	(3.221)
Interviewer too busy	(3.222)
Miscellaneous	(3.90)
4. Not eligible	(4.0)
Out of sample	(4.10)
Not an eligible respondent	(4.70)
Unregistered voter	(4.71)
Registered to vote, but did not vote	(4.72)
Quota filled.	(4.80)

Each study is asked to clearly define cut-off rates for complete and partial questionnaires, in advance. AAPOR (2000) gives a few of the most common examples of cut-off rates in Standard Definitions.

Eligible, contact or non-contact. Voters that are eligible and did not at least partially complete a questionnaire can be classified as *nonrespondents*. This category consists of three types of nonresponse:

1. Refusals and break-offs (2.10),

2. Non-contacts (2.20), and

3. Others (2.30).

AAPOR (2000) also asks that a survey provide their definition of a break-off questionnaire.

Refusals and break-offs consist of cases in which some contact has been made with a voter. The voter or another responsible person accompanying them may reject the opportunity to fill-out the questionnaire (2.11). Or, a questionnaire is returned with only a few questions completed, less than needed to qualify as a partial questionnaire, and the respondent refuses to complete it further (2.12).

Eligible respondents that were supposed to be contacted and, for some reason, were not are called *non-contacts* (2.20), which are also referred to as *missed voters*. A voter is missed when he or she does not pass the interviewer, or when an interviewer is too busy to approach the voter (Merkle and Edelman, 2000, p.73). Therefore, missed voters can be broken down into: did not pass interviewer (2.21), or interviewer too busy (2.22). However, not all missed voters are eligible respondents. If there is any question regarding a missed voter's eligibility, he or she should be categorized as a *missed voter with unknown eligibility* (3.22).

Other cases (the 2.30 subset) represent instances in which the respondent was eligible and did not refuse the questionnaire, but no interview is obtainable because:

1. A precinct is not attempted or worked (2.31)

2. The respondent is physically and/or mentally unable to complete a questionnaire (2.32)

3. Language problems (2.33)

4. Literacy (2.34) and

5. Miscellaneous other reasons (2.35).

These cases are explained below.

In AAPOR's final disposition codes for other typical data collection methods, the case of *not-attempted-or-worked* is included in the category of unknown eligibility. In telephone, household interviews, and mail surveys, the code not-attempted-or-worked refers to a household unit, while in this exit poll study, we will refer to this code as a precinct unit. In an exit poll, there is never a case where a county has an unknown precinct. Hence, this case code is considered eligible. The code *precinct-not-attempted-or-worked* (2.31) includes addresses drawn in the sample, but for which no interviewer was available. Therefore, cases were simply not assigned or attempted before the end of the field period. Cases of precinct-not-attempted-or-worked (2.31) are further separated into

1. Unable to reach/unsafe area (2.311), and

2. Unable to locate an address (2.312).

Unable-to-reach cases (2.311) include remote areas inaccessible due to weather or other causes or areas in which interviewers are not assigned because of safety concerns (e.g., bad weather, high crime, rioting, or evacuations). *Location problems* (2.312) typically involve rural residences in which the description of the precinct unit is errant (e.g., wrong street name) or inadequate to allow an interviewer to find the precinct building (e.g., the house that had been painted red to the left of where the general store used to be).

Regarding all such cases in exit polls, there is a possibility that pollsters may decide to change the selected precinct to an alternative one. If an alternative precinct is selected, the code precinct-not-attempted-or-worked is no longer applicable. Respondents who are physically or mentally unable to complete a questionnaire include both permanent conditions (e.g., senility, blindness or literacy) and temporary conditions (e.g., drunkenness) that prevail whenever attempts are made to contact the voter. Unlike other data collection methods, in an exit poll a voter with a temporary condition cannot be interviewed again, unless they are recontacted later on that same Election Day. However, there is no practical way for polling organizations to implement such an action.

Language problems (2.33) include cases of a foreign language barrier and literacy. Foreign language barriers can be subdivided in to three cases. The first case is when the respondent does not read a language in which the questionnaire is printed (2.332). A second case is when the respondent never receives a questionnaire printed in a language he or she can read (2.333). Finally, a case exists when an interviewer with appropriate language skills cannot be assigned to the polling place at the time of contact of the voter with a language barrier (2.334).

In 1975, Congress adopted the language minority provisions of the Voting Rights Act (U.S. Department of Justice, 1975). The Voting Rights Act protects voters that speak minority languages by "requiring particular jurisdictions to print ballots and other election materials in the minority language as well as in English, and to have oral translation help available at the polls where the need exists" (U.S. Department of Justice, 2000).

Literacy problems (2.34) would apply to cases in which the selected voter could speak the language in which the questionnaire was printed, but could not read it well enough to comprehend the meaning of the questions. All registered voters are entitled to vote, regardless of the impairment. "Any voter who is blind, disabled, unable to read or write, unable to read or write the English language, or is physically unable to enter a polling place" is able to receive assistance from a person of their choice that meets specific criteria (State of Utah, 2001).

The *miscellaneous* designation (2.35) includes cases involving some combination of other reasons (2.30) or special circumstances (e.g., vows of silence, lost records, faked cases invalidated later on).

Unknown eligibility, no contact. In exit polls, *unknown eligibility* cases include situations in which it is unknown whether a specific individual observed at the precinct is an eligible voter (3.20). For instance, there could be a precinct at a library where some people are at the library to check out books and others are there to vote. In such precincts, the assumption cannot be made that all people exiting the building are eligible to be in the sample. Therefore, it is unknown whether a specific individual observed at that precinct is an eligible voter without some verification of eligibility.

Cases for which it is unknown whether a specific individual observed at the precinct is an eligible voter (3.20) usually crop up because of a *failure to complete a needed screener* (3.21). Depending on the sampling design of an exit poll, a screener may or may not be needed. A screener may be needed at multi-precinct polling places, where not every precinct is a part of the sample. In exit polls without an official screener, interviewers are often expected to make some kind of voter assessment. For instance, interviewers have to assess the ages of the people at the polling place. It may be difficult to distinguish if some people are over eighteen and eligible to vote. Other cases include people who are at a polling place, but not for the purpose to vote (e.g., people who work in the building). Another type of an unknown eligible respondent is a *missed voter* (3.22), which can be further separated into *did not pass interviewer* (3.221) and *interviewer too busy* (3.222). If it is certain that the missed voter is eligible, he or she should be coded as a *non-contact, missed voter* (2.20).

Finally, a *miscellaneous* other category (3.90) should be used for highly unusual cases in which the eligibility of the precinct or potential respondent is undetermined and which do not clearly fit into one of the above designations.

Not eligible. *Not eligible* cases for an exit poll with an interviewer contact and self-administered survey include:

1. Out-of-sample voters (4.10);

2. Not an eligible respondent (4.70); and

3. Situations in which quotas have been filled (4.80).

Not eligible cases differ from unknown eligible cases in that some contact has been made with the respondent to determine that they are not eligible, or it is completely obvious that they are not eligible (e.g., children under 10 years old).

Out-of-sample cases (4.10) would include voters [who] were at a polling place, but for a purpose other than to vote. This would differ from (3.20) in that some contact and verification has been made to determine that the respondent is not eligible.

Ineligible respondents (4.70) are rare in exit polls. They mostly consist of *unregistered voters* (4.71) or *voters that are registered, but did not vote* (4.72). This category

(4.70) differs from (4.10) in that the respondents are inside-of-sample but for a different reason are ineligible, while in the former case the respondents are not properly part of the sample. These individuals might be at a precinct if they are accompanying another person or if they are at the precinct for a reason other than to vote.

Finally, in surveys that employ a quota, there may be cases in which returned questionnaires are not treated as part of the final data set because the *quota for their subgroup of respondents had already been filled* (4.80). The nature of the quotas and how they are filled must be clearly defined. While quota sampling is not specifically mentioned in current exit polling literature, there may be reasons for some pollsters to consider using quotas.

In all cases concerning final disposition codes involving ineligibility, definite evidence of the status is needed. When in doubt, a case should be presumed to be eligible or possibly eligible rather than ineligible, unless there is clear evidence leading to the latter classification.

Outcome Rates. As with the final disposition codes, the outcome rates must be modified to apply to general exit polls. The outcome rates generally referred to are response, cooperation, refusal and contact rates. Many of the definitions and formulas of the outcome rates as documented in Standard Definitions apply straightforwardly to exit polls and, therefore, will not be reproduced here, but are available on AAPOR's Web site. The modifications required to apply AAPOR's outcome rates to general exit polls are discussed below.

For purposes of calculating outcome rates, *precinct-not-attempted-or-worked* (2.31) is included in the category of other (2.30). This differs from typical data collection methods, where not-attempted-or-worked is a sub-category of unknown-if-housing-unit, abbreviated UH (3.10). In addition, *precinct-not-attempted-or-worked* is identified as a precinct-level concept, instead of a respondent-level concept. A study has the option of estimating the number of respondents in each precinct-not-attempted-or-worked or removing this code from their outcome rates, with an explanation behind the removal. In some cases, an alternative precinct is chosen, which is a valid reason to exclude precinct-not-attempted-or-worked in the calculations of the outcome rates.

This study excludes UH (3.10) from the outcome rate formulas, because there is no equivalent in an exit poll. Note that the exclusion of UH applies to all outcome rates.

In Standard Definitions, there is no provision for a miss rate, which is typical for an exit poll. This study considers one approach of applying miss rates to an exit poll and supplies the outcome rates below. Again, much of this section is parallel to Standard Definitions, and the similarities are not put into quotes.

$$
\begin{aligned}
MIS &= \text{Miss rate} \\
I &= \text{Complete Interview (1.1)} \\
P &= \text{Partial interview (1.2)} \\
NC &= \text{Non-contact/Missed voter (2.20)} \\
O &= \text{Other (2.30)}
\end{aligned}
$$

UO = Unknown if eligible respondent exists (3.20)

e = Estimated proportion of cases of unknown eligibility that are eligible

$$MIS1 = \frac{NC}{(I + P) + (R + NC + O) + (UO)} \tag{2.1}$$

Miss Rate 1 (MIS1) is the number of missed voters divided by the interviews (complete and partial) plus the nonrespondents (refusals, non-contacts, and others) plus the cases of unknown eligibility.

$$MIS2 = \frac{NC}{(I + P) + (R + NC + O) + e(UO)} \tag{2.2}$$

Miss Rate 2 (MIS2) includes estimated eligible cases among the unknown cases.

$$MIS3 = \frac{NC}{(I + P) + (R + NC + O)} \tag{2.3}$$

Miss Rate 3 (MIS3) is the number of missed voters divided by the interviews (complete and partial) plus the nonrespondents (refusals, non-contacts, and others). This case excludes the cases of unknown eligibility and estimated eligible cases among the unknown cases.

Standard Definitions of AAPOR Applied to 2000 Utah Colleges' Exit Poll. The adapted final disposition codes and outcome rates for general exit polls mentioned above will be used to evaluate the data of the 2000 Utah Colleges' exit poll (UCEP 2000). Some final disposition codes and outcome rates have been reduced and others have been excluded due to the uniqueness of UCEP 2000.

Background to the 2000 Utah Colleges' Exit Poll. In many ways, UCEP 2000 is comparable to a typical exit poll. It was designed by using a stratified, multi-stage sampling design. Initially counties within the state were appropriately stratified and, then, were selected by probability proportional to size (PPS) sampling, where the measure of size was proportional to the estimated voter turnout for that county. In the second stage, 92 polling places were selected by PPS sampling within each county to be sampled. In the third stage, interviewers selected voters according to a previously determined systematic sampling interval, within each polling place, intended to give polling places a constant workload. There were, on average, three interviewers per polling place. At the designated time every hour, one interviewer per polling place would call in the results they had collected within that hour.

UCEP 2000 used three survey forms to call the elections. The survey forms were distinguished by color, namely: white, blue, and pink. Each color contained a few questions that were common to all colors, as well as questions specific to that color. This conceptually produced three independent surveys with the same survey design.

Final disposition codes in UCEP 2000. For several reasons, not all final disposition codes for general exit polls were used in UCEP 2000. On the questionnaire, there was only a certain amount of space allotted for the nonresponse final disposition

Table 2.7   Final disposition codes and frequencies of occurrence for UCEP 2000

Question status	Disp. code	White	Blue	Pink
1. Returned questionnaire	(1.0)	2367	1070	1088
Complete	(1.1)	2335	1052	1071
Partial	(1.2)	32	18	17
2. Eligible, contact or non-contact	(2.0)	1340	612	738
Refusal and break off	(2.10)	1340	612	738
Refusal	(2.11)	1239	594	680
Break-off	(2.12)	101	86	58
Total		3707	1682	1826

codes, so they were constrained to a minimum. In addition, the final disposition codes were also reduced to minimize the complexity for the interviewers.

For example, we did not use such general exit polling options as: literacy (2.34), language (2.33), physically/mentally disabled (2.32), or missed voter (2.21). We eliminated these codes because they covered such a small part of our sampling frame. For instance, a very low percentage of registered voters in Utah are unable to speak English. In order to be a registered voter in Utah you must be a United States citizen. One requirement to gain citizenship in the United States is that an immigrant must "be able to speak, read, write and understand ordinary English words and phrases" (Immigration and Naturalization Service, 2001). Also, any voter with a foreign language barrier is able to bring an interpreter when he or she votes (U.S. Department of Justice, 1975).

We did not experience a missing rate, because in UCEP 2000 there were on average three pollsters per polling place, compared to the one pollster per precinct used in most exit polls (Mitofsky, 1991:96; Mitofsky and Edelman, 1995:82). Therefore, we did not feel like we missed any contacts.

A list of the final disposition codes used in UCEP 2000 and their frequencies are provided in Table 2.7.

Returned questionnaires. In UCEP 2000, the returned questionnaires section of the final disposition codes is uniquely defined to fit our study. The definitions for complete and partial surveys are listed below.

1. *Complete* = 100% of crucial questions and 100% of semi-crucial questions

2. *Partial* = 100% of crucial questions and 50% of semi-crucial questions

The crucial questions consisted of the five election races that we called on election night. Semi-crucial questions were the two initiatives that we also reported the night of the election.

Eligible, contact. In UCEP 2000, a *break-off* (2.12) occurred when a respondent filled-out at least one question of the crucial or semi-crucial questions, but failed to respond to 100% of crucial questions and/or less than 50% of semi-crucial questions. A *refusal* (2.11) occurred when an interviewer approached a voter, and the voter declined to fill-out the questionnaire.

Unknown eligibility, non-contact. For several reasons, it was assumed that there were no voters with unknown eligibility in UCEP 2000. Interviewers were positioned at the building exits with the largest flow of voters. In addition, the voters were sampled at the polling place level, which took care of the problem of multi-precincts. As for other cases, such as a person at a polling place but for a reason other [than] to vote, the interviewers were asked to assess a voter's eligibility.

Not eligible. There were no data collected for ineligible respondents. The interviewers rarely, if at all, encountered a respondent who was not eligible. For example, such respondents as unregistered voters are not likely to appear at their polling place on Election Day. If an ineligible respondent was identified, the interviewers were instructed to exclude him or her from their sample and select an alternative respondent.

Outcome rates. Because we did not allow for many of the final disposition codes of the general exit poll in UCEP 2000, many of the elements in the outcome rates could be reduced. This produced simplified versions of the response, refusal, and contact rates. The outcome rates that applied to UCEP 2000 are reported below.

$$RR5_s = \frac{I}{(I+P)+R} \qquad\qquad RR6_s = \frac{(I+P)}{(I+P)+R} \qquad (2.4)$$

Simplified Response Rate 5 (RR5$_s$) assumes there are no non-contact, unknown eligible, or not eligible cases. *Simplified Response Rate 6 (RR6$_s$)* does the same as Response Rate 5, but includes partials as interviews.

$$COOP3 = \frac{I}{(I+P)+R} \qquad\qquad COOP4 = \frac{(I+P)}{(I+P)+R} \qquad (2.5)$$

Cooperation Rate 3 (COOP3) assumes there are no non-contact, unknown eligible, or not eligible cases. *Cooperation Rate 4 (COOP4)* does the same as Cooperation Rate 3, but includes partials as interviews. *COOP3* is equivalent to *RR5$_s$*, and *COOP4* is equivalent to *RR6$_s$*.

$$REF_s = \frac{R}{(I+P)+R} \qquad\qquad CON3_s = \frac{(I+P)+R}{(I+P)+R} \qquad (2.6)$$

Simplified Refusal Rate 3 (REF3$_s$) assumes there are no non-contact, unknown eligible, or not eligible cases. *REF3$_s$* is equivalent to $1 - RR6_s$. *Simplified Contact Rate 3 (CON3$_s$)* assumes there are no non-contact, unknown eligible, or not eligible cases.

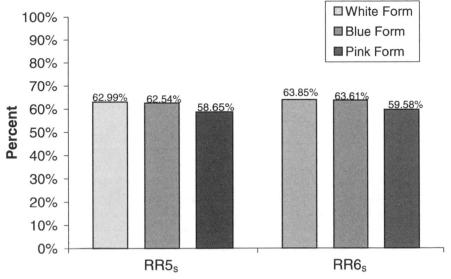

Figure 2.1   Response rates of UCEP 2000.

Results. This study does not report all of the types of outcome rates that could have been used in UCEP 2000. Note that $COOP3 = RR5_s$, $COOP4 = RR6_s$; $REF3_s = 1 - RR6_s$; and $CON3_s = 1$. Therefore, $COOP3$, $COOP4$, $REF3_s$, and $CON3_s$ were excluded in the results, because they did not add any new information. Outcome rates collected for UCEP 2000 reveal that $RR5s$ and $RR6s$ follow a similar pattern within colors (See Figure 2.1). The white form tends to have the highest response rate, closely followed by the blue form, and the pink form is quite a bit below the other two.

Summary. AAPOR's Standard Definitions provide useful final disposition codes and outcome rates for the current data collection methods employed and unique data collection methods, such as used in an exit poll. In this paper, we modified Standard Definitions to apply to general exit polls and directly applied these standards to the 2000 Utah Colleges' Exit Poll (UCEP 2000). Results from this analysis reveal an average Simplified Response Rate 5 ($RR5_S$) of 61.4%, and an average Simplified Response Rate 6 ($RR6_S$) of 62.4%, where the response rates were averaged across survey forms denoted by color. Compared to national exit polls' response rates of 57-60%, UCEP 2000's response rates were slightly higher. However, in national exit polls response rates exclude missed voters, which were not excluded in the response rates of UCEP 2000. This is due to the fact that we had, on average, three interviewers per polling place and, for this reason, we assumed that all selected voters were contacted.

References

The American Association for Public Opinion Research (1986). Certificate of Incorporation and By-Laws, Princeton, NJ: Author.

The American Association for Public Opinion Research (2000). Standard Definitions: Final Dispositions of Case Codes and Outcome Rates for Surveys, Ann Arbor, StateMI: Author, available on the Internet at: *http://www.aapor.org/ethics/stddef.html*, accessed 8-20-01.

The American Association for Public Opinion Research (2001). Best Practices for Survey and Public Opinion Research, available on the Internet at: *http://www.aapor.org/ethics/best.html*, accessed 8-20-01.

Council of American Survey Research Organizations (2001). Responsibilities in Reporting to Clients and the Public, available on the Internet at: *http://www.casro.org/codeofstandards.cfm*, accessed 8-20-01.

Forman, E.K. (1991). *Survey Sampling Principals*, New York, NY: Marcel Dekker.

Glynn, C.J;, Herbst, S.; O'Keefe, G.J.; and Sharpiro, R.Y. (1999). *Public Opinion*, Boulder, CO: Westview Press.

Immigration and Naturalization Service (2001). How to Become a United States Citizen, available on the Internet at: *http://www.aoa.gov/aoa/pages/citizen.html*, last modified 11-28-2000.

Lohr, S. (1999). *Sampling: Design and Analysis*, Pacific Grove, CA: Brooks/Cole.

Merkle, D.M. and Edelman, M.E. (2000). The Review of 1996 Voter News Service Exit Polls from a Total Survey Error Prospective, *Election Polls, the News Media, and Democracy*, Lavrakas, P.J. and Traugott, M.W. (eds.), New York, NY: Chatham House, 68-92.

Miller, P. J.; Merkle, D.M.; and Wang, P. (1991). Journalism and Footnotes: Reporting the 'Technical Details' of Polls, *Polling and Presidential Election Coverage*, Lavrakas, P.J. and Holley, J.K. (eds.), Newbury Park, CA: Sage Publications, 200-214.

Mitofsky, W. J. (1991). A Short History of Exit Polls, *Polling and Presidential Election Coverage*, Lavrakas P.J. and Holley, J.K. (eds.), Newbury Park, CA: Sage Publications,. 83-99.

Mitofsky, W. J. and Edelman, M. (1995). A Review of the 1992 VRS Exit Polls, *Presidential Polls and the News Media*, Lavrakas, P.J.; Traugott, M.W.; and Miller, P.V. (eds.), Boulder, CO: Westview Press, 81-99.

National Council for Public Polls (NCPP) (1979). Principles of Disclosure, Princeton, NJ: Author, available on the Internet at: *http://www.ncpp.org/disclosure.htm*, accessed 8-20-01.

Traugott, M. and Lavrakas, P.J. (2000). *The Voter's Guide to Election Polls*, 2nd ed., New York, NY: Chatham House.

State of Utah (2001). Utah Election Codes, Title 20A Chapter 03, 20A-3-108 Assisting Disabled, Illiterate, or Blind Voters, available on the Internet at: *http://www.le.state.ut.us/~code/TITLE20A/htm/20A03010.htm*, last modified 10-22-2001.

U.S. Department of Justice (1975). Voting Rights Act, Section 203 and Section 4(f)(4), available on the Internet at: *http://www.usdoj.gov/crtvoting/sec_203/activ_203.htm#provisions*, last accessed Sept. 3, 2002.

U.S. Department of Justice (2000). The Right to Vote: How Federal Law Protects You, Civil Rights Division, available on the Internet at: *http://www.usdoj.gov/crt/voting/misc/voterev.htm*, last accessed Sept. 3, 2002.

One data quality issue that Mitofsky talked about in his interviews had to do with variance estimation.

"In the election system these days, other than the switch to percentages, the basic models for estimating from sample precincts are quite similar to the ones we put together in the 1960s. There are still ratio estimates, still post-stratified, still computing variances the same way from Taylor expansion. I mean, we looked at all kinds of other things. We looked at whether the variances were better using a bootstrap approach and they were the same numbers. We were getting the same numbers."

"What we did discover was that the variances were being understated from the incomplete reporting. So now we developed a little fudge factor to boost the variances back up to where they belonged. When you have a sample of n and 10 percent of it reports, you are not getting the right level on the sampling error; when 20 percent reports, you're still understated. As you get more and more of the sample, the estimate of the sampling error approaches what it should be. Marty Frankel discovered this. He computed Z scores. He looked at the variances we were computing and the estimates and looked at the Z scores and, of course, they weren't what they should have been. So he took a whole bunch of historical data and figured out Z. We had been understating it for years."[30]

Of course, one way to compensate for errors in the data is to weight to population control totals. In the next article, Kaminska and Barnes (2005) examine the usefulness of party identification weighting.

2.3.2 Party Identification Weighting: Experiments to Improve Survey Quality
Olena Kaminska and Christopher Barnes

Introduction. Weighting the pre-election polls by Party Identification (PID) has become one of the most discussed and controversial issues in [the] survey research field. While most studies on the topic examine the stability of party ID and, on this basis, attempt to prove or disprove the validity of weighting by PID, it may be reasonable to weight by PID even if the distribution of PID is not completely stable. For example, a portion of change in PID may be an indicator of the change in response willingness, rather than actual change in PID. If PID is correlated with response willingness, weighting can improve the response rate error.

The purpose of this paper is to present three experiments, which evaluate if weighting by PID can improve the quality of pre-election surveys. The experiments were designed to answer four different questions: first, if weighting by PID shifts the vote

---

[30] [Bloom and Pearson (2008) report on some very current research that deals with this issue. See 7.4.2 in this volume.]

choice results closer to election outcome; second, what design for PID weights is best; third, how weighting by PID works within the likely voter model and within the all population data set; and, fourth, if weighting by PID improves the quality of the pre-election survey overall, not only vote choice variables. The study is based on 2000 pre-election polls sponsored by the following organizations: ABC News/*The Washington Post*, Gallup/CNN/*USA Today*, NBC/WSJ [*Wall Street Journal*], PSRA/Pew and *Newsweek*.

Weighting by PID

Discussion.   As noted, there has been extensive debate regarding PID weighting. Opponents of PID weighting have cited three main reasons against weighting (Blumenthal, 2004):

1. The question about PID is among the last ones in the survey

2. Weighting by PID means using four-year-old exit poll results and

3. PID is not stable.

In a typical pre-election poll the party identification question is usually asked in the demographic section at the end of the questionnaire. Because questions on political preference and issues precede the PID question, they might influence the party identification response. This problem may be exacerbated if different questions or different question order are used in the exit poll (or other data source for weighting).

Using four-year-old exit poll results is slightly different from using four-year-old Census results for demographic variables, given that PID is not a factual measure, but is an attitudinal measure. In contrast to other variables such as gender, age, race, education or income, usually used for weighting, party identification might change as a result of political or social events. Therefore using four-year-old PID distribution might not improve the accuracy of pre-election polls if the PID outcome is different in following elections.

A critical question is: are the shifts in PID that occur in pre-election polls a correct measure of the shifts in population or are they due to survey error? This question is usually answered by argument about stability of party identification. The main reasoning of researchers who decide to weight by PID is that PID is stable over years (Bowers, 2004). Opponents of weighting by PID argue that while party affiliation changes slowly over long-term periods, it can have daily or weekly shifts in the pre-election period (Traugott, 2001; Annenberg Public Policy Center, 2004).

While discussing the stability of PID is important, it is not the only possible explanation of PID shifts in polls. Before making conclusions, researchers should account for two other significant issues, which can cause shifts: response rate and likely voter models. The issue with response rate is that PID can be related to response willingness. While the shift in PID occurs in polls, it may be due to the shift in willingness of different parties' supporters to participate in surveys. Republicans or Democrats might be more willing to answer pre-election poll questions if their candidate wins the debate, as they may be more enthusiastic and have a stronger desire to express that opinion.

The problem with likely voter models is that the model selects "interested" respondents. As Rich Morin notes: ". . . more members of one party may become interested in the presidential race at any given moment (and, therefore, qualify for the likely voter pool), while members of other party become momentary bored, distracted or annoyed (and, thus, are judged less likely to vote)" (Morin, 2000). Chris Bowers (2004) has compared 12 polls, six of which are weighted by PID and six others are not. The polls took place during approximately the same time periods and were also compared to next in time polls within each company. Three major conclusions were derived from the experiment:

1. Polls weighted by PID show lower variance in vote choice over time than those not weighted by PID.

2. The shift over time within each company was higher for those companies that do not weight by PID.

3. If the not weighted by PID data sets were weighted by PID, they would look almost exactly like the six weighted polls (the vote choice results would have smaller variance).

This paper presents further research which evaluates if weighting by PID improves survey quality.

Experiment 1. The main purpose of the first experiment is to compare PID weighted vote choice results to those weighted by original weight. This experiment is conducted using a likely voter model. It also addresses the question about the best PID design, comparing two-level (Democrat vs. Republican) to three-level (Democrat vs. Republican vs. Independent) PID weights.

- **Data sets.** All three experiments were conducted, using the next five pre-election surveys: ABC News/*Washington Post* Poll # 15762: Post Presidential Debate Poll (October 12-15, 2000); Gallup/CNN/*USA Today* Poll # 38: Final Election Tracking Poll (November 5-6, 2000); NBC/*WSJ* Poll # 6010 (November 3-5, 2000); PSRA/Pew # 10MID: Mid-October 2000 Political Survey (October 18-22, 2000) and PSRA/*Newsweek* Poll # 2000-NW33: The Homestretch (October 31-November 2, 2000). The data sets were selected on the basis of two main criteria: the availability of data in The Roper Center for Public Opinion Research archive and the dates the polls were conducted (the closest available to presidential election in 2000, which was held on the 7th of November 2000).

- **Weighting.** Each data set used in this experiment has demographic-based weights and likely voter weights. The purpose of the first weight is to adjust the results according to U.S. Census demographics. The likely voter weight is used to exclude respondents who are not likely to vote. The combined weight will be called *original weight* in this paper. For each data set two different weights were created: the two-level PID weight and the three-level PID weight. They were calculated independently for each poll. PID distribution was taken from the 1996 exit poll, according to which 39% of voters identified themselves as Democrats, 35% as Republicans, and 26% as Independents. For the two-level weight only the Democrats vs. Republicans distribution was accounted for. The distribution from polls was compared to the exit poll outcome with 52.7% Democrats and 47.3% Republicans. Independents and others received a weight equal to one. For the three-level weight, the Democrat, Republican and Independent distribution was used. The "Other party" category and responses, which were coded as "Don't know" or "Refused" received weights equal to 1.

For the two-level PID weight, the highest value was 1.03 and the lowest -0.96, which means that the Democrat-Republican distribution in these polls was very close to the distribution in the exit poll and could not change the results much. The highest weight for three-level PID (step 5) was 1.19 and the lowest -0.70, which is higher than for two-level weight, but still is considerably low in comparison to typical demographic weights. The party ID-based weight was multiplied by the original weight to get the PID final weight. Thus, the PID final weight selects from the data set only likely voters, improves the distribution of demographic variables, bringing them closer to Census data, and improves the distribution of PID according to the 1996 exit poll distribution.

- **Error calculation.** There are different ways to evaluate the accuracy of pre-election polls. Detailed information about eight methods of evaluation was developed by SSRC [the Social Science Research Council] in 1949; their advantages and disadvantages can be found in "Review: Was 1996 a Worse Year for Polls than 1948?" written by Mitofsky (1998; 1999). Three of these will be used here: Method 3, Method 5, and Method 7 (it has become traditional to name the methods by numbers in the SSRC list).

The methods used in this analysis:

1. *Method 3* is the average error, calculated as the average (without regard to sign) of the percentage point deviation for each candidate between his/her estimate and the actual vote

2. *Method 5* is a margin error, which is calculated by subtracting the margin between the top two candidates in a poll from the margin between the same candidates in the election.

3. *Method 7* is the chi-square, which tests if the difference between poll and election results is statistically significant.

There is a conceptual difference between Method 3 and Method 5. Method 5 evaluates the margin between two leading candidates. In other words, it compares if the difference between two leading candidates was predicted correctly. Method 3 evaluates how close, on average, the percentage for each candidate in the poll was to the percentage in the elections. There is discussion about how many candidates should be included in Method 3. One line of reasoning, which is used here, is to include all the candidates reported by the pollster. Thus, Method 3 will take into account four candidates from the 2000 presidential elections: Al Gore, George W. Bush, Pat Buchanan, and Ralph Nader, while Method 5 will take into account the Al Gore - George W. Bush margin.

In the actual presidential election in 2000, 48.38% voted for Al Gore, 47.87% for George W. Bush, 0.42% for Pat Buchanan, and 2.74% for Ralph Nader. The poll results were compared to these percentages.

- **Results.** As was mentioned earlier, the party distribution after weighting by original weight is close to the 1996 exit poll for all polls in the experiment. While this is true, Democrats were slightly underrepresented in all surveys. This resulted in higher weights for Democrats as compared to the weights for Republicans. Thus, weighting by two-level PID has increased the percentage of Al Gore supporters and decreased the percentage of George W. Bush supporters for each poll (Table 2.8). In all the cases, the change was not higher than 1.05%. The outcome for three-level weight is very similar. According to Method 7, there is no statistically significant difference between vote choice results weighted by original weight and those weighted by two-level or three-level final PID weight for any poll.

  According to election outcome, Al Gore won the popular vote. As can be [inferred] from the Table 2.8, none of the polls predicted the right winner before weighting and two of them—the PSRA/Pew and PSRA/*Newsweek* Polls—could predict the right winner after weighting by PID. Given this, although weighting by PID has increased the margin error for the PSRA/Pew Poll (regardless of the sign), at the same time it improved the results by predicting the right winner.

  One of the main issues of interest is directional: does weighting by PID shift results closer to election outcome? Margin errors (Method 5) decreased for four out of five polls after weighting by PID. This means that all the polls, except for PSRA/Pew Poll, predict the difference between the two leading candidates better after weighting by PID. Figure 2.2 presents the results for margin errors before and after weighting by PID within each company.

  According to binomial probability theory, there is a 19% chance that four or more out of five data sets can improve the results after weighting. This is too low for rejecting the hypothesis that improving does not occur. Considering that only five data sets were used, this result suggests that further experiments should be conducted. Both two-level and three-level PID weights show very similar results and there is not enough evidence to claim that either is better.

  There was no—or almost no—change in the average error (Method 3) for all five polls after weighting by PID. The error for four polls decreased after weighting by three-level PID and—for the PSRA/Pew Poll—it slightly increased (Figure 2.3). The difference between errors for most companies is too small to make strong conclusions about the one-directional shift.

Experiment 2. The main idea of Experiment 2 is to examine if PID weight improves the outcome for all the population, including unregistered and "unlikely" to vote respondents. Thus, the likely voter model was not used for this experiment. Here,

Table 2.8 Vote choice outcome after weighting by two-level PID final weight

	ABC News/Wash Post		Gallup/CNN/USA Today		NBC/WSJ		PSRA/Pew		PSRA/Newsweek	
	original weight	PID final weight	original weight	PID final weight	original weight	PID final weight	original weight	PID final weight	original weight	PID final weight
Al Gore	44.91	44.94	45.07	45.34	44	44.36	44.84	45.55	43.28	44.33
George W. Bush	46.92	46.89	46.69	46.41	47.48	47.12	45.05	44.3	44.97	43.97
Pat Buchanan	0.57	0.57	0.65	0.66	1.56	1.55	0.62	0.62	0.12	0.12
Ralph Nader	3.6	3.6	3.74	3.74	3.54	3.55	4.09	4.1	4.94	4.96
LV counts	1472		2350		1026		663		808	

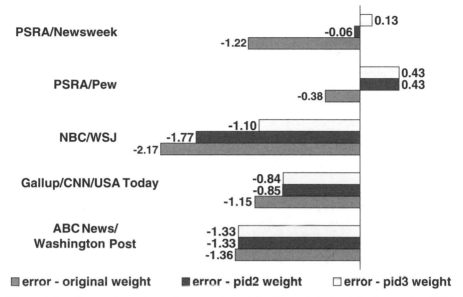

Figure 2.2 Comparison of margin errors for weighted by original weight, by two-level PID (pid2) weight and by three-level PID (pid3) weight poll results.

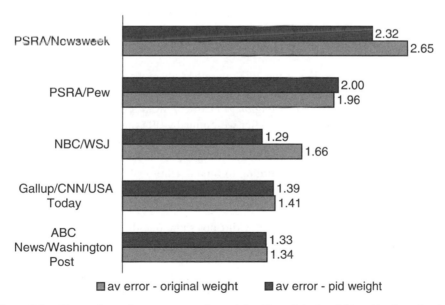

Figure 2.3 Comparison of average errors for weighted by original weight and by three-level PID weight poll results.

the vote choice outcome was calculated for unweighted, weighted only by PID, and weighted by demographic weight data sets. Poll outcomes were compared to the NES 2000 [National Election Studies] vote choice distribution. For the second and third experiments, the same data sets were used, except for the NBC/*WSJ* Poll. The reason for this was that the NBC/*WSJ* Poll did not include nonregistered respondents. Thus, comparing the results for this poll to NES would not be theoretically correct.

- **Weighting.** The vote choice frequencies were run for three following weighting designs:

  1. *Unweighted*—the data set was simply unweighted and all the cases were included.

  2. *Weighted by demographic weight*—each data set has a [demographic weight] developed by the company. Different companies might use different demographic factors for developing weights, which depended only on the company's choice.

  3. *Weighted by "PID only weight."* This weight calculation was different from the one for Experiment 1. The poll PID distribution was derived from the unweighted data set. This distribution was simply compared to the weighted NES PID distribution. Weighting in NES does not have PID factor, but accounts for demographics and selection probability (Burns, Kinder, et al., 2002). Only the first question about PID affiliation, without following questions about leaning, was used. Only the Democrat-Republican distribution was accounted [for], excluding Independents. To calculate the only PID weight, the NES outcome was divided into the outcome from the poll; both added to 100% beforehand.

- **Results.** According to the chi-square test, there was significant shift in vote choice results after weighting by PID in comparison to unweighted results for the ABC News/*Washington Post*, PSRA/Pew and Gallup/CNN/*USA Today* Polls. The shift after weighting by PID was also significant if compared to weighted by demographic weight results for PSRA/Pew and Gallup/CNN/*USA Today* Polls. As demonstrated in Figure 2.4, the margin error has decreased for all four data sets. According to binominal probability theory, there is only a 6.25% chance that weighting by PID will improve the outcome for four out of four polls.

  Error, calculated by Method 3 did not show one-directional results. For the ABC News/*Washington Post* and NBC/*WSJ* Polls the error was smaller for weighted only by PID weight, as compared to unweighted and weighted by demographic weight outcome. For the Gallup/CNN/*USA Today* Poll, the error after PID weight was in between the other two, and for the PSRA/Pew poll it was the highest. While the error differences within each poll were small and were not one-directional, no strong conclusions can be made about the influence of weighting by PID on average error.

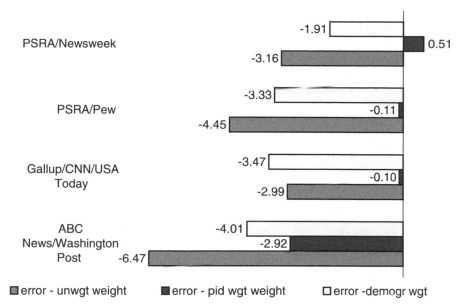

Figure 2.4   Comparison of margin error for unweighted, weighted only by PID and weighted by demographic weight polls results.

Experiment 3. The main idea of Experiment 3 is to examine if weighting by PID improves the outcome for other variables not related to vote choice. For this purpose demographic variables, such as gender, age and race, were chosen, as their distribution in the population is known from the Census.

For this experiment, the likely voter model was not used and all respondents were included. The comparison was done between the outcomes for unweighted and for weighted only by PID data. The error was calculated as a difference between poll outcome (unweighted and weighted by PID) and Census. Each variable was divided into two groups:

1. Gender—female and male

2. Age—"18 to 44 years old" and "45 years old and above"

3. Race—"White" and "Black / African-American"

- **Results.** There was no one-directional change in error for gender and age after weighting by PID. For gender, error decreased in two polls, increased for two others, and did not change in one. For age, the errors for unweighted and weighted by PID data were very close to each other. For race, in all four polls the error decreased after weighting by PID. The change in all cases was too small to legitimize any strong conclusion.

  The chi-square test showed no significant difference between unweighted and weighted by PID outcome. Thus, weighting by PID does not seem to change the results for demographic variables. It does not seem to improve constantly the demographic distribution, but importantly it does not seem to violate it [either].

Conclusion.   Weighting by [party identification] is one of the most controversial topics in survey methodology. In this paper we would like to shift the debate from the stability of PID, as reasoning for weighting (or not weighting) by PID, to experimental research designed to measure the performance of PID weighting. The results of these initial experiments indicate that there is potential for PID weighting to be an addition to likely voter models as aids to election predictions. The results are not definitive, by any means; rather , [they are] suggestive of the need for further exploration.

Four main conclusions can be made from our three experiments:

1. Weighting by PID seems to improve prediction for the margin between two leading candidates. This is true for likely voter data (both two-level and three-level PID weights), the same as for the whole population data sets.

2. Weighting by PID does not seem to have one-directional influence on average error for four presidential candidates. It does not seem to improve, but it also does not violate, the overall vote choice results.

3. There is no substantial difference between two-level PID and three-level PID weight designs.

4. The PID weight does not seem to change the demographic variable distribution.

Attention should be paid to the fact that this paper's experiments are based on the polls from one year and one presidential election—November 2000. The exit poll PID distribution in 1996 and 2000 years were very close: 39% Democrats, 35% Republican and 26% in 1996 (27% in 2000) Independent. Careful consideration and other research should be conducted before the results can be extrapolated on other time points.

References

Annenberg Public Policy Center of the University of Pennsylvania (2004). Party Identification Shifts toward Republicans, but Democrats Still Lead, Annenberg Data Show, press release, November 19, 2004, available on the Internet at: *http://www.annenbergpublicpolicycenter.org/naes/*.

Blumenthal, Mark (2004). Weighting by Party, September 23, 2004, available on the Internet at: *http://mysterypollster.typepad.com/main/2004/09/weighting_by_pa.html*.

Bowers, Chris (2004). Rapid Poll Movement Is a General Election Myth, September 18, 2004, available on the Internet at: *http://www.mydd.com/story/2004/9/18/454/00913*.

Burns, Nancy; Kinder, Donald R.; Rosenstone, Steven J.; Sapiro, Virginia; and the National Election Studies (2001). American National Election Study, 2000: Pre- and Post-election Survey (Computer file), 2nd ICPSR [Inter-university Consortium for Political and Social Research] version, Ann Arbor, MI: University of Michigan, Center for Political Studies (producer); Ann Arbor, MI: ICPSR (distributor), 2002.

Mitofsky, Warren J. (1998). Review: Was 1996 a Worse Year for Polls Than 1948? *Public Opinion Quarterly*, 62 (2), Summer, 230-249.

Mitofsky, Warren J. (1999). Reply to Panagakis, *Public Opinion Quarterly*, 63 (2), Summer, 282-284.

Morin, Richard (2000). Unconventional Wisdom; New Facts and Hot Stats from the Social Sciences, *The Washington Post*, Washington D.C., October 15, 2000, B 5.

Traugott, Michael W. (2001). Assessing Poll Performance in the 2000 Campaign, *Public Opinion Quarterly*, 65 (3), Fall, 389-419.

2.4 SUMMARY OBSERVATIONS

The 2000 U.S. presidential election—in which Al Gore won the popular vote, while George Bush captured the electoral majority and the presidency—not only generated interesting headlines during the post-election season, but also raised interest in election polls—and exit polls—serving as an impetus for much future research. The papers in this chapter are only a glimpse at the efforts that followed, but they provide a brief example of the breadth of study that resulted. This "early" research also led to additional improvements in methods and quality in subsequent elections, as the later chapters in this book demonstrate.

Editors' Additional References

Adams, Greg (2001). Voting Irregularities in Palm Beach, Florida, *Chance*, 14 (1), 22-24.

Basson, Danna (2008). The Impact of Accessible Political Knowledge on Voters' Candidate Evaluations, Issue Positions, and Issue Consistency, a paper presented at the Washington, DC, chapter meeting of the American Association for Public Opinion Research, February 20, 2008.

Bloom, Joel David and Pearson, Jennie Elizabeth (2008). Reliable Compared to What? A Probability-theory-based Test of the Reliability of Election Polls, unpublished paper.

Dillman, Don A. (2000). Statement on Palm Beach County Florida November 2000 Ballot, available on the Internet at: *http://www.sesrc.wsu.edu/dillman/palmbeach_statement.htm*.

Fan, David (2001). Predicting the Bush/Gore Election from the Press: The Cost to George W. Bush of his November Surprise of an Old Arrest, *American Statistical Association Proceedings*, AAPOR-ASA Section on Survey Research Methods.

Finklestein, Michael D. and Levin, Bruce (2003). Bush v Gore: Two Neglected Lessons from a Statistical Perspective, *Chance*, 16(3), 30-36.

Frankovic, Kathleen A. and McDermott, Monika L. (2001). Public Opinion in the 2000 Election: the Ambivalent Electorate, *The Election of 2000*, Pomper, Gerald M. et al. (eds.), New York, NY: Chatham House Publishers, Seven Bridges Press, 73-91.

Grier, David Alan (2002). Predicting the Unpredictable Election, *Chance*, 15 (2), 30-35.

Kaminska, Olena and Christopher Barnes (2005). Party Identification Weighting: Experiments to Improve Survey Quality, *American Statistical Association Proceedings*, AAPOR-ASA Section on Survey Research Methods, 3882-3886.

Moore, David W. (2006). *How to Steal an Election! The Inside Story of How George Bush's Brother and FOX Network Miscalled the 2000 Election and Changed the Course of History*, New York, NY: Nation Books, Avalon Publishing Group.

Pomper, Gerald M. (2001). The Presidential Election, *The Election of 2000*, Pomper, Gerald M. et al. (eds.), New York, NY: Chatham House Publishers, Seven Bridges Press, 125-154.

Slater, Melaney and Christensen, Howard (2002). Applying AAPOR's Final Disposition Codes and Outcome Rates to the 2000 Utah College's Exit Poll, *American Statistical Association Proceedings*, AAPOR-ASA Section on Survey Research Methods, 540-545.

Wolter, Kirk; Jergovic, Diana; Moore, Whitney; Murphy, Joe; and O'Muircheartaigh, Colm (2003). Reliability of the Uncertified Ballots in the 2000 Presidential Election in Florida, *The American Statistician*, 57 (1), 1-14.

CHAPTER 3

2004: FRAUDULENT ELECTION?

3.1 INTRODUCTION

In 2004, the U.S. electorate was strongly polarized. Much of the country represented "red" states—voting for George W. Bush—but key states were solidly "blue," favoring John Kerry for President. Because much of the population distribution was concentrated in the blue states, the 2004 presidential election eventually came down to a handful of battleground states. This time the deciding voters were in Ohio, not Florida, as in the 2000 presidential election. The candidates' electoral tallies were close enough that the winner in Ohio would win the election.

Below, Warren Mitofsky recalls some of the problems he encountered in conducting exit polls during the 2004 presidential elections. The rest of the chapter is divided into three main sections—the first deals with the controversial discrepancy between exit polling and official vote counts in Ohio and presents three papers that capture much of the discussion: an excerpt from the Edison-Mitofsky report evaluating the

Editors' Note: Text in gray shading is provided by the editors of this volume. Boxed-in text represents excerpts from interviews with Warren Mitofsky. The balance of the chapter is made up of papers prepared by contributing authors. [Text in square brackets within the selected readings was added by the editors.]

Elections and Exit Polling. Edited by Fritz Scheuren and Wendy Alvey
Copyright © 2008 John Wiley & Sons, Inc.

problems that resulted in inaccurate exit poll results; a description of some issues that pointed to possible fraud; and a paper that rebuts the fraud argument. The next section presents an excerpt of an article that addresses confidentiality issues. The final section explores some further research, looking at the electoral contest in Ohio to better understand the reasons for the outcome in 2004.

We begin with Mitofsky's memories about the 2004 election. Notice how he takes this opportunity to improve exit polling. He even tried out some ideas in preparation for the 2008 election. (See Chapter 6 (6.2.1) in this volume.)

"In 2004 I did an interview with *The Chicago Tribune* and I said the AP [Associated Press] did not have any statistical quality control in its county vote processing and, to protect the projection system from using bad data, we built quality control of the county data into the projection system. Of course, we could not go back to the data source, but we could keep questionable data out of the county-based projection calculations. The typical news executive's thoughts about quality control were to ask at the end of the night whether the final numbers were correct. I argued over the years that that is not good enough. We are presenting these numbers all night long and they could be full of bad votes. If there were errors, you won't know it..."

"When networks formed the first pool we were doing [election research] over the telephone [and we had to collapse precincts to ensure consistent boundaries]. I was concerned about not resolving the boundary changes properly. All I was left with was a quality control scheme I told you about earlier. [See Chapter 2.] Not having the size of the precinct was giving me a problem, as well."

"So I said, 'If I convert votes to percentages for the precincts, I'm not going to have this problem. If they do lousy field work over the telephone, I'm not going to get penalized, because they did bad field work. Now, it may not make a good ratio estimate if they get the right place, or if they get the right place and it's grown a lot and changed in character and so on. But at least I'm not going to get hurt on the size of the place.' So, I decided I'm not going to estimate votes any more; I'm going to estimate proportions. I'm going to convert all of my data to proportions and I'm going to leave the models in terms of proportions. Now, I'm going to give up something, because now I have a biased estimate, even without the ratio estimate, because I'm not reflecting the variation in turnout from polling place to polling place."

"I said, 'Well I can make up a little of that' and I made some estimate of current data with stratum size. Joe [Lenski] and Murray [Edelman] fought with me not to do this. They didn't want to use proportions, because of the introduction of the bias. I didn't agree. The problem was far more serious from the field work end than from the ramifications of having the wrong size. There were all kinds of reasons why that was going to hurt. And this would take care of that problem and I could make up some of it by making a current estimate of the size of the strata. Anyway, I switched everything to estimating proportions when Joe and I took this over in 2004."

"We've been using percentages now since the 2004 primaries. It's working out just fine. The estimates in the primaries in 2004 were probably the best primary estimates I've ever made. The 2004 general election was not that good. It wasn't the approach, because when you put the vote counts in, the estimates were very good; [the problems then were related to] the exit polling."

"I mean the errors in those exits polls were awful. The networks [tend to be] really resistant. We told them changes we wanted to make [to polling procedures] and they didn't do anything. They're talking about the legal problems. In the states that were pushing [the interviewers] back, you have got to get us closer to the door. We [are allowed to] work at election hearing distances to do an exit poll. So, now they're talking about approaching the lawyers. I said, 'That's what you did in 2004 and none of them listened to you. It's a civil rights suit, you get your money back. We've won eight suits so far, we've lost none. What are you hesitating about?' I said, 'What I'd do is pick the easiest states, you know the ones the lawyers say are most winnable, and go after those and use those to pressure the others. When they get the bill from the lawyer, they'll soften.'"

"We told them to take the logo off of the exit poll questionnaire. You know the names of five networks and the AP is not helping us. The conservatives see CNN, the liberals see FOX, and the next thing you know, we have a problem. They say, 'Unless you can prove the logos are hurting us, we're not going to take them off.' I said, 'How about you say it the other way around—unless you can prove its helping you, we take them off?'"

3.2 THE GREAT EL ECTION DEBATE

The closeness of the 2004 vote led to some controversy about the outcome of the election. This debate took place not only in the media, but it was also the subject of papers and discussions in the statistical community. In 2004, researchers examined battleground states—in particular, Ohio, Pennsylvania, and Florida—where exit polling suggested that Kerry would win, but in which official vote tallies yielded electoral votes for Bush. Some suggested that since these discrepancies were not due to chance or random error, they may be due to voting count fraud or problems with the exit polling methodology and called for further investigation to explain the differences.[31] In January 2005 Edison Media Research and Mitofsky International (Edison/Mitofsky, 2005) released an evaluation report that addressed the problems they had encountered during the 2004 election. The Executive Summary of that report is provided here.

---

[31] See, for example, Freeman, Steven F. (2004).

3.2.1   Excerpt from: Evaluation of Edison/Mitofsky Election System 2004[32]

Edison Media Research and Mitofsky International

Executive Summary.   On November 2, 2004, the election system created by Edison Media Research and Mitofsky International for the National Election Pool (NEP) produced election estimates and exit poll data for analysis in 120 races in all 50 states and the District of Columbia. In addition, between January and March 2004, Edison and Mitofsky conducted exit polls for 23 Democratic Primaries and Caucuses. For every election, the system delivered on its main goals: there were no incorrect NEP winner projections, and the exit poll data produced on Election Day were used on-air and in print by the six members of the NEP (AP, ABC, CBS, CNN, FOX, and NBC) as well as several dozen media organizations that subscribed to those data. However, the estimates produced by the exit poll data on November 2nd were not as accurate as we have produced with previous exit polls.

Our investigation of the differences between the exit poll estimates and the actual vote count point to one primary reason: in a number of precincts a higher than average Within Precinct Error most likely due to Kerry voters participating in the exit polls at a higher rate than Bush voters. There have been partisan overstatements in previous elections, more often overstating the Democratic vote, but occasionally overstating the Republicans. While the size of the average exit poll error has varied, it was higher in 2004 than in previous years for which we have data. This report measures the errors in the exit poll estimates and attempts to identify the factors that contributed to these errors.

The body of this report contains the details of our analysis of the performance of the exit polls and the election system. In addition to the information included in this report, exit poll data from this election is being archived at The Roper Center at the University of Connecticut and at the Institute for Social Research [ISR] at the University of Michigan and is available there for review and further analysis. This is the procedure that we have followed for all previous exit polls, which are also available at The Roper Center and ISR. The description of the methodology of the exit polls and the completion rates nationally and by state have already been posted on our Web site along with all questionnaires used on Election Day.

Here is a brief summary of our findings:

Exit poll estimates.   The exit poll estimates in the 2004 general election overstated John Kerry's share of the vote nationally and in many states. There were 26 states in which the estimates produced by the exit poll data overstated the vote for John Kerry by more than one standard error, and there were four states in which the exit poll estimates overstated the vote for George W. Bush by more than one standard error. The inaccuracies in the exit poll estimates were not due to the sample selection of the polling locations at which the exit polls were conducted. We have not discovered

---

any systematic problem in how the exit poll data were collected and processed. Exit polls do not support the allegations of fraud due to rigging of voting equipment. Our analysis of the difference between the vote count and the exit poll at each polling location in our sample has found no systematic differences for precincts using touch screen and optical scan voting equipment. We say this because these differences are similar to the differences for punch card voting equipment, and less than the difference for mechanical voting equipment.

Our detailed analysis by polling location and by interviewer has identified several factors that may have contributed to the size of the Within Precinct Error that led to the inaccuracies in the exit poll estimates. Some of these factors are within our control, while others are not. It is difficult to pinpoint precisely the reasons that, in general, Kerry voters were more likely to participate in the exit polls than Bush voters. There were certainly motivational factors that are impossible to quantify, but which led to Kerry voters being less likely than Bush voters to refuse to take the survey. In addition, there are interactions between respondents and interviewers that can contribute to differential nonresponse rates. We can identify some factors that appear to have contributed, even in a small way, to the discrepancy. These include:

- Distance restrictions imposed upon our interviewers by election officials at the state and local level
- Weather conditions which lowered completion rates at certain polling locations
- Multiple precincts voting at the same location as the precinct in our sample
- Polling locations with a large number of total voters, where a smaller portion of voters was selected to be asked to fill out questionnaires
- Interviewer characteristics, such as age, which were more often related to precinct error this year than in past elections.

We plan further analysis on the following factors:

- Interviewer training and Election Day procedures
- Interviewing rate calculations
- Interviewer characteristics
- Precinct characteristics
- Questionnaire length and design.

We also suggest the following changes for future exit polls:

- Working to improve cooperation with state and local election officials
- Improvements in interviewing training procedures
- Changes in our procedures for hiring, recruiting, and monitoring interviewers.

Even with these improvements, differences in response rates between Democratic and Republican voters may still occur in future elections. However, we believe that these steps will help to minimize the discrepancies.

It is also important to note that the exit poll estimates did not lead to a single incorrect NEP winner projection on election night. The Election Night System does not rely solely on exit polls in its computations and estimates. After voting is completed, reported vote totals are entered into the system. Edison/Mitofsky and the NEP members do not project the outcome of close races until a significant number of actual votes are counted.

As in past elections, the final exit poll data used for analysis in 2004 were adjusted to match the actual vote returns by geographic region within each state. Thus, the discrepancy due to differing response rates was minimized and did not significantly affect the analysis of the vote. The exit polls reliably describe the composition of the electorate and how certain demographic subgroups voted.

Survey weighting. Early in the afternoon on November 2nd, preliminary weightings for the national exit poll overstated the proportion of women in the electorate. This problem was caused by a programming error involving the gender composition that was being used for the absentee/early voter portion of the national exit poll. This error was discovered after the first two sets of weighting; subsequent weightings were corrected. This adjustment was made before NEP members and subscribers used exit poll results on-air or in print.

After Election Day, we adjusted the exit poll analysis data in three states (Tennessee, Texas, and Washington) to more accurately reflect the proportion of absentee ballots that came from each geographic region in those states. We have implemented a change to the survey weighting program to take into account the geographic distribution of the absentee votes in the future.

Technical performance. While the computer system performed well for most of the night, a database server problem led to NEP member and subscriber screens "freezing up" shortly after 10:35 p.m. [Eastern Standard Time (EST)] election night. This problem caused disruptions in the system until shortly after midnight, when we switched to a backup server for the rest of the night. There was a second occurrence of this problem at approximately 2:45 a.m. EST. Details of the data server problems and other technical issues are outlined in the technical performance report being distributed to the NEP Technical Committee. We have isolated the reasons behind the database server problem and list several recommended technical changes in this report to help avoid a repeat of this problem in future elections.

As soon as discrepancies between the exit poll data and official vote counts were reported, a host of observers tried to explain the differences. Debates followed regarding the likelihood that the 2004 presidential election was "stolen" from Kerry.

Among the prominent spokespeople who suspected corruption were Steve Free-man[33] and Kathy Dopp.[34] Unfortunately, we were unable to include a paper that represented this side of the debates, so we will attempt, below, to provide a flavor of the concerns raised. To fully understand the arguments and review the evidence, readers are encouraged to look at the references cited.

3.2.2   The Argument for Fraud
Wendy Alvey

Noting that the shift in support favored Bush in all but one of the battleground states, Steve Freeman set about to understand why. Using National Election Pool (NEP) data that were leaked to the public prematurely, he examined the distribution of votes nationwide—and particularly in Florida, Ohio, and Pennsylvania—and concluded from those exit poll data that "it is impossible that the discrepancies between pre-dicted and actual vote counts in the three critical battleground states of the 2004 election could have been due to chance or random error." (Freeman, 2004:7) He then proceeded to systematically examine explanations for the differences in poll results and the official vote counts, eventually ruling out arguments that pointed to skewed survey data, under-representative samples, and disproportionate nonresponse rates. He concluded that the "exit poll disparity doesn't tell the whole story" (Freeman and Mitteldorf, 2005:4). He felt that the potential for undetected technological ma-nipulation;[35] the concentration of Bush votes in Republican strongholds, where the likelihood of detection is smaller; and a high incidence of reported irregularities at the polls—from voting machine shortages to purged voter rolls and lost absentee ballots—suggested that a more concentrated effort to direct the outcome could be in play.

Freeman's main focus regarding the polling data centered on what Edison-Mitofsky (2005) called Within-Precinct Error (WPE), but which he called Precinct Level Dis-parity (PLD). "PLD is the difference between the official vote tally in just those precincts where exit polls were conducted and the exit poll numbers in those very same precincts. In other words, it is the difference between whom people said they voted for as they walked out of the voting booth and the way those votes were of-ficially recorded" (Freeman, 2005:3). Freeman points out that while the margin of error for the exit polls was quite small—less than 1 percent—and vote counts showed that Bush won by a slim margin, the PLD indicated a clear win for Kerry. Edison-Mitofsky concluded that this was due to the fact that Kerry voters were more likely to respond to the polls than Bush voters and attributed that to interviewer error. Freeman acknowledges that interviewer characteristics may have had an impact on poll results

---

[33] See, for example, Freeman (2004) or Freeman and Bleifuss (2006).

[34] Baiman, R.; Dopp, K.; et al. (2005).

[35] Without an adequate paper trail to verify that votes are being recorded for the correct candidate, there is no way to prove or disprove that the data had been manipulated.

(though not always the impact the pollsters suggested), but feels that they are not the sole reason for such high PLDs. In fact, the high PLDs might be evidence of fraud occurring.

Contending that the data do not support the Edison-Mitofsky argument—and that weighting corrections used by the pollsters further distorted the situation—Freeman considered another possibility: that the Kerry votes were not fully counted, while Bush votes were. Freeman's examination of response rates, participation rates, and PLDs contradicts the Edison-Mitofsky findings.[36] According to his analysis, there was "improbability after improbability in the official story; neglected correlation after neglected correlation pointing toward fraud." (Freeman, 2005: 12). Added to that was Freeman's concern about the fact that Edison-Mitofsky refused to release raw data for the research community to examine. He intimated that this failure to permit scholarly peer review may be indicative of problems, as well.

Freeman's arguments are further supported by the U.S. Counts' National Election Data Archive Project. In their report, Baiman, Dopp and colleagues considered the problems that Edison-Mitofsky and Freeman had examined and wondered whether these issues were "collectively of sufficient magnitude to reverse the outcome of the election, or were they isolated incidents, procedurally disturbing but of little overall consequence?" (Baiman, Dopp et al., 2005:4). They addressed three main areas of inquiry—sampling error; polling error; and voter system error—and concluded that "If the discrepancies between exit poll and election results cannot be explained by random sampling error; the 'Reluctant Bush Responder' hypothesis is inconsistent with the data; and other exit polling errors are insufficient to explain the large exit polling discrepancies, then the only remaining explanation—that the official vote count was corrupted—must be seriously considered." (Baiman, Dopp, et al., 2005:18). In the end, they determined that, "The absence of any statistically-plausible explanation for the discrepancy between Edison/Mitofsky's exit poll data and the official presidential vote tally is an unanswered question of vital national importance that demands a thorough and unflinching investigation" (Baiman, Dopp et al., 2005:22).

Note: The author expresses her apologies to Steve Freeman and Baiman, Dopp et al. for any misrepresentation of their positions.

References

Baiman, Ron; Dopp, Kathy, et al. (2005). Analysis of the 2004 Presidential Election Exit Poll Discrepancies, U.S. Count's National Election Data Archive Project, available on the Internet at: *http://electionarchive. org/ucvAnalysis/US/Exit_Polls_2004_Mitofsky-Edison.pdf* .

Edison Media Research and Mitofsky International (E/M) (2005). *Evaluation of Edison/Mitofsky Election System 2004,* January 19, 2005, available on the Internet at: *http://www.vote.caltech.edu/media/ documents/EvaluationJan192005.pdf*

Freeman, Steven F. (2004). The Unexplained Exit Poll Discrepancy, available on the Internet at: *http: //center.grad.upenn.edu/get.cgi?item=exitpollp*

Freeman, Steven F. (2005). Polling Bias or Corrupted Count? Accepted Improbabilities and Neglected Correlations in 2004 U.S. Presidential Exit Poll Data, paper presented to the American Statistical Associa-

[36] See, for example, Freeman and Bleifuss (2006).

tion Philadelphia Chapter, October 14, 2005; available on the Internet at: *http://www.appliedresearch. us/sf/Documents/ASAP-Improbabilities051014.pdf*

Freeman, Steven F. and Bleifuss, Joel (2006). *Was the 2004 Presidential Election Stolen? Exit Polls, Election Fraud, and the Official Count*, New York, NY: Seven Stories Press.

Freeman, Steven F. and Mitteldorf, Josh (2005). A Corrupted Election: Despite What you May Have Heard, the Exit Polls Were Right, available on the Internet at: *http://www.appliedresearch.us/sf/Documents/ ASAP-Improbabilities051014.pdf*, February 15, 2005.

Mark Lindeman, on the other hand, systematically and thoroughly reviews Freeman's argument that exit poll results provide evidence of massive fraud in the battleground states—particularly in Ohio (Lindeman, 2006). Instead, he proposes that the discrepancies between exit polling and official vote counts are due primarily to polling error. Later in this chapter, Kyle et al. (2005) address the confidentiality issue that Freeman and his colleagues raised—see 3.3.1 in this volume.

3.2.3   Excerpt from: Beyond Exit Poll Fundamentalism: Surveying the 2004 Election Debate
Mark Lindeman

Many sources seem simply to assume that exit polls should be, or have been, accurate within sampling error. Some observers cite Warren Mitofsky's reputation for making accurate *calls*—apparently not understanding that Mitofsky and other exit pollsters incorporate official vote counts before making predictions in competitive races.

Relatively little systematic information has been available about exit poll accuracy (Lindeman and Brady (2006) surveys much of the available literature). The 2004 controversy, itself, has shaken loose some facts that do not support confidence in U.S. exit polls' extreme accuracy. The Edison/Mitofsky [E/M] evaluation report[37] gives average *Within Precinct Error* (WPE) values for the largest exit polls in the last five elections (including the CBS/*New York Times* exit poll in 1988). (Within Precinct Error or WPE is, simply put, the difference between percentage margins in the exit poll sample and the vote count.) In *all five cases*, the exit polls, on average, overstated the Democratic share of the official vote, a result coded as a negative average WPE. The average WPE was −6.5 points in 2004, −5.0 points in 1992, and close to −2 points in 1988, 1996, and 2000 (E/M, 2005: 34).[38]

... My questions *about the exit polls* are: Do the exit polls support the hypothesis of massive, widespread fraud? And secondarily, Do the exit polls support or suggest other hypotheses of fraud or miscount of whatever scope and scale? I do not argue that these are the most important questions about the 2004 election or about election integrity in general.

---

[37]For the full report, see Edison Media Research and Mitofsky International (E/M) (2005).
[38][See, also, Lindeman, Liddle, and Brady (2005) in Chapter 7 of this volume (7.4.1).]

Machine Type and Red Shift.   The day after the election, a colleague came to my office and suggested a simple, obvious empirical inquiry: whether exit poll discrepancies were greater in precincts using Diebold machines than in other precincts. This line of reasoning can readily be extended to other tests: Are exit poll discrepancies greater on Direct Recording Electronic (DRE) machines, in general, (not necessarily limited to Diebold) than on other equipment? Are they greater on optical-scan equipment, which some people regard as vulnerable to hacking? The answers appear to be *No*, and *No*, but these answers have had remarkably little impact on the "debate."

Charles Stewart (2004) reported that, at the state level, there existed no significant relationship between types of equipment used and state-level exit poll error. In fact, red shift[39] was (insignificantly) *negatively* correlated with use of DREs and optical scanners, and *positively* correlated with use of other technologies, including paper.... There is no evidence that exit poll red shift was significantly smaller in paper-ballot precincts than in other precincts, nor that red shift was larger in DRE and optical-scan precincts...

Nonresponse Bias.   ...[We] can frame hypotheses regarding precinct-level variables that are likely to covary with nonresponse and/or selection bias (cf. Edelman and Merkle, 1995). (By "selection bias" I mean that some interviewers may have had different propensities to *approach* Bush and Kerry voters, whereas nonresponse bias refers to the voters' propensities to complete the survey if approached.) Indeed, as Freeman admits, the E/M evaluation report offers support for several such hypotheses.[40]

First, if voters vary in propensity to be interviewed, factors that influence how easily voters can simply avoid interviewers are likely to affect the magnitude of nonresponse bias. Hence, we can conjecture that bias would be larger in large precincts, and in precincts where interviewers were far from the polling place (see... Edelman and Merkle, 1995: Table 7, reporting *absolute* error). These hypotheses receive strong support. Precincts where interviewers were instructed to approach every, or every other, voter (indicating relatively small polling places) had a mean WPE [Within Precinct Error] of approximately −3.6; precincts where interviewers approached every 10th voter (large polling places) had mean WPE of −10.5 (E/M, 2005: 36). Precincts where the interviewer stood inside the polling place had mean WPE of −5.3; precincts where the interviewer stood more than 100 feet away had mean WPE of −12.3 (E/M 2005, 37).

Moreover, if voters are (for instance) reluctant to engage with interviewers they perceive as likely to disagree with them, then nonresponse bias may vary according to interviewer characteristics. Contrariwise, interviewer characteristics are unlikely to affect the magnitude of fraud. The E/M report documents, *inter alia*, that mean WPE was −7.7 points among interviewers under 35, but approximately −5.5 points

[39] [Red shift refers to the fact that states with a majority of Democrats reported voting Republican instead. Blue shift refers to the reverse—states that have more registered Republicans reported more votes for the Democratic candidate.]

[40] [See Freeman and Mitteldorff (2005).]

among interviewers 35 and over (E/M, 2005: 43). Moreover, mean WPE was −3.9 points among interviewers with a high school education or less, but −7.3 points among interviewers with one to three years of college (45). Combining these results, it seems eminently plausible that some Bush voters were predisposed to steer around college-age interviewers (and/or vice versa).... Nonresponse bias, by definition, is not directly measurable, and there is no *a priori* reason to expect measured variables to account statistically for *all* the red shift. Nevertheless, the evidence in the E/M report does support the conclusion that at least part of the red shift is explained by nonresponse (and/or selection) bias.

Does Bush Do Surprisingly Well in Red-Shift States?   If variations in red shift reflect variations in the magnitude of vote miscount favoring Bush, then we would expect Bush to do better than expected in states and precincts with large red shift.... If red shift and "surprise" [variation from pre-election predictions] both reflect fraud, then we would expect a positive correlation. In fact, the observed correlation is *negative* and statistically significant ($r = -0.293$, $p = 0.04$). That is, red shift tended to be largest ... in states where Kerry did *better*—or at least not much worse—than expected.

So, Bush does not do surprisingly well in red shift states, at least not by comparison with other states. If anything, the opposite is true.

Does Bush Improve on Past Performance where Red Shift Is High?   Overall, Bush's vote margin was about 3 points larger in 2004 than in 2000 (+2.5 versus −0.5). If red shift evinces vote miscount, we would expect—all else being equal— that Bush's margin would improve more in states with large red shift than in other states. In practice, there is no discernible relationship between red shift and election-to-election swing (change in margin) at the state level. This result, or non-result, is difficult to square with fraud, unless fraud was concentrated in states where Bush otherwise would have done much worse in 2004 than in 2000.

... Happily, we now know the result of extending shift-swing analysis to the entire country. Warren Mitofsky first presented these results in a joint appearance with Steve Freeman in October 2005; Lindeman (2005) reproduces (with permission) two crucial slides. These slides, again, evince *almost exactly zero correlation between red shift and swing* at the precinct level.

How can this be? I think that in order to reconcile this result with massive, widespread fraud, one almost has to adopt one or both of two heroic auxiliary hypotheses. The first would be that fraud was targeted to counterbalance "real" variations in swing, which seems to entail complete control over vote counts in almost all precincts (at least exit poll precincts) nationwide. On this hypothesis, one would almost imagine that Freeman's various "neglected correlations" (discussed below) were planted as red herrings. And so we can offer a second auxiliary hypothesis: perhaps the entire exit poll was rigged, soup to nuts, to frustrate any attempt to reconcile their results with a coherent account of massive, widespread fraud.

I believe that all these findings constitute important evidence *against* massive, widespread fraud. Certainly they argue against the view that the exit polls offer strong evidence of such fraud.

Assessing Arguments for Vote Miscount. But what of all the exit poll evidence that purportedly *does* point to vote miscount? ... Even if the exit polls do not support the inference of massive, widespread fraud, they may support narrower inferences or hypotheses about vote miscount or other important political phenomena.

Do response rates inform the nonresponse bias hypothesis? U.S. Count Votes and others[41] have argued that the reported response (completion) rates inform the Edison/Mitofsky hypothesis that Bush voters participated at a lower rate.... The observed response rates are slightly higher in the strong-Bush precincts (those that voted 80%+ for Bush) than in the strong-Kerry precincts: 56% versus 53% (E/M, 2005: 37). However, this fact seems inconsequential for several reasons. E/M reports that the difference in response rates is statistically insignificant (which seems evident). Indeed, the precinct-level response rates are statistically consistent with a slight negative slope—which is only to say that the "true" relationship, if any, *might* be slightly positive or slightly negative.

... Freeman describes the expectation of a higher response rate in strong-Kerry precincts as "practically a mathematical necessity" (Freeman, 2005: 6), but it is not.... It is perfectly plausible that both Kerry voters and Bush voters might participate at higher rates in Republican than in Democratic precincts (suppose, for instance, that the Democratic precincts are disproportionately urban, crowded, and chaotic), while Bush voters might participate at higher rates than Kerry voters within precincts across the partisan spectrum. All in all, the response rate argument—modestly suggestive when it was first offered in the spring of 2005—has little force.

What about the excess of 2000 Bush voters? One popular argument that the exit polls evince fraud hinges on the reported distribution of 2000 votes. In the final results weighted to the official vote totals, 43% of respondents reported that they voted for George W. Bush in 2000, and 37% that they voted for Al Gore.... Bush received about 50.5 million votes in 2000, and perhaps 2 million or so of these voters may have died before the 2004 election. Yet 43% of the 2004 electorate would yield 51.7 million people who had voted for Bush in 2000.... [I]sn't it apparent that the exit pollsters essentially had to invent several million Bush 2000 voters in order to account for Bush's re-election?

Actually not.... In the 2000 VNS [Voter News Service] exit poll, 45.6% of respondents reported having voted for Bill Clinton in 1996, only 31.3% for Bob Dole—although Clinton won in the official returns by just 8.5 points. Moreover, 45.6% of the 2000 electorate would equal about 48.1 million Clinton voters, although Clinton received only 47.4 million votes in 1996. Did exit pollsters invent millions of Clinton

---

[41][See, for example, Dopp (2005).]

1996 voters in 2000, in effect concealing evidence of massive vote miscount favoring Al Gore?

But we need not stop in 2000. I examined ten exit polls in eight elections dating back to 1976. In *all ten polls*, the previous winner's reported vote share (among people who reported having voted four years earlier) was larger than his actual vote share had been—about 11 points larger on average. This evidence, and much other evidence, indicates that some respondents falsely report having voted for the incumbent. (An extensive discussion appears in Lindeman, 2006.)

The swing states. Freeman seizes on E/M's observation that precinct-level error (Within Precinct Error) is slightly larger in precincts in the 11 swing states (-7.9 points on average) than in other precincts (-6.1 points).... Freeman [argues]: "The first priority was, of course, to win states. Thus, it would make sense that votes would be most vigorously coveted in the swing states..." (Freeman, 2005: 7).

... One might expect that votes would be stolen *only* in swing states, but perhaps that would be too obvious. Well, then, if... Bush supporters had the wherewithal to steal votes in states all over the country, why would they not try to steal roughly equal proportions of votes in swing and non-swing states, so as to throw the hounds off their trail? And can one really say that it "would make sense" that the malefactors managed to steal just enough votes in Pennsylvania to lose by 2 points instead of double digits—and to match the pre-election polls...? Any statistical "swing state effect" seems to be far outweighed by a correlation Freeman does not mention: exit poll discrepancies favoring Kerry tended to be larger in states where he did better (not worse!).

The GOP strongholds. Freeman reproduces a table of mean WPE by range of "precinct partisanship." He summarizes, "...[T]he stronger Bush's support, the greater the [average exit poll] disparity" (Freeman, 2005: 7). Freeman elaborates, "If fraud were afoot, it would make sense that the President's men would steal votes in GOP [Republican] strongholds, where they control the machinery of government...." As it happens, the mean red shift is 2.4 points *smaller* (less negative) in what E/M dubs the "Moderate Republican" precincts, where Kerry got only 20% to 40% of the vote, than in the "Even" precincts (40% to 60% Kerry). It is true that the mean WPE is close to 0 in the high-Kerry precincts, and that the mean is highest in the high-Bush precincts—although the median in the high-Bush precincts is actually smaller than in the rest of the sample. It might be true that the high-Bush precincts are generally controlled by "the President's men," although there are only 40 such precincts in the exit poll sample (about 3% of the total)—somewhat meager pickings for fraudsters afoot. One also wonders, again, what to make of the fact that red shift is greater on average in *Kerry* states.

Do "GOP stronghold" precincts hold the key to interpreting the exit poll data?... It is apparent... that large red shift values—and, for that matter, large blue shift values—are not concentrated in any particular part of the vote share spectrum.... [I]t is hard to imagine how these data support a "Men in Strongholds" fraud narrative.

Republican governors. In a similar vein, Freeman hypothesizes that where "Republicans controlled the state government, there might have been greater opportunity to steal votes" (Freeman, 2005: 8). [He] reports that the mean WPE was −6.7 in states with Republican governors, but −5.0 in states with Democratic governors... [T]he causal reasoning is not so compelling, for rarely do governors control elections. Most often, secretaries of state and/or nonpartisan election boards do, in concert—and/or in conflict—with local elections officials (usually at the county level). This caveat turns out to be non-trivial. Of the five states with the largest red-shift WPE, one (Delaware) had a Democratic governor who appointed the commissioner of elections; three (Connecticut, Vermont, and New Hampshire) had Republican governors but Democratic secretaries of state who administer elections; and the last (New York) had a Republican governor and secretary of state, but elections are administered by a bipartisan board. It seems inapt to construe any of these states as Republican-controlled for purposes of assessing the likelihood of fraud....

African American voters. Freeman notes that African Americans have been disenfranchised in a host of ways, including felony disenfranchisement and voting machine shortages. "And is [WPE] higher in states with more blacks? Yes, it is. There is no apparent polling explanation; so one must at least consider the possibility that this higher [WPE] indicates that blacks are being disenfranchised by yet other means, as well" (Freeman, 2005: 8).

Differential disenfranchisement of blacks is a real and distressing fact in 2004, as in other elections (e.g., Klinkner, n.d.; Mebane, 2006), and one hesitates to complain that Freeman's argument here verges on *non sequitur*. Freeman has not actually proposed any mechanism by which fraud would induce a relationship between the African American population and exit poll red shift.... [T]he mechanisms he enumerates are unlikely to explain the outcome, because blacks who are prevented from casting ballots will rarely, if ever, participate in the exit poll. One possible explanation is that black voters disproportionately cast provisional ballots that were ultimately not counted. However, this explanation is unlikely to account for much red shift in most states, because the proportions of uncounted provisional ballots are simply too small....

Another explanation is that black voters experienced greater ballot spoilage or residual vote (overvote and undervote) rates, as county-level analysis suggests was true for Florida in 2000 (Klinkner, n.d.). However, given that residual vote rates decreased substantially in several states after the adoption of electronic voting machines (Stewart, 2005), it seems unlikely that residual votes can account for any substantial portion of a large *increase* in exit poll red shift, no matter how racially disproportionate the "new" residual vote pattern might be....

But what about all the other evidence? But what about all the other evidence that I have ignored in this paper? The Warren County (OH) lockdown, the punch card switches in Cleveland, the fact that an obscure black judge got more votes than John Kerry in several Ohio counties, the returns in Florida where heavily Democratic counties voted for Bush, the complaints of vote-flipping, the analysis of Snohomish County (WA), the divergence between election day and early/absentee voting in North

Carolina, the undervotes in New Mexico, the drastic misallocation of voting machines in Franklin County (OH)....

With apologies to readers who have little idea what the preceding sentence was about, I will not attempt to enlighten them here. For what it is worth, I have sifted through quite a bit of evidence about the 2004 election apart from the exit polls. Although many of the anomalies cited do not strike me as robust evidence of anything whatsoever, some evince serious flaws in the conduct of the election (for instance, on voting machine allocation in Franklin County, see Mebane (2006); Highton (2006)[42]). Some of the evidence could possibly point to fraud. Isn't it irresponsible of me not to present all that evidence? Or rather, some would say, shouldn't I focus on presenting the evidence that raises the most questions about questionable election systems?

I can only appeal to readers' capacity to make basic intellectual distinctions. This paper does not address the possibility of "fraud" in general; it addresses specific arguments that the exit polls evince massive, widespread fraud. Some observers seem to believe that any evidence (however tenuous) for any form of fraud in 2004 constitutes supporting evidence for all other forms of fraud, but I do not. Nor does offering evidence against one form of fraud entail denying the existence of any fraud, misfeasance, or systemic vulnerability. We should not be afraid to think clearly.

Lessons for Survey Researchers. ... The 2004 NEP [National Election Pool] researchers and sponsors could be forgiven for concluding that when it comes to "transparency" and disclosure, no good deed goes unpunished. Nevertheless, disclosure was wise and constructive. For instance, *CNN.com* posted preliminary tabulations almost immediately as the polls closed (accompanied by a lucid methodological explanation which few people read)—thus setting in train endless accusations of conspiracy. Would it have been better not to release the initial results at all? It might have *delayed* the accusations, but when the tabulations inevitably leaked, the outcome would have been far worse. The January 2005 Edison/Mitofsky evaluation report, which offered an unprecedented level of detail about methods and results, seems to have been mined to support prior assumptions about fraud even as it was criticized for continuing the cover-up. Could E/M have starved the controversy by withholding these results? Surely not: various fraud theorists have proven very resourceful in finding support for their priors. On the contrary, the evaluation report provided useful evidence of nonresponse bias, as well as many other details of interest to independent researchers.

Given [the] intense scrutiny of election returns, and of exit polls as a *de facto* election audit mechanism, public discourse would likely benefit from making more information available faster—not primarily on Election Day, but before and after. I will not push the argument for transparency very far: no plausible degree of openness can serve as a universal solvent for suspicion.... [A]ny careful reader of the Edison/Mitofsky evaluation report can recognize some analytical gaps that could usefully be filled, potentially shedding light well beyond the "exit poll debate." I

[42][See, also, Mulrow and Scheuren (2007) in Chapter 6 of this volume (6.5.2).]

will not opine here on what the disclosure policies should be, but formulating and explaining such policies in advance would help to mitigate the perception that data and analytical results were released selectively and capriciously.

I harbor no illusions—certainly not after the last year—that if information is simply made available, everyone will become informed. I do think that some of Warren Mitofsky's remarks 17 years ago, in his AAPOR [American Association for Public Opinion Research] presidential address, can be usefully revisited and extended in this context. Invoking Julian Woodward's 1946 description of polls as a "public utility," Mitofsky criticized AAPOR's contemporaneous *Standard for Minimal Disclosure*:

> "That information is not informative enough for most consumers of survey research, certainly not for most members of the news media and other members of the public. It is barely useful to trained survey researchers. What we have done with our minimal disclosure is to place the burden on our users for making sense of the limitations of our surveys. . . ."

> "That is wrong. It is the responsibility of the researcher to spell out the limitations of his or her own survey when there are limitations. . . . Let me be even clearer about what I want to see disclosed. When I say limitations, I want to know how the design affects conclusions based on that survey. Specific statements about what can and cannot be concluded should be part of the disclosure. . . . [For instance, if] the sampling error is not reported, then the disclosure statement should [explain why]. . . . If there are question biases, I want to know what they are. In surveys about racial attitudes, I want to know whether there is an interaction between the race of interviewers and respondents and whether the conclusions are affected by this interaction."

> "To me, this type of disclosure would come closer to "truth in packaging" of survey research. What we have now does not do it." (Mitofsky, 1989: 451–452)

Make no mistake: in many respects Edison/Mitofsky's and the NEP sponsors' disclosure has been not minimal, but exemplary, and I do not question their ethics—in fact, I have vehemently defended them. In no way do I think that any of these parties oversold the accuracy of the exit poll interviews. Their critics did that, and much else besides. Nevertheless, if we construe the public at large as having some legitimate interest—however misinformed—in the exit poll interview results, I do not believe that we have done enough to explain the limitations. Better information and understanding of exit polls' potential pitfalls will benefit activists who are exploring the potential of better exit polls, or independent exit polls, or "parallel elections" in election auditing (see, for instance, *studycaliforniaballots.org*). I have serious reservations about this line of research, but people who pursue it should be able to benefit from our best knowledge and questions about exit poll research.

And Finally. In the course of the exit poll controversy, I have often recalled Walter Lippmann's words in *Public Opinion*:

> "If [a social scientist] is really critical and saturated in the scientific spirit, he cannot be doctrinaire, and go to Armageddon. . . for a theory of which he is not sure. If you are going to Armageddon, you have to battle for the Lord, but

the political scientist is always a little doubtful whether the Lord called him."
(Lippmann, 1922: 234-235)

Some observers have seemed in a great hurry to go to Armageddon. But I will not end with the apocalypse. I have been privileged to talk and to correspond with many people who have legitimate concerns about American electoral processes, and who are working hard and well to address the faults and vulnerabilities that they see.

References

Edelman, Murray and Merkle, Daniel M. (1995). The Impact of Interviewer Characteristics and Election Day Factors on Exit Poll Data Quality, paper presented at the conference of the American Association for Public Opinion Research, Fort Lauderdale, FL, May 18–21, 1995.

Edison Media Research and Mitofsky International (E/M) (2005). *Evaluation of Edison/Mitofsky Election System 2004*, January 19, 2005, available on the Internet at: *http://www.vote.caltech.edu/media/documents/EvaluationJan192005.pdf*

Freeman, Steve (2005). Polling Bias or Corrupted Count? Accepted Improbabilities and Neglected Correlations in 2004 US Presidential Exit Poll Data, presentation to the Fall Meeting of the American Statistical Association, Philadelphia Chapter, October 14, 2005, available on the Internet at: *http://www.appliedresearch.us/sf/Documents/ASAP-Improbabilities051014.pdf*, (last accessed 5/6/06).

Highton, Benjamin (2006). Long Lines, Voting Machine Availability, and Turnout: The Case of Franklin County, Ohio in the 2004 Presidential Election, *PS: Political Science and Politics*, 39 (1), January 2006, 65-68.

Klinkner, Philip A. (n.d.). Whose Votes Don't Count?: An Analysis of Spoiled Ballots in the 2000 Florida Election, available on the Internet at: *http://www.hamilton.edu/news/florida/Klinkner%20Analysis.pdf* (last accessed 5/5/06).

Lindeman, Mark (2005). Exit Polls, Vote Fraud, and 'ESI's Hypothesis': A Response to Kathy Dopp's 'Mathematical Proof,' working paper, November 10, 2005 (original version 11/9/05), available on the Internet at: *http://inside.bard.edu/~lindeman/doppresponse.pdf* (last accessed 5/6/06).

Lindeman, Mark (2006). Too Many Bush Voters? False Vote Recall and the 2004 Exit Poll, paper prepared for presentation at the Annual Meeting of the Midwest Political Science Association, Chicago, IL, April 20-23, 2006.

Lindeman, Mark; and Brady, Rick (2006). Behind the Controversy: A Primer on U.S. Presidential Exit Polls, *Public Opinion Pros*, Internet journal, available on the Internet at: *http://publicopinionpros.com/from_field/2006/jan/lindeman_1.asp* (subscription required), January 2006.

Lippmann, Walter (1922). *Public Opinion*, 234-235, available on the Internet at: *http://xroads.virginia.edu/~Hyper/Lippman/contents.html*, (last accessed 11/28/07).

Mebane, Walter R., Jr. (2006). Voting Machine Allocation in Franklin County, Ohio, 2004: Response to U.S. Department of Justice Letter of June 29, 2005, February 11, 2006 (original draft July 7, 2005), available on the Internet at: *http://macht.arts.cornell.edu/wrm1/franklin2.pdf* (last accessed 5/5/06).

Mitofsky, Warren (1989). Presidential Address: Methods and Standards: A Challenge for Change, *Public Opinion Quarterly* 53 (3), Autumn 1989, 446-453.

Stewart, Charles, III (2004). Addendum to Voting Machines and the Underestimate of the Bush Vote, available on the Internet at: *http://www.vote.caltech.edu/media/documents/Addendum_Voting_Machines_Bush_Vote.pdf*, December 5, 2004 (last accessed 4/30/06).

Stewart, Charles, III. (2005). Residual Vote in the 2004 Election, Caltech/MIT Voting Technology Project working paper, available on the Internet at: *http://www.vote.caltech.edu/media/documents/vtp_wp21v2.3.pdf*, February 2005, version 2.3, (last accessed 5/7/06).

Another paper that focuses on these controversies and examines the problems that occurred is Traugott, Highton and Brady's report for the National Research Commission on Elections and Voting.[43] Also, Chapter 7 (7.4.1) of this volume covers several alternatives that might be used to test the *fitness for use* of an election result.[44]

3.3   CONFIDENTIALITY—ANOTHER SIDE OF THE CONTROVERSY

One major complaint of Freeman and his colleagues was that the Edison/Mitofsky evaluation was based on data that the National Election Pool consortium refused to release for external analysis. Kyle et al. (2007) examine Ohio exit poll and vote counts and describe a disclosure avoidance approach that permits them to analyze the evaluation conclusions without access to the actual raw data by precinct. Alternatives that permit restricted access to identifiable precinct data while still maintaining respondent confidentiality do exist,[45] but new problems may arise from these approaches.[46] In the end, it is the need to honor the pollsters' pledge of confidentiality that must take precedence. Below, the authors go on to show that the disclosure protected data are, indeed, adequate for their analysis.

3.3.1   Excerpt from: Explaining Discrepancies between Official Votes and Exit Polls in the 2004 Presidential Election[47]
Susan Kyle, Douglas A. Samuelson, Fritz Scheuren, and Nicole Vicinanza

Preliminary exit poll data on the [U.S.] presidential race of November 2, 2004, leaked early on the Internet, showed Democrat John Kerry in the lead. However, projections after the polls closed showed Republican George W. Bush the winner, which the official results confirmed. Many Kerry supporters believed, and continue to believe, that the discrepancy between the national exit polls and official results is evidence the election may have been "stolen."

In response to this belief, the organizations that conducted the national exit polls, Edison Media Research and Mitofsky International (Edison-Mitofsky) released an analysis that attributed the discrepancy to differences in response rates between Bush and Kerry voters [(Edison-Mitofsky, 2005)]. The Edison-Mitofsky explanation was rejected by critics, who called for a full release of all data, including identification of the actual precincts sampled.[48]

---

[43] See Traugott, Highton and Brady (2005).
[44] See Lindeman, Liddle, and Brady (2005).
[45] See, for example, Sekhon (2007).
[46] Scheuren (2007).
[47] Excerpt from: Explaining Discrepancies between Official Votes and Exit Polls in the 2004 Presidential Election, by Susan Kyle, Douglas A. Samuelson, Fritz Scheuren, and Nicole Vicinanza, reprinted with permission from *Chance*. Copyright ©2007: American Statistical Association. All rights reserved
[48] [See, for example, Freeman (2004).]

Edison-Mitofsky released a subsample of the full sample as a public-use file, but not with identifiable precincts. Critics regarded even this as inadequate and continued to pressure for full disclosure. But Edison-Mitofsky remained adamant that to release identifiable precinct data would violate the confidentiality promises made to exit poll respondents. The Edison-Mitofsky concern, in fact, had merit because names, addresses, and other contact information about each voter are a matter of public record by precinct. Thus, in precincts where a voter's party affiliation is in an extreme minority—for example, a Republican in heavily Democratic Washington, DC—identification of an individual voter could conceivably happen. To buttress their case, Warren Mitofsky contacted the American Statistical Association [ASA] and obtained a ruling from the chair of the ASA Privacy and Confidentiality Committee that the Edison-Mitofsky position was, indeed, wholly appropriate and supportable. There, for a time, matters were left.

Method of Aligning Vote Definitions. According to the Ohio statewide figures, Kerry received 48.83% of the official final vote, and Bush received 50.81%. Third-party candidates accounted for the remaining 0.36%. Because the exit polls focused on only Kerry and Bush, the two major-party candidates' vote percentages were recalculated using only these candidates. According to the adjusted statewide totals published on the official Ohio Web page, Kerry obtained 48.94% of the vote statewide and Bush 51.06%. To compare official Ohio votes to the exit poll estimates, we recalculated again, using only the 49 precincts included in the exit-polling sample, as Edison-Mitofsky also did. Our independent analyses indicate the sample size is more than adequate for its intended purpose of predicting the election outcome and profiling each candidate's voters. This means we can rule out random variation due to sampling as more than a minor cause of the discrepancy between the Ohio official vote and the exit poll.

What about the impact of absentee and provisional votes? These have been included by implication in the overall final results. But assessing the impact of absentee and provisional voting on a precinct-by-precinct basis requires further steps. To make these precinct-by-precinct comparisons in the sample precincts, we proceeded as follows:

Absentee voters. Absentee voters were not, by definition, included in the exit polls. However, data from counties where absentee votes are tallied separately indicate absentees appear to vote much the same (within 1 to 2 percentage points) as voters in the counties from which they come. This allows us to assume absentees would have little or no impact on the Kerry percentage, particularly as they are a small fraction of the Ohio vote.

Provisional voters. Provisional voters were included in the exit polls. People whose provisional ballots were rejected would not know it at the time they answered the exit polls, however, so this could be a source of potential difference. In three counties where provisional ballot counts were available by precinct—Cuyahoga, Delaware, and Montgomery (which account for 18% of the total Ohio vote)—we adjusted the percent vote for each candidate for the proportion of provisional ballots rejected. To do this, we took the change in votes for each candidate between the unofficial and official canvass; divided it by the number of votes added between canvasses—which, in theory, should be accepted only if they are provisional or military ballots (of which there were few)—to generate the proportion of added votes for each candidate; multiplied this proportion for Kerry by the number of rejected provisional ballots; added this adjustment to the official precinct total for Kerry; added the total number of ballots between canvasses to the total of the *Bush* + *Kerry* votes for that precinct; and divided to create a new Kerry percent. As these adjustments tended to be small, for counties where precinct-by-precinct provisional ballot data were not available, we used the unadjusted Kerry average.

In summary, the effect of absentee and provisional votes in the Ohio sampled precincts was minor. As with sampling error, it is not a major source of concern in assessing reasons for the difference between the exit poll and official results.

The Blurring Method of Protecting Respondent Confidentiality.   Our goal in using the statistical procedure known as blurring was to mask, with minimal information loss, the identification of the actual precincts drawn in the Ohio exit poll sample. In ordinary random rounding, the same expected value is always added to or subtracted from the observations being obscured. This would offer little or no protection, however, if the observations were more widely spaced than the expected value of the amount being added or subtracted.

Blurring is a form of conditional random rounding, where the amount added or subtracted to the actual value depends on where in the distribution the observation is found. If the points around the observation to be obscured are close, then the amount added or subtracted is small. If the values are far apart, as can occur in heavy-tailed distributions, larger changes are used to obscure an individual point.

In our Ohio application, the plan was to take the official Kerry percent for each precinct (after being adjusted), sort it by size, and replace the official values in groups of 20 by changes that used to obscure an individual point. Three steps are involved in implementing blurring:

- Choose the blurring interval.

- Sort the individual observations from lowest to highest and group them according to the chosen blurring interval.

- Replace the actual value for every observation with the average for the group into which it falls.

This can be done for as many variables as desired, but protection is greatly reduced when more than one or two values are blurred.[49]

In our Ohio application, the plan was to take the official Kerry percent for each precinct (after being adjusted), sort it by size, and replace the official values in groups of 20 by the 20-group average in which the precinct fell. This would keep the overall

---

[49][For more information on disclosure limitation procedures, see the Federal Committee on Statistical Methodology (2005).]

vote unchanged, while making it impossible to identify the actual exit-polling precinct. In all the precincts selected for exit polling, the difference between the actual and blurred values turned out to be less than 0.5%—mostly 0.1% or less. Thus, all we had to do to protect the identity of every sampled precinct was to round Kerry's proportion of the vote to the nearest whole percent.

Blurring. Blurring is one technique for disguising the exact percentages of individual precincts. A hypothetical example illustrates the approach. Suppose there are 12 precincts with official Kerry vote percentages, as follows (after ordering from smallest to largest): 20.3%, 20.9%, 35.6%, 36.4%, 40.4%, 41.2%, 52.4%, 53.8%, 60.4%, 60.6%, 72.2%, and 74.2%. Let's use blurring with a group of two. The blurred percentages are the averages of the successive pairs of percentages. That is, we replace 20.3% and 20.9% with 20.6% and 20.6%; we replace 35.6% and 36.4% with 36.0% and 36.0%, and so on. The final blurred data using a group of two is 20.6%, 20.6%, 36.0%, 36.0%, 40.8%, 40.8%, 53.1%, 53.1%, 60.5%, 60.5%, 73.2%, and 73.2%.

A blurring with a group of three would result in 25.6%, 25.6%, 25.6%, 39.3%, 39.3%, 39.3%, 56.8%, 56.8%, 56.8%, 69%, 69%, and 69%. If someone can correctly place a target precinct in a grouping of size k, that person has a $1/k$ chance of identifying the target precinct, assuming there is no available information about the precincts other than the percentages. Hence, the larger k, the greater the protection. On the flip-side, the larger k, the greater the differences are between the released and actual values. Big perturbations negatively impact the usefulness of the released data.

Blurring is related to random rounding, which is mentioned by the authors. To illustrate random rounding, let's randomly round 20.3% and 53.8% to the nearest 1%. We round 20.3% by flipping an unfair coin: heads we round to 20.0% and tails we round to 21.0%. The probability for heads is $(.203 - .20)/.01 = 30\%$, and the probability for tails is $(.21 - .203)/.01 = 70\%$. For 53.8%, we flip a coin that yields an 80% chance of rounding to 53% and a 20% chance of rounding to 54%.

Protecting data confidentiality is an active area of statistical research. For an introduction to data protection techniques and the statistical issues of evaluating the effectiveness of these techniques, see the Summer 2004 issue of *Chance*.

Overall Comparison of Exit Poll with Official Ohio Vote. We now turn to the exit poll results, themselves, comparing them with the official votes and displaying the range of possible statewide estimates for the Kerry percentage that could arise, given different models of nonresponse. We examine three possible adjustment models for the voters who refused to respond to the exit poll. These three models span the range of possible outcomes:

- Model 1 divides the refusals in the same proportion as those who did respond. This *split the difference* approach is essentially what was done when the original

Table 3.1   Exit poll and vote results by precinct for Ohio, November 2, 2004

Mitofsky precinct number	Official vote	Exit Poll			Difference: Fraction of range
		Minimum	Original	Maximum	
48	22%	11%	38%	81%	0.223
14	24%	18%	28%	54%	0.121
7	25%	25%	34%	53%	0.334
23	28%	9%	31%	79%	0.037
26	28%	12%	26%	67%	−0.04
2	30%	20%	41%	71%	0.213
3	30%	22%	41%	69%	0.246
37	30%	19%	32%	59%	0.048
29	32%	18%	30%	59%	−0.05
47	32%	36%	43%	53%	0.648
21	34%	31%	44%	61%	0.339
28	34%	20%	43%	73%	0.166
6	36%	13%	53%	88%	0.227
15	37%	24%	39%	63%	0.048
43	37%	30%	49%	68%	0.313
17	38%	12%	48%	87%	0.132
19	38%	28%	45%	66%	0.178
27	38%	11%	67%	95%	0.342
30	39%	21%	50%	79%	0.192
25	40%	42%	68%	80%	0.73
18	42%	18%	46%	79%	0.066
1	43%	19%	50%	81%	0.113
40	43%	25%	39%	61%	−0.12
11	45%	8%	41%	88%	−0.06
46	45%	40%	47%	55%	0.132

continued next page

exit poll estimates were made. Under this assumption, Kerry comes out ahead in 40 out of 49 Ohio precincts and wins an estimated 53% of the vote. (See Table 3.1.) This is essentially the pattern that led to accusations of fraud or error. However, as Edison-Mitofsky pointed out, the alternative of a differential response rate between Kerry and Bush voters cannot be ruled out.

• Model 2 assumes all the refusals were Bush voters. In this case, Kerry would have received only an estimated 27% of the vote.

• Model 3 assumes all the refusals were Kerry voters, in which case Kerry would have received an estimated 76% of the vote.

Table 3.1   Exit poll and vote results by precinct for Ohio, November 2, 2004
(*continued from previous page*)

Mitofsky precinct number	Official vote	Exit Poll			Difference: Fraction of range
		Minimum	Original	Maximum	
39	46%	32%	54%	73%	0.207
13	47%	24%	47%	74%	0.009
22	47%	30%	41%	58%	−0.2
5	48%	17%	41%	76%	−0.12
34	48%	19%	54%	84%	0.096
16	51%	47%	57%	64%	0.328
36	52%	44%	66%	77%	0.425
50	52%	43%	58%	69%	0.229
20	54%	40%	69%	82%	0.358
42	54%	28%	66%	85%	0.212
49	54%	14%	58%	90%	0.05
4	55%	28%	70%	88%	0.254
44	55%	11%	55%	91%	−0.01
31	57%	28%	68%	87%	0.191
38	57%	11%	41%	84%	−0.22
35	62%	53%	75%	82%	0.443
9	64%	29%	67%	86%	0.047
41	66%	33%	57%	75%	−0.21
12	70%	33%	68%	85%	−0.03
32	71%	55%	82%	88%	0.337
8	80%	41%	90%	95%	0.189
33	81%	19%	68%	91%	−0.19
24	85%	64%	87%	90%	0.07
10	96%	40%	96%	99%	0.007
Overall	47%	27%	53%	76%	0.093

Thus, the three models yield a wide range of possible outcomes. Obviously, what one assumes about the nature of missingness is crucial. Indeed, an assumption failure could explain virtually all the difference between the exit poll and official Kerry vote.

How do we know which assumption to accept? Frankly, we don't. In many settings, the usual split the difference approach to handling refusals works well enough. But, every case is unique, and the split the difference approach does not always work. For example, in the Virginia gubernatorial race of 1989, the "refusals/don't knows" broke heavily against Doug Wilder, who had a big lead in pre-election polls. Apparently, the refusals/don't knows did not want to vote for Wilder, but did not want to tell that to the interviewer. Wilder is African-American, and some voters seem to have been reluctant to state a preference because of racial implications. In the case of the 2004

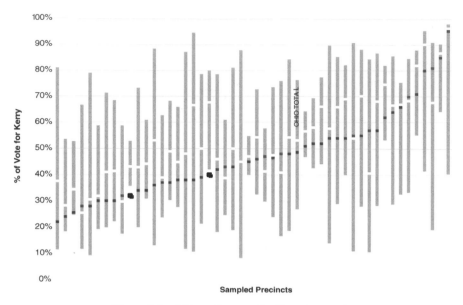

Figure 3.1   Ohio poll ranges for Kerry by precinct.

presidential election, the split the difference assumption led to an estimated 53% for Kerry, as noted. But, what reason do we have to suppose this is the correct assumption?

Basic Precinct-by-Precinct Comparisons of Exit and Official Vote. We consider again the three models of attributing refusals: attribute refusals in the same proportion as respondents, attribute all refusals to Bush, and attribute all refusals to Kerry. Figure 3.1 shows the estimates for the Kerry exit poll vote using each of these three models, as well as the official vote (blurred) for the 49 precincts in Ohio where exit polls were conducted.

The vertical axis represents the percentage of votes for Kerry; the horizontal axis represents the precincts sampled, in ascending order of the precinct's official vote for Kerry. The grey bar represents the exit poll estimates, ranging from attributing all refusals to Bush (lowest Kerry percentage) to attributing all refusals to Kerry (highest Kerry percentage). The gap, or white block, in the grey bar represents the statistical estimate of the exit poll results using nonresponse Model 1 (i.e., the model that attributes exit poll refusals in the same Figure 3.1 Ohio poll ranges for Kerry by precinct proportion as respondents). The colored block in the graph represents the official (blurred) result for each precinct. It is colored red where the official result fell below the estimate (i.e., exit polls overestimated the Kerry result) and blue where the official result fell above the estimate (i.e., exit polls underestimated the Kerry result). The dominance of red in the graph supports Edison-Mitofsky's hypothesis that Bush voters were more likely than Kerry voters to refuse to answer the exit poll.

Note that, for two precincts (*outliers*), the official blurred result lies outside the interval formed by the range of attribution models. For the other 47 precincts, Figure 3.1 shows that differences in attribution of refusals would be sufficient to account for the differences between the exit poll estimate and the blurred official result. This analysis, therefore, indicates higher exit poll refusal rates among Bush voters could constitute at least a partial plausible explanation for the observed discrepancy.

The two outlier precincts indicate that attribution of refusals is insufficient to account for all observed discrepancies, however. The hypothesis of different refusal rates, therefore, is not entirely satisfying, although the amount by which the outliers lie outside the range of possible refusal rate bias is not large.

Patterns that Might Be Expected if Systematic Irregularities Did Occur in Ohio's Vote-Counting Process. Two additional comparisons at the precinct level for the 49-precinct Ohio exit poll sample help shed light on the likelihood of substantial and systematic irregularities in the Ohio vote counting process. Both involve examining the extent to which the sample precincts voted similarly in the 2000 and 2004 presidential elections. Such comparisons are informative when considering possible irregularities, because if there were substantial vote counting fraud or error at the precinct level, it would be unlikely to occur to a similar extent in all precincts or over both elections. Differences between 2000 and 2004 in vigilance by election officials, partisan poll-watchers, and the press, compounded by differences in the demographics and partisan composition of precincts, can be confidently expected to produce variations among precincts in the extent to which fraud or errors would go undetected. And this would almost certainly remain true even if there were substantial similar irregularities in both the 2000 and 2004 presidential elections. For both reasons, a close association between voting patterns in the 2000 and 2004 presidential elections would be further evidence against the hypothesis of widespread fraud or systematic error.

We also would expect—if vote-counting irregularities (purposeful or not) favoring the Republicans occurred in 2004, but not in 2000—that precincts in which Bush did much better in 2004 also would be precincts in which the exit poll more seriously overestimated Kerry's proportion of the official vote. Put simply, if Bush received a substantial number of "extra" votes in a precinct in 2004, we would expect that the exit poll would overstate Kerry's proportion of the official vote by a large amount in that precinct.

Figure 3.2 plots the proportion of the vote Bush received in each of the sample precincts in 2004 (vertical axis) and the proportion he received in the same precincts in 2000 (horizontal axis). The circle symbol indicates that the 2004 exit poll for the precinct overstated Kerry; the triangle symbol indicates that the 2004 exit poll for the precinct overestimated the proportion for Bush. It is evident from the plot that all precincts voted similarly in 2000 and 2004 and that the direction of error in the exit poll estimates is not strongly related to how large a proportion of the official vote Bush received in the precinct.

Next, we consider whether the difference between the proportions of the vote Bush received in these precincts in 2000 and 2004 (vertical axis in Figure 3.3) is associated with the difference between the official and exit poll results for these precincts (horizontal axis in Figure 3.3). Again, if systematic irregularities in vote

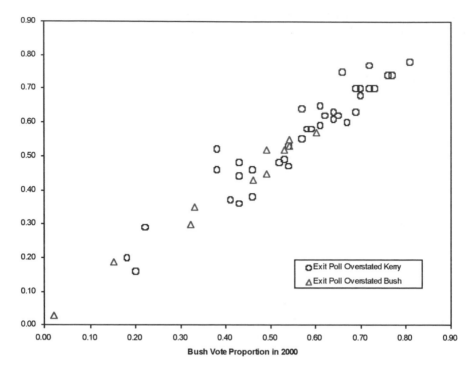

Figure 3.2   Bush vote proportion in 2000 and 2004 for the Ohio sample precincts.

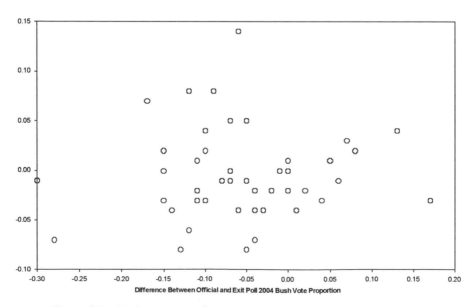

Figure 3.3   Bush vote proportion comparison for the Ohio sample precincts.

counting in some precincts occurred in 2004, but not in 2000, we would expect Bush to do significantly better in those precincts in 2004 and for larger exit poll differences to occur in those precincts. As is apparent in Figure 3.3, these two sets of differences are not associated; their correlation coefficient (0.032) is about as close to zero as real data can achieve.

Summary of Ohio Results and Recommendations for Further Analysis and Future Exit Polls. While it is impossible to prove that there were no attempts to "steal" votes in the counting process in Ohio in 2004, the available evidence from the exit polls does not support the conclusion that any such attempt succeeded on a scale sufficient to alter the outcome of the election or that innocent error could have altered the outcome. In addition, while there have been reports of scattered procedural irregularities, no credible direct evidence of widespread wrongdoing has so far come to light. The anecdotal irregularities reported include such items as controversies over voters being ruled ineligible to vote or voters giving up because of long wait times. This analysis of the Ohio exit polls sheds no light on the extent to which these events occurred.

The observed discrepancy between the official vote and the exit polls provides an insufficient scientific basis for finding a fraudulent result. Indeed, the scientific evidence supports alternative explanations for the discrepancy between the official results and the exit polls, chiefly the difference in response rates between Kerry and Bush voters.

Readers who feel they need more data to reach a firm conclusion should support the expansion of this approach to other states. This certainly can be done if funds are made available. The focus for those who feel the election was stolen should shift to the voting experience in 2004 directly, rather than using exit polls as a measure of how well the count was done. The main goal in revisiting and auditing the 2004 election is not to overturn the outcome, but to learn enough to make the election for 2008 more transparent and auditable. The difficulty of auditing election results inevitably compounds uncertainty. And, ironically, 'improving' the electronic technology of voting, even if it produces a paper trail, could conceivably make the assessment of wrongdoing only harder to resolve. For example, subtle changes in programming code for electronic voting stations could be developed by potential wrongdoers that would randomly change votes, vastly complicating the task of detecting fraud.

Note: The paper on which this article is based was commissioned by the Election Science Institute.

References

Edison Media Research and Mitofsky International (2005). Evaluation of Edison/Mitofsky Election System 2004, available on the Internet at: *http://www.vote.caltech.edu/media/documents/EvaluationJan192005.pdf*.

Freeman, Steven F. (2004). Unexplained Exit Poll Discrepancy, University of Pennsylvania, Graduate Division, School of Arts and Sciences Center for Organizational Dynamics, available on the Internet at: *http://center.grad.upenn.edu/center/get.cgi?item=exitpollp*.

Further Reading

Hayes Phillips, Richard. (2004). Deposition to Ohio Supreme Court, available on the Internet at: *http: //web.northnet.org/minstrel/alpage.htm*.

Miller, Anita (ed) (2005). *What Went Wrong in Ohio: The Conyers Report on the 2004 Presidential Election,* Chicago: Academy Chicago Publisher.

Strudler, Michael.; Oh, H. Lock; and Scheuren, Fritz (1986). Protection of Taxpayer Confidentiality with Respect to the Tax Model, *1986 Proceedings of the American Statistical Association,* Section on Survey Research Methods, 375–381.

Pavía-Miralles, Jose Manuel (2005). Forecasts from Nonrandom Samples: The Election Night Case, *Journal of the American Statistical Association,* 100 (472), 1113–1122.

3.4   FURTHER RESEARCH

As occurred after the controversies of the 2000 U.S. presidential election, the discrepancies that were discovered after the 2004 election also led to research efforts to improve pre-election and exit polling.

One area of research worth mentioning here involved a dual frame approach in Oregon, where traditional exit polls could not be carried out because voting is done by mail (Mitofsky et al., 2005). As elections become more than one-day events, a dual frame approach will likely become increasingly important. Data from early voters are often picked up using telephone surveys. In 2008, perhaps a third or more of all voters will use nontraditional means and the measurement of voter preferences will have to adapt accordingly. Another case where a dual frame approach was used is in Green and Gerber (2006), to assess the usefulness of a registration-based sample to predict election outcomes. Like Mitofsky et al., their experience was favorable, as well.

In another attempt to understand what happened in Ohio, Batcher, Liu, and Scheuren (2005) conducted additional analyses with the intent of improving methods for the future. In their paper, presented below, they conducted trend analysis and grouping analysis to see if the 2004 election results in Ohio were consistent with expected results from past elections. They found outliers (in both directions) in the three largest counties in Ohio, but were unable to conclude whether the results were due to changes in voter preferences, precinct demographics, voter turnout, voting mechanics, or the general voting process. Subsequently, Mulrow and Scheuren (2007) reported on both an exit poll and an independent telephone survey in Franklin County, Ohio, to examine if voting problems in the 2004 election persisted in the 2006 congressional election. The results of this 2006 test were then used to develop statistical methods for studying the upcoming 2008 election.[50]

---

[50] See Mulrow and Scheuren (2007) in Chapter 6 (6.5.2) in this volume.

3.4.1 Ohio Election Analysis
Mary Batcher, Yan Liu and Fritz Scheuren

Overview. In response to speculation that election results in some Ohio counties were unexpectedly biased towards the Republican presidential candidate, official Ohio election results from 1984–2004 were used to assess the reasonableness of the 2004 election results in light of past election results. The underlying premise was that a shift in the 2004 results from what would have been expected given past results would be present if the election results had been manipulated. Such a shift could also occur naturally through a change in voter preferences or demographics. In counties where a shift occurred, further analysis should be done to explore the reasons for the shift.

Two approaches were taken. The first was a look at the trend of presidential results over time. The trend of the presidential years from 1984 to 2000 was plotted [by county] and a line fitting the trend was generated and used to predict the 2004 results. This method is referred to below as trend analysis. The second approach was to group counties in two ways:

1. By size and previous voting pattern, and

2. By race/ethnicity, and use the 2000 results to predict 2004.

This method is referred to below as grouping analysis.

Trend Analysis. The Democratic votes were graphed against the total vote for both parties for the presidential election years 1984, 1988, 1992, 1996, and 2000, and a trend determined for each county. That trend was used to predict the 2004 outcome for the counties in Ohio. The county predictions were compared to the actual 2004 results and are plotted in Figure 3.4.

The difference between the actual number of individuals voting Democratic and the predicted number voting Democratic in each county are plotted. The total vote of both parties for the counties in Ohio is shown on the horizontal axis and the size and direction of the difference between actual and predicted votes are on the vertical axis. Thus, on the graph, a positive difference (i.e., a point marked above the zero line in the middle) indicates that the Democratic vote for the county represented by that point was higher than would be expected based on the trend. Similarly, marks below the zero line indicate that the actual Democratic vote was lower than would be expected from the trend. It is important to note the scale differences on the plot. The total vote is approximately 10 times the size of the differences. Thus, even the largest difference of approximately 50,000 votes is small relative to the total vote. It is necessary to retain the scale differences to be able to see these small differences.

The general pattern shows that for smaller counties (those with fewer votes in the presidential elections since 1984) the differences cluster around zero. However, starting around 200,000 total votes, the difference tends to be positive, indicating more Democratic votes than would have been expected based on previous presidential voting patterns over time. Cuyahoga County has been the focus of many of the questions about the election in Ohio. ... As it is a very large county, [it] can be seen to be one

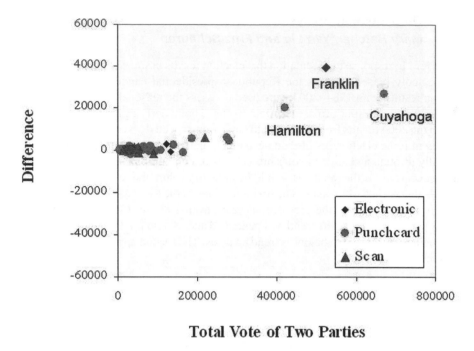

Total Vote of Two Parties

Figure 3.4   Scatter plot of difference between actual and predicted vs. the total vote—trending prediction for 2004 Democratic vote.

of the more extreme points on the graph. Based on [just] the trend analysis, Democratic candidates actually did better than the trends predicted in all large counties and most small ones. [This is counter to concerns about Republican vote manipulation. Of course, like other evidence, it is not, by itself, conclusive.]

Grouping Analysis. Trending over time allows us to compare the actual vote with what would have been expected if this election followed the pattern of several recent presidential elections. However, over time, real changes in voter preference can occur. We, therefore, conducted a grouping analysis to compare the actual vote to what would have been expected based on the 2000 presidential election and

1. County size and

2. County race/ethnicity measured as percent white.

In the grouping analyses, we first divided the 88 Ohio counties into "similar" groups by certain characteristics. Then, the adjusted Democratic votes of the 2000 election were used to predict the 2004 Democratic votes for each county within its group. The adjusted 2000 Democratic vote for each county was obtained by multiplying the 2004 total vote of both parties by the percent of the 2000 Democratic vote. The county results were combined to obtain the overall projection for the state of Ohio. We

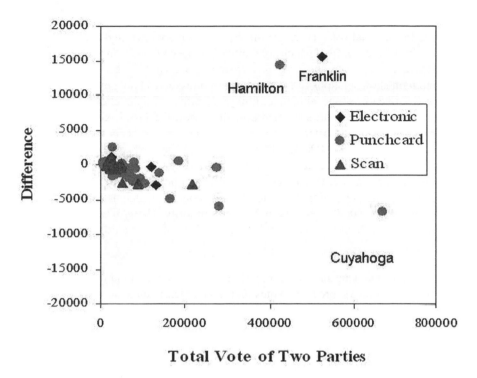

Total Vote of Two Parties

Figure 3.5   Scatter plot of difference between actual and predicted vs. the total vote—grouping prediction for 2004 Democratic vote grouping by county size and voting pattern.

used two grouping criteria: [counties were grouped into two size classes] and, within those, [arrayed] according to their voting patterns in the 2000 presidential election; the [second approach was] to group counties by the percent of white residents. The county predictions were compared to the actual 2004 results and are plotted in Figure 3.5.

[As with Figure 3.4,] the total vote for the counties in Ohio is [again] on the horizontal axis and, [as before] the size and direction of the difference are on the vertical axis. Thus, for the predicted Democratic vote, a positive difference indicates that, for the county plotted at that difference, the Democratic vote was higher than would be expected based on the voting patterns in the 2000 election. Similarly, negative differences indicate that the actual Democratic vote was lower than would be expected. As in the trending plot, the reader should note the relatively small differences compared to the total vote. We have retained these scale difference to allow easier viewing of the points. If both axes were on the same scale, [all] the points would be collapsed close to the horizontal axis.

The general pattern observed in Figure 3.5 is for the differences to cluster around zero for the smallest counties but to begin moving downward as size of county increases and then generally level off, [centered] around a line slightly below zero, indicating slightly fewer Democratic votes than would have been expected based on the 2000 Presidential vote. There are three points widely separated from the rest in terms of the difference, all of which are relatively large counties: Cuyahoga, Franklin, and Hamilton counties are the three most extreme points in Figure 3.5, with Franklin and Hamilton showing a positive difference from predicted and Cuyahoga showing a slight negative difference, although falling generally in the direction of the trend.

The three counties that are separated from the rest are the three largest and also contain proportionately the most minorities. Cuyahoga is the largest county in Ohio, with a population of over 1,300,000. It is largely urban, and has 36% minority population, compared to a statewide average of 17%. Cleveland is its major city. Franklin County contains the state capitol, Columbus, and has a total population of over 1,000,000, with 27% minority population. Hamilton County contains Cincinnati, and has a total population of over 600,000, of whom 28% are minority. These three counties combined represent 29% of the total population and 55% of the minority population of Ohio.

Outlier Analysis. Our analysis, [above] does not show the kind of unidirectional trend in actual votes vs. predicted votes that would clearly indicate voting irregularities. However, we can see possible outliers under both the trending and grouping methods of prediction. [At the county level, these methods differ, as was noted in the text. This suggests that patterns may be random (i.e., no evidence of tampering).] It is useful to identify common outliers under both methods, so they can be looked at more closely by conducting the same type of graphical analysis at a precinct or ward level and, then, looking more closely at specific features of the county or precincts (e.g., the location of the precinct, the availability and type of voting equipment, specific problems or issues reported, etc.). This type of precinct level analysis can help identify why outliers may occur and whether the trends were more pronounced in certain precincts or areas of the county.

The next section presents a precinct level analysis for Cuyahoga County.

Precinct Analysis for Cuyahoga County. Cuyahoga County was selected because it was identified as a possible outlier in the county level analysis and also because it was among the Ohio counties identified as having election irregularities.

The precincts were first grouped based on their geographic locations. Then each precinct's vote was predicted using a model built on data from the rest of the precincts within its group. Finally, predicted votes and actual votes were compared to identify possible outliers. This comparison can be performed for either Democratic votes or for Republican votes. Here, we focus on the Democratic vote.

Linear regression models were used for prediction, where the dependent variable y is the official 2004 Democratic vote and the prediction variable x is the adjusted 2000 Democratic vote. The adjusted 2000 Democratic vote is the Democratic percent of the 2000 vote multiplied by the 2004 total vote. The purpose of the adjustment is to avoid any effect due to changes in precinct size from 2000 to 2004.

The actual prediction models were developed using all precincts within a geographic group, after dropping the one to be predicted. The 2004 Democratic vote in each precinct is predicted using the precinct's group prediction model. Precincts with large differences between the actual and predicted vote are flagged as potential outliers. We also flagged precincts with large percent differences as potential outliers.

The 1,391 precincts in Cuyahoga were divided into groups. The last group— '999'—includes precincts that have fewer than 10 total votes for both major parties, big changes in the percent of votes of the two parties from 2000 to 2004, or cannot be put in a group of at least 4 precincts.

The model fit within group is good for all groups, with an at least 0.93. These models are not exactly the same as the models actually used for prediction, but are very close to the prediction models. The actual prediction models are fit for each precinct using all precincts after dropping the one to be predicted. In other words, there is a different prediction model for each precinct, but each precinct's model is very similar to its group model.

When we looked at the difference between actual and predicted Democratic votes vs. the total votes of both parties, we found that the positive differences indicated that the Democratic vote was larger than predicted and negative differences indicated a smaller Democratic vote than predicted. Based on visual review of the scatter plots, precincts with a vote difference of more than 80 or with a percent difference greater than 40% were flagged as potential outliers. There were 32 such precincts.

The identification of outlier precincts is based on the difference in the number of predicted and actual Democratic votes. It flags the largest differences without consideration of precinct size. While there are both positive and negative differences, the positive outliers are concentrated primarily in the middle size precincts, while the negative outliers tend to be in the midsize-to-large precincts.

When we looked at the same data in terms of percent of predicted relative to actual, we saw that the positive differences become smaller as a percent of the actual vote and the larger differences relative to the actual precinct vote were all negative, indicating a worse Democratic performance than predicted in those 11 precincts. This is an interesting finding, suggesting further investigation into the specifics of changes in voter preferences, precinct demographics, voter turnout, voting mechanics, and the general voting process in those precincts. There are many reasons for a precinct to look different from its prediction. Possible explanations include everything from a real shift in voter preference to the full range of potential explanatory factors speculated about for Ohio.

In addition to these readings, see Chapter 7 for a more methodological paper regarding error analysis. The paper by Lindeman, Liddle and Brady[51] might be particularly important for those wishing to explore differences between exit poll results and official vote counts by looking at Within- Precinct Error, as it proposes some alternatives measures to assess the discrepancies.

---

[51]Lindeman, Liddle, and Brady (2005), 7.4.1 in this volume.

3.5 SUMMARY OBSERVATIONS

So, what is the bottom line? Was the 2004 U.S. presidential election "stolen" from John Kerry? While these articles certainly raised a host of issues about voter processes and the strengths and limitations of exit poll results and predictions, the sense of the researchers is that purposeful fraud is, as yet, unproven. Nonetheless, that sentiment is not universal. More importantly, these efforts opened up a lot of areas for future research and improvement to address before the 2006 and 2008 elections.

Some of that work was already under way.[52] In addition to analysis of mid-term elections—which present some differences from the national presidential votes—the next chapter in this book also provides some specifics on the testing that grew out of the 2004 controversies. These efforts set the stage for changes and enhancements to current methods and procedures for 2008.

Editors' Additional References

Baiman, Ron; Dopp, Kathy, et al. (2005). Analysis of the 2004 Presidential Election Exit Poll Discrepancies, U.S. Count's National Election Data Archive Project, available on the Internet at: *http://electionarchive. org/ucvAnalysis/US/Exit_Polls_2004_Mitofsky-Edison.pdf*.

Batcher, Mary; Liu, Yan; and Scheuren, Fritz (2005). Ohio Election Analysis, *American Statistical Association Proceedings,* Social Statistics Section, 1891-1896.

Dopp, Kathy (2005). Response to Lindeman's Response to 'Mathematical Proof that Election Science Institute's Test to Rule Out Vote Fraud is Logically Invalid,' available on the Internet at: *http:// electionarchive.org/ucvAnalysis/US/exit-polls/ESI/Mark-Lindeman-Response.pdf*, December 1, 2005 (last accessed 12/31/05).

Edison Media Research and Mitofsky International (E/M) (2005). *Evaluation of Edison/Mitofsky Election System 2004*, January 19, 2005, available on the Internet at: *http://www.vote.caltech.edu/media/ documents/EvaluationJan192005.pdf*.

Federal Committee on Statistical Methodology (2005). Report on Statistical Disclosure Limitation Methodology, available on the Internet at: *http://www.fcsm.gov/working-papers/spwp22.html* (last accessed January 2008).

Freeman, Steven F. (2004). The Unexplained Exit Poll Discrepancy, available on the Internet at: *http: //center.grad.upenn.edu/get.cgi?item=exitpollp.*

Freeman, Steven F. and Bleifuss, Joel (2006). *Was the 2004 Presidential Election Stolen? Exit Polls, Election Fraud, and the Official Count*, New York, NY: Seven Stories Press.

Freeman, Steven F. and Mitteldorf, Josh (2005). A Corrupted Election: Despite What you May Have Heard, the Exit Polls Were Right, available on the Internet at: *http://www.appliedresearch.us/sf/Documents/ ASAP-Improbabilities051014.pdf*, February 15, 2005.

Green, Donald P. and Gerber, Alan S. (2006). Can Registration-based Sampling Improve the Accuracy of Mid-term Election Forecasts? *Public Opinion Quarterly*, 70(2), 197-223.

Lindeman, Mark (2006). Beyond Exit Poll Fundamentalism: Surveying the 2004 Election Debate, paper presented at the 61st Annual Conference of the American Association for Public Opinion Research, Montreal, Canada, May 19, 2006.

Lindeman, Mark; Liddle, Elizabeth; and Brady, Richard (2005). Investigating Causes of Within Precinct Error in Exit Polls: Confounds and Controversies, *Proceedings of the American Statistical Association*, Biometrics Section, 282-292.

Kyle, Susan; Samuelson, Douglas A.; Scheuren, Fritz; and Vicinanza, Nicole (2007). Explaining Discrepancies between Official Votes and Exit Polls in the 2004 Presidential Election, *Chance*, 20 (2), 36-43.

---

[52] For example, Liddle and colleagues conducted work similar to that of Kyle et al. (2007) on a national basis. An unpublished paper currently documents that effort.

Mitofsky, Warren; Bloom, Joel; Lenski, Joseph; Dingman, Scott; Agiesta, Jennifer (2005). A Dual Frame RDD/Registration-Based Sample Design: Lessons from Oregon's 2004 National Election Pool Survey, *Proceedings of the American Statistical Association,* Section on Survey Research Methods, 3929-3936. (Note: Look for an update of this paper to be forthcoming in *Public Opinion Quarterly*.)

Mulrow, Edward and Scheuren, Fritz (2007). Producing a 2008 National Election Voters Scorecard, paper prepared for this volume, 6.5.2.

Sekhon, Jasjeet S. (2007). Comment: Data Troubles: Explaining Discrepancies between Official Votes and Exit Polls in the 2004 Presidential Election, *Chance*, 20 (2), 43-45.

Scheuren, Fritz (2007). Rejoinder: Explaining Discrepancies between Official Votes and Exit Polls in the 2004 Presidential Election, *Chance*, 20 (2), 45-47.

Slater, Melaney and Christensen, Howard (2002). Applying AAPOR's Final Disposition Codes and Outcome Rates to the 2000 Utah College's Exit Poll, *American Statistical Association Proceedings,* AAPOR-ASA Section on Survey Research Methods, 540-545.

Traugott, Michael; Highton, Benjamin; and Brady, Henry (2005). *A Review of Recent Controversies Concerning the 2004 Presidential Election Exit Polls*, report for the National Research Commission on Elections and Voting, New York, NY: Social Science Research Council.

"IF THE ELECTION WERE HELD TODAY, WHICH CANDIDATE WOULD I BE WRONG ABOUT YOU VOTING FOR?"

CHAPTER 4

MID-TERM ELECTIONS: 2006

4.1 INTRODUCTION

Chapter 4 focuses on the U.S. elections in 2006. In addition to presidential elections, which are held in the U.S. every four years, each state also holds mid-term elections in off-years, to elect some Senators and Representatives to Congress, Governors, and selected state and local officials. Some states also add referenda to their ballots in election years to obtain direction from the citizenry. Like the presidential elections, polls are used in these mid-term contests to predict the winners and support for ballot issues; the experiences of the past elections often serve as the basis for improved methods for making projections.

We begin with three papers that are based on research of pre-election polls. The first considers the impacts of campaign dynamics on the survey outcomes; the second looks at means for predicting likely voters; and the third presents a model for

Editors' Note: Text in gray shading is provided by the editors of this volume. Boxed-in text represents excerpts from interviews with Warren Mitofsky. The balance of the chapter is made up of papers prepared by contributing authors. [Text in square brackets was added by the editors.]

forecasting election outcomes in the U.S. House of Representatives. Then, the chapter concludes with a reading that draws on results that explore the impact of voters' attitudes on ballot initiatives.

First, though, we provide some comments from Warren Mitofsky about projections and some of the factors that go into making those projections.

"Exit polls are only one way to make projections; we make projections many different ways. [Mistakenly,] everything that goes into a projection in the media [has been] referred to as an exit poll, but in truth they are not… In 2004 in the United States, there were 124 races projected on election night: 51 of them were presidential contests in each of the states and in the District of Columbia. There were 34 Senate races, 11 Governors races, 7 statewide House races (otherwise we would skip the House races altogether), and 20 referenda. We also did projections for 23 Democratic primaries in the winter of 2004. So there were a lot of projections on television in 2004. This involved sampling from 50 states, interviews from the exit polls with 60–100 voters in almost 1,500 precincts… There were 5,800 'absentee voters' interviewed in 13 states… Early voting is where they set-up polling places in a state prior to Election Day and, instead of waiting until Election Day to vote, you can go in and vote early. So I used the word 'absentee' as a generic description, but it's really both [early voting and absentee voting]. We also collected the actual vote returns, not [just] the exit poll interviews with the voters, in almost 3,000 precincts and we received county vote totals."

"I said there were different projections or estimates coming out. The first source of estimates is from the exit polls. The exit polls are interviews with voters throughout Election Day at a sample of polling places. We're talking about probability sampling (the precincts are a probability sample of the state) and the voters (every nth voter exiting the polling place); and the interviews from those voters are what give us the proportion of the vote for each of those candidates."

"It's all stratified random sampling. You have ratio estimates, simple unbiased estimates, post-stratified estimates, and variances computed for each estimator. The variance estimator is not $\left(\frac{pq}{n}\right)$. The only characteristics that are being used are either the past vote or the geographic portion of the voting states. I don't have the luxury of sending interviewers out there and giving them a random start. So, if the interval is 4, I get them to start 1, 2, 3, 4. I confess there is no random start, but they're still taking every nth voter. And the problem that occurs—the bias that comes into these exit polls—is when they don't follow directions and take every nth voter, as we instructed them, but start freelancing."

"There is bias because, when they don't take every nth voter and [instead] they take people that they think will respond, the people are more like them [the interviewers] and, since many of these people [conducting the interviews] are young, there are too many young [respondents]; if they are Democrats, there are too many Democratic voters. It's a never ending cycle. The biggest source of bias in the exit polls has to do with the nonrandom selection of the voters, [because the interviewers are] not following our directions."

"There are. . . three sources of data for projections: the exit polls are available first and reported at poll closings. The sample precincts returns are used for the closer races and that's after the polls close and the vote has a chance to be tabulated and reported for the different polling places. . . And the third source is the county data, which take anything up to a couple of hours [to be released]. It may take a half hour to start and several hours to get to 100 percent. But the closest elections are projected from models that are based on county votes."

"On election night [in 2004] with 124 races, in spite of what you may have read in the paper, all the projections were correct. Not only the ones that we made (we being Edison Media Research and my company, Mitofsky International), but also our six clients, the five television networks and the Associated Press. All those projections were correct. Now this was in contrast to the stuff that was on the Web saying that Kerry was going to win in a landslide. That was based on leaked data, which were being leaked by people who didn't know what they were leaking and they were reaching conclusions that we were never reaching. There was an article in *Newsweek* about Bob Schrum going to Kerry and saying, 'Let me be the first one to call you Mr. President.' That was in the afternoon. There was a story about Karen Hughes going to Bush and telling him to prepare his congratulations to Kerry for winning. This is what was going on at the [political] parties, based on the leaked exit polls."

"We think there won't be leaks in 2006. The networks have now agreed that we don't have to give them the data . . . until 5 o'clock in the evening. And if they don't [receive the data early], it's going to be very hard for the data to leak and the leaks are what caused all the problems."

4.2   PRE-ELECTION POLLS

Pre-election polls are conducted between elections and often used to assess the feasibility of a particular candidate's run for office. In addition to other questions, they often ask respondents, "If the election were held today, who would you vote for in the upcoming primary (or election)?" Many pollsters conduct successive pre-election polls, building up to the election, to examine potential voting trends and how they are impacted by current events. Harrison (2007) examines pre-election polls for the 2006 mid-term elections in an electoral context, to determine which factors are most likely to impact results.

4.2.1   Poll and Races: The Impact of Methods and Conditions on the Accuracy of 2006 Pre-Election Polls
Chase Harrison

This paper studies the accuracy of pre-election polls in different types of elections. Using data from the 2006 mid-term congressional elections, I conduct an analysis that accounts for factors related to both survey methods and factors involving the dynamics of elections. Pre-election polls represent an important institution in contemporary politics, journalism, and campaigns. During the course of an election, the accuracy of electoral forecasts, or the predictive accuracy of a pre-election poll, represents an item of substantive concern. In a different capacity, survey researchers and methodologists look to pre-election polls as a rich and valuable source of information about survey methods. Political polls are, after all, surveys. In contrast with many surveys, pre-election polls—at least those close to an election—which attempt to forecast a vote can be externally validated by actual election results.

Though the study of the relationship between survey methods and the accuracy of pre-election polls dates, at least informally, to the early days of newspaper polling, attention was formally focused on measures of the accuracy of pre-election polls after newspapers incorrectly predicted the winner of the 1948 presidential election. A distinguished group of social scientists established by the Social Science Research Council (SSRC) produced an extensive evaluation of pre-election survey methodologies and sources of error. (Mosteller et. al., 1949). Since the Mosteller report, a number of authors have examined the accuracy of pre-election polls. (Traugott, 2001; Traugott, 2005; Mitofsky, 1998). Additional research has also focused on the impact of survey methods on poll accuracy. (e.g., Visser et al., 2000; Gelman and King, 1993; Crespi, 1988; Lau, 1994; and DeSart and Holbrook, 2003).

Crespi (1988), Lau (1994), and DeSart and Holbrook (2003) conduct analyses to understand the impact of different methodological decisions on the overall accuracy of pre-election polls. Crespi, using data from a number of statewide polls, finds that sample size, the importance of accuracy to the researcher, and the number of days before an election that a poll is conducted all influence the accuracy of the survey. Lau (1994) examines national pre-election polls conducted during the 2000 presidential campaign and considers a wide range of variables. He finds that the number of days a poll is in the field and whether the poll includes weekend calling both have an impact on poll accuracy. Interestingly, in Lau's analysis, sample size does not have an impact on the accuracy of survey results, nor does using a likely voter screen. DeSart and Holbrook (2003), however, replicate Lau's analysis using statewide presidential trial-heat polls over three elections: 1992, 1996, and 2000 and come to different conclusions. Specifically, they find that sample size and using likely voter models *do* matter for the accuracy of these polls, while the number of days in the field does not. They find significantly more complex dynamics involving the length of time prior to an election that the poll was conducted and that this relationship changes form in different election years.

This paper seeks to understand the reasons for these differing results by placing polls in an electoral context. The primary methodological studies involving pre-election polls typically focus on individual elections or a small number of similar elections. With the exception of Crespi, who discusses factors such as turnout and incumbency, the electoral context of pre-election polls is largely ignored or treated descriptively. This paper is an attempt to develop a framework to understand the extent that factors involving campaign dynamics might systematically interact with survey methods and whether including electoral contexts into the study of pre-election polls might be useful.

Data. The data for this study are based on publicly released pre-election surveys conducted during the mid-term elections of 2006. The unit of analysis for this study is the pre-election forecast, taken as the publicly released data measuring the winner of a campaign "trial heat" survey question. Data were obtained from databases compiled by *Pollster.com*, (*http://www.pollster.com/polls/*) and The Polling Report (*http://www.pollingreport.com/*). Data from different types of races—House, Senate, and gubernatorial races were combined to form a single data set.[53] This analysis includes only polls conducted between Labor Day and the mid-term elections in 2006. It specifically excludes Internet polls and a small number of mail polls, but includes automated telephone polls conducted using Interactive Voice Response (IVR) technologies. Due to peculiar and possibly unique circumstances, the Connecticut Senate race is also excluded from these data. In total, 1,031 electoral forecasts are included in this study, spread across 158 separate electoral contests or races. This includes 93 House races, 34 gubernatorial races, and 31 Senate races.

Model. The dependent variable used in this model is conceptualized as the predictive accuracy of the poll or the difference between the predicted results of an election and the actual result. A variety of different measures have been discussed by various authors. (Mosteller et al., 1949; Mitofsky, 1998; Martin, Traugott, and Kennedy, 2005.) This analysis uses the Martin-Traugott-Kennedy measure as a dependent variable. This measure is signed to evaluate both the magnitude of error, as well as whether error is systematically biased toward one party. The measure, itself, compares the odds-ratios of Republican votes versus Democratic votes for both an individual survey and final election results. The actual measure is based on the natural log of the product of these two ratios, and consequently will be 0 when the two ratios are the same (i.e., there is no bias in the poll); negative values indicate a poll over-represents the Democratic vote, while a positive value signifies that the poll over-represents Republican votes.

This paper will replicate models which evaluate the impact of methodological features of surveys on accuracy and will conduct the analysis separately in several electoral contexts. Five variables are included in the model to measure different methodological decisions made by pre-election researchers. Following Lau and DeSart and Holbrook, I include the duration of the field period, measured as the number of days a survey was conducted[54], days before the election, the number of interviews conducted, and whether the survey screens for or predicts likely voters. In addition, a variable to measure whether the survey is conducted by IVR (*IVR = 1*) is included.

---

[53] Most of these data are from separate surveys. In a small number of cases, questions about Senate and gubernatorial races were asked in the same survey. In these cases, we take the unit of analysis to be the survey question or electoral prediction.

[54] This can be taken as an implicit measure of nonresponse.

Table 4.1   Impact of survey methods on accuracy of pre-election pools

Variable	Coef.	Robust Std. Err.	Sig
Days before Election	0.00137	0.00039	***
Days in Field	−0.00792	0.00451	*
Likely Voter Screen	0.00670	0.01660	
Number of Interviews	−0.00001	0.00004	
IVR	−0.02566	0.01935	
Constant	0.01726	0.03017	
Adjusted R-Square	0.025		
N	1,030		

Sig *<=.10; ** <=.05; ***<=.01

Three fundamental types of electoral context will be examined in this paper: the political office contested in the election, the party of the candidate, and the incumbency of the candidate. Incumbency and party are evaluated simultaneously.

Table 4.1 displays the results of an ordinary least square regression predicting survey bias by the five survey methodology variables. To control for correlated errors due to the clustering of the surveys within individual races, robust standard errors are calculated in the manner suggested by Huber (1967) and White (1980). Since this regression is predicting signed error or bias in polls, statistically significant items will be those which produce errors in polls. This analysis finds two of these variables produce bias: days before election and (marginally) number of days in the field. In 2006, surveys conducted further away from the election were less accurate than those conducted closer, with distance from the election being positively correlated with overestimating the Republican vote. Though the impact of these variables is muted in such a heterogeneous data set, the findings are consistent with those of Lau (1994). This effect is also not entirely surprising in an election which may have seen a genuine swing in electoral intentions during the course of campaigns, with voters less likely to choose Republican candidates as the election drew near.

Table 4.2 replicates the methodological model separately for gubernatorial, Senate, and House political races. Since the Senate and Governor races are both statewide, while most congressional districts are smaller units of a state, there may be other methodological factors involving the accuracy of these races than those accounted for in this model. Table 4.2 only finds one statistically significant effect: that of distance from the election for House races. In other types of races, there are no statistically significant predictors of survey error or bias, and no other factors emerge within House races.

Table 4.2 Impact of survey methods on accuracy of pre-election polls by type of race

Variable	House			Senate			Governor		
	Coef	Robust Std. Err.	Sig.	Coef	Robust Std. Err.	Sig.	Coef.	Robust Std. Err.	Sig
Days before Election	0.0025	0.0009	***	0.0008	0.0005		0.0009	0.0006	
Days in Field	−0.0116	0.0073		−0.0046	0.0040		−0.0038	0.0100	
Likely Voter Screen	0.0476	0.0389		0.0101	0.0191		−0.0294	0.0322	
Number of Interviews	0.0000	0.0001		0.0001	0.0001		0.0001	0.0001	
IVR	−0.0588	0.0514		−0.0072	0.0254		0.0475	0.0306	
Constant	−0.0364	0.0649		−0.0053	0.0486		−0.0239	0.0492	
Adjusted R-Square	0.067			0.020			0.032		
N	367			330			333		

Sig *<=.10; ** <=.05; ***<=.01

Table 4.3 examines this methodological model for three types of elections: elections with a Republican incumbent, those with a Democratic incumbent, and those with no incumbent office-holder. In contrast to Table 4.2, which suggested no methodological bias or artifacts within pre-election polls, models which examine incumbency and party seem full of systematic methodological influence.

When controlling for other factors, the length of time before an election has a statistical impact on survey-based electoral forecasts within races with a Republican incumbent, but not in other types of races. The further away from Election Day a poll was conducted, the more likely it was to overestimate the vote for Republican incumbents. Interestingly, surveys with a large sample size actually have a slight Republican bias in Democratic races, but not others.[55] IVR surveys overstate the Republican vote in surveys with incumbents, but not others[56].

Perhaps the most important and problematic finding in this study is the negative impact that long field periods seem to have in races with an incumbent. With both Democratic and Republican incumbents, surveys which had longer field periods produced at least marginally significant bias. In both cases, the results suggest that these surveys with longer field periods, and presumably higher response rates, selected or included respondents who over-report the likelihood that they will vote for a Democratic candidate. Interestingly, this effect is diminished in races with no incumbent.

Discussion.   From a survey research standpoint, these findings suggest both good news and a cause for further investigation. On the one hand, the only factor which had a negative impact on poll results, factoring everything together, is potentially a real measure as opposed to a methodological artifact. Across House, Senate, and gubernatorial races, there is little reason presented here for systematic concern. On the other hand, the strong relationships between several variables and poll accuracy in models which look at incumbency separately could raise real concern. In particular, the potential Democratic bias in surveys with long field periods suggests the possibility of further exploration. For example, is it possible that, in some conditions, attempts to reduce nonresponse error may increase error in pre-election polls? Or is it possible that short field periods do a better job of selecting likely voters than survey questions or statistical models?[57]

This paper has presented a framework which might prove to be a useful contribution to our understanding of the accuracy of election polls. Specifically, it examines pre-election forecasts across a wide range of races and finds that, in some cases, survey methods work differently within different electoral contexts. At the same time, the data here are limited to elections held in a single and potentially unusual year. Consequently, attempts to develop models which are generalizable across elections will require additional data and further analysis.

---

[55] Although interesting and potentially important, it is not clear whether this result is real or the result of relatively smaller sample sizes in these races.

[56] IVR and Internet surveys are also relatively more likely to conduct surveys in noncompetitive districts. Consequently, the significance of this result is also less clear.

[57] [For example, Newport (2006) reported on the relationship that timing has on registered voters vs. likely voters.]

Table 4.3 Impact of survey methods or accuracy of pre-election polls by incumbency

Variable	Democratic Incumbent			Republican Incumbent			No Incumbent		
	Coef.	Robust Std. Err.	Sig.	Coef.	Robust Std. Err.	Sig.	Coef.	Robust Std. Err.	Sig.
Days before Election	0.0005	0.0006		0.0019	0.0006	***	0.0010	0.0008	
Days in Field	−0.0161	0.0071	**	−0.0078	0.0041	*	0.0003	0.0109	
Likely voter Screen	0.0197	0.0268		0.0122	0.0253		−0.0449	0.0318	
Number of Interviews	0.0002	0.0001	**	−0.0793	0.0216	***	0.0148	0.0578	
Constant	−0.0958	0.0500	*	0.0698	0.0414	*	−0.0148	0.0578	
Adjusted R Square	0.159			0.163			0.013		
N	267			500			263		

Sig * <=.10; ** <=.05; ***<=.01

References

Crespi, Irving (1988). *Pre-Election Polling; Sources of Accuracy and Error*, New York, NY: Russell Sage Foundation.

DeSart, Jay, and Holbrook, Thomas (2003). Campaigns, Polls, and the States: Assessing the Accuracy of Statewide Presidential Trial-heat Polls, *Political Research Quarterly*, 56 (4), December, 431–439.

Gelman, Andrew, and King, Gary (1993). Why Are American Presidential Election Campaign Polls so Variable When Votes Are so Predictable? *British Journal of Political Science*, 23 (4), October, 409–451.

Huber, P. J. (1967). The Behavior of Maximum Likelihood Estimates under Nonstandard Conditions, *Proceedings of the Fifth Berkeley Symposium on Mathematical Statistics and Probability*, Berkeley, CA: University of California Press, 1, 221–223.

Lau, Richard R. (1994). An Analysis of the Accuracy of "Trial Heat" Polls during the 1992 Presidential Election, *The Public Opinion Quarterly*, 58 (1), Spring, 2–20.

Martin, Elizabeth A; Traugott, Michael W.; and Kennedy, Courtney (2005). A Review and Proposal for a New Measure of Poll Accuracy, *Public Opinion Quarterly*, 69 (3), January 1, 342–369.

Mitofsky, Warren J.(1998). Review: Was 1996 a Worse Year for Polls than 1948? *The Public Opinion Quarterly*, 62 (2), Summer, 230–249.

Mosteller, Frederick; Hyman, Herbert; McCarthy, Philip J.; Marks, Eli S.; and Truman, David B. (1949) *The Pre-election Polls of 1948*, New York, NY: Social Science Research Council.

Traugott, Michael W. (2001). Trends: Assessing Poll Performance in the 2000 Campaign, *The Public Opinion Quarterly*, 65 (3), Autumn, 389–419.

Traugott, Michael W. (2005). The Accuracy of the National Pre-election Polls in the 2004 Presidential Election, *Public Opinion Quarterly*, 69 (5), January 1, 642–654.

Visser, P. S. ; Krosnick, J. A. ; Marquette, J. ; and Curtin, M. (2000). Improving Election Forecasting: Allocation of Unlikely Respondents, Identification of Likely Voters, and Response Order Effects, *Election Polls, the News Media, and Democracy*. Lavrakas, P. and Traugott, M. (eds.), New York, NY: Chatham House (Seven Bridges), 224-260.

White, H. (1980). A Heteroskedasticity-consistent Covariance Matrix Estimator and a Direct Test for Heteroskedasticity, *Econometrica*, 48, 817–830.

One of the big challenges in conducting pre-election polls is identification of likely voters. If one assumes that nonvoters differ significantly from voters, then accurate pre-election polls require that most respondents be persons who are likely to vote. In the excerpt below, from Murray, Riley, and Scime (2007), the authors describe the results of data mining research they conducted to address this critical issue using decision trees. The methodology they used is described in Chapter 7: Technical Appendix (7.3).

4.2.2   Excerpt from: A New Age Solution for an Age-Old Problem: Mining Data for Likely Voters

Gregg R. Murray, Chris Riley, and Anthony Scime

Introduction.   The difficulty of conducting pre-election surveys that provide accurate electoral predictions is an ongoing problem of note, at least since the spectacularly flawed *Literary Digest* poll in 1936 and Gallup's and others' errant estimates in the Truman-Dewey contest in 1948. While pre-election polling has improved dramatically since its earliest days, there is continuing public concern and criticism about differences between contemporaneous pre-election surveys and, more importantly, between pre-election surveys and Election Day results. One often-noted culprit is the difficulty of identifying likely voters—those citizens whose attitudes we need to evaluate because they actually vote on Election Day (Dimock, Keeter, Schulman, and Miller, 2001; Erikson, Panagopoulos, and Wlezien, 2004).

Given the high stakes of pre-election polling, both in terms of costs and reputations, a number of researchers and practitioners have created likely voter screens (Voss, Gelman, and King, 1995; Blumenthal, 2004). For example, Traugott and Tucker (1984) built a three-variable model that correctly classified 77% of respondents in 1980 as voters or non-voters using measures of voter registration, previous presidential voting, and interest in the campaign. While Erikson and his co-authors (2004) expressed concern about the invariant nature of voter registration, Petrocik (1991) argued that these three questions are unduly subject to the effects of social desirability. As an alternative, he proposed a 10-variable model that relies heavily on demographics, which presumably are less vulnerable to social desirability, and correctly classified 78% of respondents in the 1980–1988 presidential elections. Freedman and Goldstein (1996) also rely heavily on demographics in their two-stage screen. The first stage consists of a single question, intention to vote in the upcoming election, which alone is 75% accurate. The second stage consists of nine questions that improved accuracy to 78% in the 1988 election by primarily capturing respondent demographics. Dimock and his co-authors accurately classified 73% of respondents in Philadelphia's 1999 mayoral election by employing the Pew Research Center's eight item index, which ignores demographics and includes voter interest and intentions, as well as self-reported past voting behavior (Dimock et al., 2001).

Our intent is to build a more efficient likely voter model. With this objective in mind, we use "new age" data mining techniques[58] and data from the American National Election Studies (ANES) to identify a number of survey questions that can be used to classify likely voters while maintaining or surpassing the accuracy rates of other models. In undertaking this endeavor, [in this excerpt,] we follow by . . . describing and analyzing the likely voter model that results [from applying data mining techniques to the ANES data]. We end by suggesting that our findings offer a new approach to identifying and evaluating likely voters that may increase accuracy without also increasing cost.

Results: The Classification Decision Tree Model. In this section we review the results of the Iterative-Expert-Data-Mining (IEDM) modeling process.[59] These results include

1. identification of sets of input variables and the accuracy of the decision trees constructed before domain expertise-driven adjustments,

2. an analysis of domain expertise-driven adjustments to the set of variables, and

---

[58][Data mining is a process of inductively analyzing data to find interesting patterns and previously unknown relationships.]

[59][See the description of the model in Chapter 7 (7.2.1) for methodological details.]

Table 4.4   Input variable identification (top 10)

Rank	Input variable	$\chi^{2}*$	Cumulative accuracy
1	Vote Intent	857.9	77%
2	Previous Vote	835.9	79%
3	Primary Vote	305.4	80%
4	Political Knowledge	263.0	80%
5	Interest: Public Affairs	259.2	80%
6	Interest: Election	248.3	80%
7	Salience: Major Party Presidential Candidates	206.2	81%
8	Salience: Major Parties	190.7	81%
9	Salience: Republican Presidential Candidate	185.2	81%
10	Party ID	169.9	81%

* $p<0.001$ for all input variables.

3. identification of the final decision tree model to be used to classify individuals as voters or non-voters.

Following the IEDM methodology, we first ordered the input variables by strength of association with validated voter turnout, as indicated by χ^2 value. Table 4.4 presents the 10 input variables most strongly associated with voter turnout in order from highest to lowest χ^2 value. The χ^2 estimates indicate that the most strongly associated input variables are intention to vote and self-reported previous presidential vote. These are followed by self-reported voting in the primary election, political knowledge as represented by knowledge of partisan control of the U.S. House, and interest in public affairs. The remaining variables include a combination of measures of political attitudes and assessments. Absent from the list of top predictors are the demographic measures commonly identified in the likely voter and turnout literature (e.g., Freedman and Goldstein, 1996; Petrocik, 1991; Rosenstone and Hansen, 1993; Wolfinger and Rosenstone, 1980). While age (11th ranked), homeownership (13th ranked), and education (15th ranked) appear in the top 20 input variables, others—such as church attendance, income, marital status, race, region, and residential tenure—are statistically superseded by a large number of other variables. This result indicates that demographic variables are less salient when forced to compete with other predictors. It suggests that a heavy reliance on relatively fixed indicators, such as demographic characteristics, may be undermining the identification of likely voters.

Continuing with the IEDM methodology, we constructed a series of decision tree models using the Chi-squared Automatic Interaction Detection (CHAID) algorithm and progressively eliminating input variables by weakest strength of association with the outcome variable. That is, the first tree was constructed from all 146 input variables,

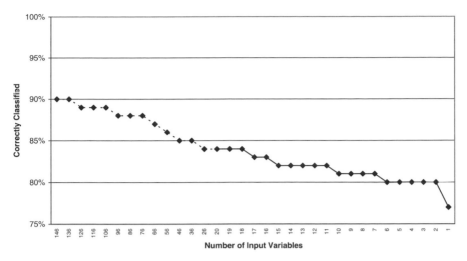

Figure 4.1   Decision trees' predictive accuracy by number of input variables.

the second tree was constructed with the 136 input variables most strongly associated with vote turnout, and the last tree was constructed from the one input variable most strongly associated with vote turnout. It should be noted that, following Freedman and Goldstein (1996), we automatically classified respondents as non-voters if they said they did not intend to vote. Figure 4.1 reports the accuracy of prediction of a series of trees as represented by the proportion of respondents correctly classified as a voter or non-voter by each tree. As one would generally expect, more input variables produce greater predictive accuracy. For example, the full set of 146 input variables generates a tree that predicts turnout with 90% accuracy, and the top 76 construct a tree with 88% accuracy, but the top 16 input variables produce a tree that predicts turnout with 83% accuracy. Of course, given the costs and other resource constraints faced when fielding a survey, the goal is to balance accuracy with number of questions. Figure 4.1 indicates that accuracy ultimately declines to and stabilizes around 80%, a rate similar to other approaches (Freedman and Goldstein, 1996; Petrocik, 1991; Traugott and Tucker, 1984). The top 10 input variables reported in Table 4.4 together construct a tree with 81% predictive accuracy, the top five achieve 80% accuracy, and the top input variable achieves 77% accuracy.

The next step in the IEDM methodology is to identify a set of variables and associated decision tree model to evaluate and then to adjust that model using domain expertise. Given the trivial differences between the top trees in predictive accuracy, we focus primarily on number of input variables. Traugott and Tucker (1984) use three variables to identify likely voters, Dimock and co-authors (2001) use eight, and Freedman and Goldstein (1996) and Petrocik (1991) use 10. In this case, we choose three input variables, in an attempt to provide at least an alternative to Traugott and Tucker, but with the desire to improve upon their model. As implied by Table 4.4,

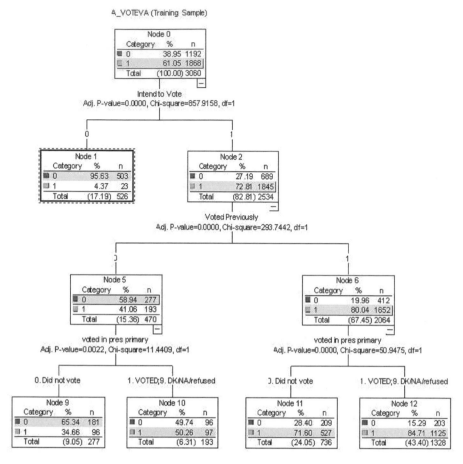

Figure 4.2   Final decision tree using IEDM-LV3 predictors.

our initial model employs three input variables to predict validated vote: intention to vote, previous presidential vote, and voted in the primary election.

Figure 4.2 presents the three-variable IEDM decision tree. The text below each node indicates the input variable name and the statistical relationship of that variable with the outcome variable for that subdivision of cases. It is important to note that the relationship indicated at each node is statistically significant. Node 10 demonstrates how CHAID merges statistically indistinguishable categories of an input variable by merging the categories of "voted" and "don't know" for this group of respondents in the primary vote input variable.

Figure 4.2 also indicates that the tree consists of a root node (Node 0) and eight child nodes, of which five are leaf nodes, which terminate each branch and present a probability of classification of the outcome variable. The leaf nodes are the outcome variable nodes with values of either voted (Nodes 10, 11, and 12) or not voted (1 and 9). The IEDM process identified vote intent as the root node, or most salient node. Following Freedman and Goldstein (1996), we manually pruned that branch at Node 1, which eliminated Nodes 3, 4, 7, and 8. That is, we believed respondents who said they did not intend to vote and classified them as non-voters. This is the only branch we pruned in such a manner. The root node is followed at its first child node by self-reported previous presidential vote and, then, self-reported vote in the primary election. These three variables capture previous voting behavior, which is also employed by Traugott and Tucker (1984) and Dimock et al. (2001), as well as vote intent, which is used by each of the other models except Traugott and Tucker (1984).

The next step in the IEDM process is to apply theory-driven domain expertise to evaluate and to adjust the initial set of three input variables. The demographic measures commonly identified in the likely voter and turnout literature—church attendance, education, income, marital status, race, region, and residential tenure (e.g., Freedman and Goldstein, 1996; Petrocik, 1991; Rosenstone and Hansen, 1993; Wolfinger and Rosenstone, 1980)—stand out as likely candidates and were forced back into the set of input variables for further analysis. This entire set together trivially improved the predictive accuracy of the top three input variables from 79.5% to 81.2%, while five measures individually improved the tree's accuracy: age (to 79.8%), education (to 79.9%), marital status (to 79.7%), race (to 80.3%), and region (to 80.2%). Together, these five variables increased the tree's predictive accuracy to 81.0%. The results indicate that including these conventional demographic input variables as a set, and in combinations with the top three input variables, provides only trivial improvement to the model's predictive accuracy. Given the trivial gain and additional costs that result from increasing the total number of survey questions, the variables are not added to [the] top three input variables and resulting tree. As such, the final tree produced by the IEDM process includes the top three input variables, which we now identify as the "Iterative Expert Data Mining-Likely Voter 3" (IEDM-LV3) model. The IEDM-LV3 model correctly classifies respondents as either a voter or non-voter in 80% of the cases.

Of course, the final decision tree model may be specific to the data from which it was constructed and, therefore, lack generalizability. In order to test for model-to-data over-fit and generalizability, we used the IEDM-LV3 model to classify the test data that we reserved for this purpose. For the test set, the model correctly predicted vote choice in 79.8% of the cases (versus 79.5% for the training set). We also estimated the relationships using logit regression. The results, which are not reported here but are available from the authors upon request, indicate that the probability of voting increases with intent to vote, previous presidential voting, and voting in the primary election. Additionally, each of the relationships easily achieves conventional levels of statistical significance. These two tests suggest that the IEDM-LV3 model is generalizable.

Applying the Decision Tree Model.   Table 4.5 presents the IEDM-LV3 rules as a decision table. The rule numbers appear in the first column, with the questions and relevant responses following in the next three columns. Each rule in the table is read

Table 4.5   Decision table from IEDM-LV3 decision tree model

Rule/ Leaf Node	Vote Intent	Previous Vote	Primary Vote	Outcome Value	% Correctly Classified	% of Total Cases
1	No	–	–	Non-voter	95.6	17.2
9	Yes	No	No	Non-voter	65.3	9.1
10	Yes	No	*Yes	Voter	50.3	6.3
11	Yes	Yes	No	Voter	71.6	24.1
12	Yes	Yes	*Yes	Voter	84.7	43.4

* Includes "don't know" and refused.

left to right to classify whether a person is predicted to vote or not, which appears in the column following the responses. Rule numbers are assigned by decision tree leaf node number in Figure 4.2 to simplify interpretation. For example, Rule 12 is derived from the branch ending with Node 12. Respondents in a leaf node are classified as voters when 50% or more of the respondents in that node actually vote, and they are classified as non-voters when less than 50% of the respondents in that node actually vote. For example, Rule 12 indicates that respondents with the following characteristics will vote 84.7% of the time and, therefore, are classified as voters:

Rule 12: IF intent is to vote AND voted previously AND voted in primary election THEN individuals classified as VOTERS.

A rule should be evaluated in terms of efficiency as indicated by the proportion of cases to which it applies and its accuracy. For example, a rule that is 100% accurate but applies to only a handful of cases must be evaluated relative to a rule that is 60% accurate but applies to a large proportion of the cases. In this instance, Rule 12 applies to more than 40% of cases and correctly classifies more than eight out of 10 applicable respondents. Rule 11 applies to almost 25% of cases and correctly classifies seven out of 10 applicable respondents. In comparison, Rule 10 applies to only 6% of cases and correctly classifies half of the applicable respondents. In all, Table 4.5 indicates that about 94% of respondents are subject to rules that correctly classify at least 65% of applicable respondents (i.e., Rules 1, 9, 11, and 12) and that about 85% of respondents are subject to rules that correctly classify at least 71% of applicable respondents (i.e., Rules 1, 11, and 12). This suggests that the IEDM-LV3 rules are efficient.

Validating the Model: Predictions.   A valid likely voter model should correctly classify individual respondents as a voter or non-voter, as well as be predictive of the other actual outcomes, such as aggregate turnout and election results. Table 4.6 details the accuracy of the IEDM-LV3 model at the individual-respondent and aggregate level for the pooled data and each election year. It indicates that 80% of the pooled respondents are correctly classified as either a voter or non-voter, with the

Table 4.6 Measures of individual- and aggregate-level accuracy, by year and pooled

	INDIVIDUAL LEVEL				AGGREGATE LEVEL		
Year	% Correctly Classified: All	% Correctly Classified: Voter	% Correctly Classified: Non-voter	PRE*	Projected Turnout	Turnout %: House Clerk†	Turnout %: Census‡
Pooled	80	78	85	.47	–	–	–
1976	79	75	94	.43	62	54	59
1980	79	76	85	.48	59	53	59
1988	80	79	81	.51	59	50	57

* Proportional Reduction in Error (PRE) = (% correctly classified—% modal category)/(1—% modal category). In this case the modal category is validated vote.

† (U.S. Census Bureau, 2007).
‡ (U.S. Census Bureau, 2002).

Table 4.7 Election outcomes: comparative two-party vote share, by year

	1976		1980		1988	
	Dem. %	Repub. %	Dem. %	Repub. %	Dem. %	Repub. %
Actual*	51	49	45	55	46	54
IEDM-LV3	49	51	42	58	47	53

* U.S. Census Bureau (2007).

individual years tightly clustered around the pooled value in a range of 79% to 80%. The substantial proportional reduction in error (PRE) for each entry indicates that the model predicts voting and non-voting much more effectively than simply choosing the modal response category.

Table 4.6 also reports aggregate turnout as projected by the model by year and, as a rough means of comparison, turnout as reported by two authoritative sources: the Current Population Reports (U.S. Census Bureau, 2002) and the U.S. House of Representatives Office of the Clerk (U.S. Census Bureau, 2007). Although projecting aggregate turnout is not our primary objective—and reports of "true" turnout are variant—the estimate is a useful indication of the validity of the approach. Similar to Freedman and Goldstein (1996), we projected aggregate turnout by applying a weighted probability of voting, which is derived from each decision tree rule as reported in Table 4.5, to respondents who indicate that they intend to vote (i.e., "intenders"). The non-intenders were assigned a probability of voting of *0*. In all, the results suggest that the model slightly over-estimates aggregate turnout, but not to a degree that undermines the validity of the model.

Finally, a valid model should also generate results that reflect the actual outcome of the election. Table 4.7 reports the projected two-party vote share for each candidate by year. We estimated vote share using the weighted sum of intenders by party of presidential vote. The table indicates that the IEDM-LV3 model overestimated Republican Ford's vote share by 2 percentage points in 1976, overestimated Republican Reagan's vote share by 3 percentage points in 1980, and underestimated Republican Bush's vote share by 1 percentage point in 1988. The results, again, support the validity of the model. In all, the predictions reported here and the results of classifying respondents in the test set as reported previously suggest that the IEDM-LV3 model is valid.

Discussion and Conclusion.   We employed a "new age" data mining technique to address a lingering question in election polling. Specifically, we used the Iterative-Expert-Data Mining process to identify a small number of survey questions that together can be asked to classify likely voters while maintaining or surpassing the accuracy rates of other models. [Using the] three items, the IEDM-LV3 correctly classify 80% of respondents. This accuracy and efficiency compare favorably with the likely voter models constructed by our predecessors. The three survey items proposed here are vote intent, previous presidential vote, and vote in the primary election. We conclude that the IEDM-LV3 model compares favorably to these other models, because it captures important elements of each. These three variables include previous voting behavior, which is employed by Traugott and Tucker (1984) and Dimock et al. (2001), as well as vote intent, which is used by three of the four competing models. Absent from our predictors are the demographic measures frequently identified in the likely voter and turnout literature (e.g., Freedman and Goldstein, 1996; Petrocik, 1991; Rosenstone and Hansen, 1993; Wolfinger and Rosenstone, 1980). Our findings indicate that demographic variables are less salient when forced to compete with other indicators. This suggests that reliance on relatively invariant indicators of individual behavior, such as demographic characteristics, may not efficiently reflect dynamic political considerations.

In all, the results present a new approach to identifying and evaluating likely voters. The findings suggest an alternative method that may increase accuracy without also increasing cost. This is hopeful news for the competitive world of election polling and public opinion research.

References

Blumenthal, Mark (2004). How Do Pollsters Select 'Likely Voters?' available on the Internet at: *http://www.mysterypollster.com/main/2004/09/how_do_pollster_1.html*

Dimock, Michael; Keeter, Scott; Schulman, Mark; and Miller, Carolyn (2001). Screening for Likely Voters in Pre-election Surveys, paper presented at the meeting of the American Association for Public Opinion Research, Montreal, Quebec, Canada.

Erikson, Robert S.; Panagopoulos, Costas; and Wlezien, Christopher (2004). Likely (and Unlikely) Voters and the Assessment of Campaign Dynamics, *Public Opinion Quarterly,* 68(4), 588-601.

Freedman, Paul and Goldstein, Ken (1996). Building a Probable Electorate from Pre-election Polls: A Two-stage Approach, *Public Opinion Quarterly,* 60, 574-587.

Petrocik, John R. (1991). An Algorithm for Estimating Turnout as a Guide to Predicting Elections, *Public Opinion Quarterly,* 55, 643-647.

Rosenstone, Steven J. and Hansen, John Mark (1993). *Mobilization, Participation, and Democracy in America*, New York, NY: MacMillan.

Traugott, Michael W. and Tucker, Clyde (1984). Strategies for Predicting Whether a Citizen Will Vote and Estimation of Electoral Outcomes, *Public Opinion Quarterly,* 48(1), 330-343.

U.S. Census Bureau (2002). Voting and Registration in the Election of November 2000: Population Characteristics, *Current Population Reports*, P20-542, Washington, DC: U.S. Census Bureau.

U.S. Census Bureau (2007). *2007 Statistical Abstract*, Washington, DC: U.S. Census Bureau. (Reported by the U.S. House of Representatives, Office of the Clerk, Statistics of the Presidential and Congressional Election, biennial.)

Voss, D. Stephen; Gelman, Andrew; and King, Gary (1995). Pre-election Survey Methodology: Details from Eight Polling Organizations, 1988 and 1992, *Public Opinion Quarterly,* 59, 98-132.

Wolfinger, Raymond E. and Rosenstone, Steven J. (1980). *Who Votes?* New Haven, CT: Yale University Press.

In the excerpt of Bafumi, Erikson, and Wlezien (2006), the authors describe a statistical model that predicts the outcome of upcoming Congressional elections based on generic polling data. They describe their model and present the forecast results for the 2006 mid-term elections. They then compare their projections to the actual results directly after the election was held. A companion paper—Bafumi, Erikson, and Wlezien (2007)—examines the tendency for the non-incumbent party to win during mid-term elections. The 2006 election, indeed, did result in a shift of power, due to an increase in Democrats in both the Senate and the House of Representatives.

4.2.3   Excerpt from: Forecasting House Seats from Generic Congressional Polls

Joseph Bafumi, Robert S. Erikson, and Christopher Wlezien

Introduction. On October 24, 2006, we released a forecast of the 2006 mid-term election based on a statistical model that estimated both the national partisan tide and the vote in contested districts. The forecast paper was published on many politically oriented blogs and portions of it were covered in mainstream print media. The forecast turned out to perform extraordinarily well. The original paper release is re-printed here. It is followed by some post-election analysis as well as evidence that our prediction could have been made very early in the election year. The utility of the generic polls (when properly analyzed) in predicting congressional election outcomes is highlighted by our work.

Forecasts from Generic Polls. According to the frequent polling on the generic ballot for Congress, the Democrats hold a large advantage leading up to the vote on November 7. But does this Democratic edge mean that the Democrats will win a majority of House seats? Doubts are often expressed about the accuracy of the generic ballot polls. And even if the polls are correct in indicating a majority of *votes* going to Democratic candidates, further doubts are expressed about whether the Democrats' vote margin will be sufficient to win the most seats.[60]

This paper is intended to provide some guidance for translating the results of generic congressional polls into the election outcome.[61] Via computer simulation based on statistical analysis of historical data, we show how generic vote polls can be used to forecast the election outcome. We convert the results of generic vote polls into a projection of the actual national vote for Congress and ultimately into the partisan division of seats in the House of Representatives. Our model allows both a point forecast—our expectation of the seat division between Republicans and Democrats—and an estimate of the probability of partisan control. *Based on current generic ballot*

[60] See, for example, Kastellec, Gelman, and Chandler, (2006).

[61] Respondents typically are asked which party they plan to vote for (or who they want to win) "if the election were being held today," though there is variation in question wording. For instance, some organizations use the wording "Looking ahead to the Congressional elections in November..." Other organizations use "Thinking about the next election for U.S. Congress..."

polls, we forecast an expected Democratic gain of 32 seats with Democratic control (a gain of 15 seats or more) a near certainty. The details follow.

The easy part is forecasting the vote from the generic polls. To properly interpret the generic polls, we estimate a regression equation predicting the vote in the 15 most recent mid-term elections, 1946-2002, from the average generic poll result during the last 30 days of each campaign. (Details are shown in the Appendix [in Chapter 7 (7.4)].) Based on this analysis, we can confidently offer the following rule of thumb for predicting the national vote based on polls over the last 30 days before the election:

1. Convert the percentage point lead, e.g., Democrats 51% Republicans 41%, in the generic poll to a percent Democratic of the two party vote, e.g., 51-41 converts to 55% Democratic or 5% more Democratic than 50-50;

2. If the poll is based on registered voters rather than "likely" voters, subtract 1.5 percentage points—thus, a 56%–44% Democratic lead in a registered voter poll converts to a narrower 54.5%–45.5% lead in terms of likely voters;[62]

3. Cut this lead in half; and

4. Add a percentage point to the Democrats as a reward for being the non-presidential party.

From the regression analysis, our 95% confidence interval for the forecast using this formula is $+/-3.7$ percentage points.[63]

Now, consider the polls over the final thirty days of the 2006 campaign. As of early October 24, *PollingReport.com* listed the results of six likely-voter generic ballot polls conducted during the final 30 days of the campaign, by CNN (2), *USA Today*/Gallup, ABC/*Washington Post*, Fox/Opinion Dynamics, and *Newsweek*. The average Democratic two-party share in these polls is 57.7%. Applying our formula as described above, the Democrats should win 55% of the two-party vote with a confidence interval from 51.3 to 58.7 percent, implying that the Democrats almost certainly would win a majority of the votes cast.

But would this mean that the Democrats would also win the most seats? If the Democrats were to win 55% of the vote, this would represent a 6.4 percentage point swing from 2004, when they received 48.6%. If Democrats were to win exactly 6.4% more of the 2006 vote in every district than they won in 2004, they would win 228 seats. However, an average swing of 6.4% percentage points will be spread unevenly—sometimes more than 6.4% and sometimes less. We must take the degree of uniformity into account. Moreover, as we already discussed, the prediction that the average vote swing will be 6.4% is, itself, subject to error. We take these considerations

---

[62] The registered voter correction represents the average adjustment necessary from generic polls over past mid-terms. Democratic registrants in the past have turned out with slightly greater frequency than today. There is no certainty that this correction will hold exactly in 2006.

[63] The exact equation is: Dem Vote Share = 24.38 + 0.51 * Dem Poll Share − 1.09 * Presidential Party Adjusted R squared = 0.75; Root MSE = 1.90,where Presidential Party takes the value "1" under a Democratic President and "−1" under a Republican. For more detail and analysis, see Bafumi, Erikson, and Wlezien (2007).

into account by a set of simulations described below. The simulations suggest that a predicted national vote surge of 6.4 percentage points would yield the Democrats 235 seats, for a 32-seat gain. This is 7 seats more than with uniform swing.

The simulations are constructed as follows. For each possible generic ballot integer value from 50% Democratic to 60% Democratic, we compute 1,000 simulations of the 435 seat outcomes. Each simulation includes:

1. A random draw from the density of the possible vote outcomes from our generic poll regression equation, based on the predict from the generic poll plus forecast error; and

2. A set of 435 random draws of district level predictions conditional on the 2006 national shock (from 1) plus district-level characteristics and shocks based on a regression model from the 2004 election.[64]

Details are presented in Chapter 7 [see 7.2.3 in this volume].

For each generic ballot integer value from 50% Democratic to 60% Democratic, we have conducted 1,000 simulations of the seat division as per the methodology discussed in the previous paragraph. For each of the possible generic poll outcomes (50% Democratic, 51% Democratic, etc), we provide both a seat forecast (as an expectation) plus a probability estimate regarding whether the Democrats will win a majority (218 seats or more) of the seats.

Figure 4.3 shows the translation of the generic vote into a likely seat outcome with an accompanying probability of a Democratic majority of seats. As can be seen, the threshold on the generic vote where the Democrats are favored to win the seat majority is 51 percentage points (in likely voter surveys) or between 52 and 53 percent (in registered voter surveys). In terms of point spreads, the cut point where the odds of partisan control are even is a spread of about 2 points in likely voter surveys (or about 5 points in registered voter surveys).

Thus, we see that the Democrats need more than the plurality of the generic vote in order to expect to control the House. In the range where the Democrats have between 50 and 52 percent of the "two-party" generic vote, majority control is up for grabs. But if the Democrats have 53% or more of the vote by the generic ballot question, as they do as of this writing, they are the heavy favorites (probability > .90) to win the House of Representatives.[65]

---

[64] Actually, the number of random draws is 374. The remaining 61 districts are assigned automatically to the current party holding the seat, due to unopposed candidacies in 2004.

[65] Readers conditioned to the idea that their districting advantage would allow the Republicans to govern with a minority of votes cast might be surprised that the threshold in terms of the national vote at which control is likely to revert to the Democrats is only slightly greater than 50%–about 51% of the generic vote and about 50.5 of the national vote. (This threshold is at least one percentage point higher, however, in terms of the *mean* district vote.) The explanation is the partisan asymmetry in 2006 retirements. Among retirees who had faced major-party competition in 2004, 19 were Republicans and only 6 were Democrats. Strategic Republican retirements in anticipation of a Democratic wave would cause an electoral ripple even if the larger wave does not arrive. Our calculations are that if there is no vote swing whatsoever from 2004 to 2006, the Democrats would pick up 5 or more seats just from the greater number of Republican than Democratic retirements.

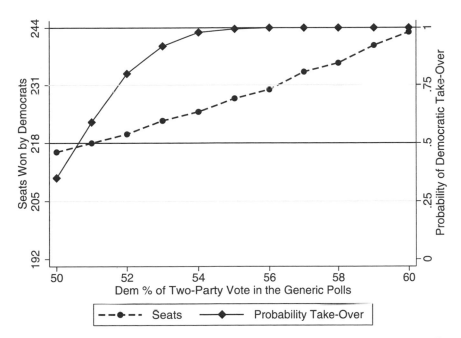

Figure 4.3   Summary of simulated election outcomes predicted from varying generic ballot poll results.

Our forecast based on current generic polls is a Democratic gain in the range of 32 Democratic seats, an amount that exceeds most current forecasts. Do we exaggerate Democratic prospects? A useful reality check is to consult available district level polls and compare them with our predictions for the same districts from our simulations, assuming a 58%-42% split in the generic ballot polls. Averaged across 32 Republican-held districts with October polling, the average district level poll margin is 51.7% Democratic, 48.7% Republican. For these same 32 districts, our average prediction is 50.3% Democratic, 49.7% Republican. The two sets of numbers match nicely. If anything, our simulations might underestimate Democratic strength in the sampled districts.[66]

Of course, if the poll numbers to the generic ballot question shift as the election nears, the forecast should be revised according to the weight of new polling information. Figure 4.3 provides the guidance. If current trends in the congressional generic ballot polling persist (which they have in past election campaigns), the Democrats are near certain to win control of the House. But this assumes a continued Democratic lead of 8 or more points among likely voters in the generic ballot. If the lead dips below this level, the Republicans can rekindle their hope of holding the House.

[66]Like with the generic polls, we should discount the district leads in the polls by about half. Pollsters are most likely to sample preferences in the districts thought to be competitive. Thus, sampled districts might have larger than average Democratic gains.

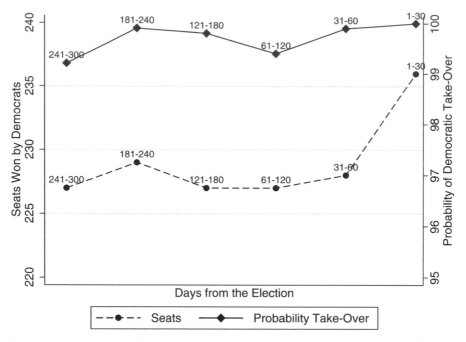

Figure 4.4 Summary of simulated election outcomes predicted from generic ballot poll results over intervals during the election year.

After the Election. The Republicans did not rekindle their hopes in the weeks leading up to the election and the result was very much as we forecasted. Although the distribution of seats is not yet final, we do know that the Democrats will hold somewhere between 231 and 237, pending the counting in six undecided districts. Regardless of how these districts sort out, the final result will be very close to our forecast of 235 Democratic seats. This result confirms what we knew in advance of the election, namely, that the generic polls do contain meaningful information about the national vote swing.

It is important to add that generic polls predict well not only when conducted near the end of the election cycle, as was done in our forecasting exercise. They also predict well early in the election year. Indeed, we could have forecasted the vote throughout the election year using available polls at the time. Figure 4.4 shows the seat outcome and the probability of a Democratic take-over we would have predicted at different intervals throughout the election year. The results are derived as explained above, except that the rule for converting the Democratic generic poll result to a national Democratic swing changes over the intervals studied. Primarily, the amount that the Democrats are rewarded for being the out-party increases in intervals further from the election; that is, since impending balancing sentiments are increasingly incorporated into the polls as the election draws near.[67] While we can detect a bit of wiggle from

---

[67] See Bafumi, Erikson and Wlezien (2007) for more information regarding this conversion.

period to period, the expected verdict was essentially the same from the beginning of the year through to the very end. That is, we could see the Democratic tide coming well in advance.

References

Bafumi, Joseph; Erikson Robert S.; and Wlezien, Christopher (2007). Ideological Balancing, Generic Polls and Mid-term Congressional Elections, available on the Internet at: *http://www.temple.edu/ipa/workingPapers/*.

Kastellec, Jonathan; Gelman, Andrew; and Chandler, Jamie, (2006). Seeking 50% of Seats, Needing More than 50% of Votes: Predicting the Seats-Votes Curve in the 2006 Elections, available on the Internet at: *http://www.stat.columbia.edu/~gelman/research/unpublished/house2006.pdf*.

4.3 REFERENDA ON THE BALLOT

The final paper in this chapter looks at a study of referenda issues in recent past elections—the 2006 election, in particular. While most of the discussion of polling has focused on prediction of winning candidates—especially in presidential elections—polls have also proven useful to assess public sentiment regarding ballot initiatives. In the next reading, Baldassare and Katz (2007) uses survey polls to explore voters' attitudes towards ballot referenda and the public policymaking process as a means of governance.

4.3.1 The Role of Direct Democracy in California Governance: Public Opinion on Making Policy at the Ballot Box
Mark Baldassare and Cheryl Katz

Introduction. In recent years, direct democracy has become an increasingly significant force in state and local policymaking, siphoning power away from elected representatives and toward a "parallel legislature" of governance by citizen lawmaking (McCuan and Stambough, 2005). Notable actions taken through direct democracy include enacting term limits for elected leaders, outlawing the use of racial preferences in hiring and university admissions, requiring a "supermajority" vote for tax increases, prohibiting marriage between same-sex partners, and removing a sitting Governor from office.

Chief among the tools of direct democracy is the initiative, which gives citizens the ability to draft legislation and place it on the ballot through a public petition. Twenty-four states now allow for statewide citizens' initiatives. Other direct democracy provisions include a mechanism to recall elected officials, and the ability to overturn decisions of state lawmakers through a public referendum. While most of the states with provisions for direct democracy added them to their constitutions in the late 19th and early 20th century, the modern era of increasing reliance on ballot initiatives began in 1978, with voters' passage of California's historic tax-limiting

Table 4.8   California statewide propositions 2000–2006

	Number on Ballot	Number Approved
2000	28	17
2001	–	–
2002	13	10
2003	2	–
2004	20	12
2005	8	–
2006	15	7
Total 2000–2006	86	46

Source: California Secretary of State.

measure, Proposition 13 (Waters, 2003). Since then, more than 800 initiatives have appeared on statewide ballots throughout the nation (National Conference of State Legislatures, 2006).

California is at the forefront of the recent surge in direct legislation, having used the initiative process an average of 10 times per election cycle since the passage of Proposition 13 (see Baldassare, 2000). Its initiative use has accelerated in the current century, with a record 86 propositions appearing on the state ballot between 2000 and 2006, including 58 from 2002 on (see Table 4.8). California's initiative use so far this decade has already surpassed the number of statewide measures during the entire 1990s (69), which was the state's previous record. What is more, the rate of passage is rising—from an average of 35 percent throughout the last century (Waters, 2003) to 53 percent from 2000 to 2006 (Baldassare and Katz, 2007). A major factor in this increasing use is the growing trend of special interest groups, such as business, labor, industry and consumer organizations, as well as office-holders themselves, to use the initiative process to enact their policy agendas (California Commission on Campaign Financing, 1992).

That trend passed a new milestone in 2003, with the use of another of California's direct democracy provisions—the recall—to remove Governor Gray Davis from office and elect Arnold Schwarzenegger as his replacement. Since his election, Governor Schwarzenegger has utilized the initiative process more than any other Governor in the nation's history: using either ballot measures or the threat of them to enact virtually his entire policy agenda in his first year in office; calling voters to a special election on his government reform proposals in 2005; and convincing the public to pass a massive package of bond measures in 2006 to finance his plans to rebuild the state's infrastructure. With Schwarzenegger's re-election in 2006, the prominence of direct democracy in the state's policy-making process is all but certain to continue. And, with California and its Governor playing a prominent role on the national political stage, we could see a growing use of ballot initiatives in other states.

Against this backdrop of direct democracy playing an increasing role in California, and with many other states also having provisions for citizen legislation, it is important to examine the public's opinions about the process. Because initiative-driven politics are a direct expression of public opinion, a great deal of scholarly research has focused on voters' perceptions, attitudes, knowledge and information sources on ballot measures, and how and why they make their choices (see, for example, Abrams, 2002; Allswang, 2000; Baldassare, 2000, 2002; Bowler et al., 1998; Bowler and Donovan, 2000; Citrin, 1996; Cronin, 1999; Magleby, 1984; McCuan and Stambough, 2005; Smith and Tolbert, 2004). This paper seeks to contribute to the public opinion literature on direct democracy by examining Californians' changing attitudes toward citizens' initiatives and the factors that influence voters' satisfaction with making policy at the ballot box.

Four Enabling Factors in Direct Democracy. In previous work, we focused attention on the effects of populism, voter distrust, partisanship, and special interests on the use of the citizens' initiative process (Baldassare and Katz, 2007). These four factors were driving forces behind the call for direct democracy during the populist and progressive eras of the late 19th and early 20th centuries, when most of the states that currently allow citizens' initiatives adopted the provision. The time was characterized by public skepticism about big government, widespread mistrust of politicians, and fears that the needs of ordinary people were being ignored (see Cronin, 1999). A series of scandals involving machine politics and corrupt governments under the influence of wealthy special interests fueled a desire for direct democracy as a way to limit the power of politicians and make them more accountable to the people (Cain and Miller, 2001; Cronin, 1999; Hofstadter, 1955; Kazin, 1995).

While the nation's interest in direct democracy tapered off during much of the 20th century, the resurgence of initiative use in the wake of Proposition 13 has largely coincided with increasing distrust in government, concerns about the role of powerful special interests, dissatisfaction with partisan gridlock in the legislative arena, and calls for greater citizen involvement to counter the public's lack of confidence in elected officials (Baldassare, 2000, 2002).

In the California context of the unprecedented use of the ballot box to make policy in this decade, the emergence of a new and more prominent role for ballot initiatives has coincided with the resurgence of the four enabling factors for direct democracy—the perceived effects of partisanship and [of] special interests on state politics, voters' distrust in their elected leaders, and populist attitudes among the public. Specific examples of the trends in these four factors include increasing numbers of voters registering as independents and rejecting the major parties, record levels of expenditures in gubernatorial campaigns raising concerns about the role of special interests, record levels of voters voicing distrust in state government, and strong majorities of Californians consistently favoring direct democracy over representative government as the vehicle for public policy. All of these factors have reached a new level of prominence on the California political scene in the events surrounding the 2003 recall election and efforts to address a wide range of policy issues in its aftermath (Baldassare and Katz, 2007; Baldassare et al., 2004; Baldassare, 2006; Gerston and Christensen, 2004; Lubenow, 2003).

In this paper, we examine whether voters' general attitudes toward direct democracy have changed over time as initiative use increased. Specifically, we compare voters' views in 2000 and 2006 on whether the initiative provision is a good thing or a bad thing, and whether public policy decisions made through the initiative process are generally better or worse than those made by the Governor and legislature. Next, we consider opinions on the use of initiatives in the context of the 2006 California general election, when voters faced 13 measures on the state ballot. Election voters were asked about their satisfaction with the way the initiative process was working in California and whether they felt changes were needed. We explore the effect of the four enabling factors—populism, voter distrust, partisanship, and concerns about special interest influence—on attitudes toward the initiative process. Since Governor Schwarzenegger has raised awareness and use of the initiative process in state governance during his time in office, we also examine whether voters' attitudes about the initiative process are influenced by their approval ratings of the Governor. Using multiple regression analysis, we measure the extent to which the four enabling factors and the Governor's approval ratings contribute to voters' satisfaction with making public policy through ballot measures, as well as their perceived need for changes to the initiative system. We conclude with a discussion of voters' perceptions of specific flaws in the initiative process, and their support for a variety of initiative reforms.

Data and Method.   The public opinion data in this study are derived from the Public Policy Institute of California (PPIC) Statewide Survey, directed by Mark Baldassare. The PPIC Statewide Surveys have been conducted by RDD [Random-Digit Dialing] telephone methodology several times a year since 1998. For this study, we gauge general attitudes and consider trends over time through comparison of surveys conducted in October 2000 and August 2006. A similar methodology was used in each of the surveys, with respondents selected at random from a computer-generated sample of listed and unlisted telephone numbers. Interviewing was conducted by Schulman, Ronca & Bucuvalas, Inc. Respondent interviews took about twenty minutes to complete. Each survey included interviews with 2,000 adults, from which likely voters were selected based on their responses to survey questions about their voting history and status. Interviews were conducted in English or Spanish, according to the respondent's preference. The surveys included the following questions gauging general attitudes toward the initiative process: "Generally speaking, do you think it is a good thing or a bad thing that a majority of voters can make laws and change public policies by passing initiatives?" and "Overall, do you think public policy decisions made through the initiative process by California voters are probably better or probably worse than public policy decisions made by the Governor and legislature?" These questions were also used as measures of populism in analyses of the August 2006 survey results.

PPIC Statewide Survey questions on the use of initiatives in the context of the November 2006 election were taken from a post-election survey of 2,000 voters conducted in the days immediately after the November 2006 general election. The survey followed the same methodology as the other PPIC surveys, but included a question screening for voters who had participated in the recent election. Questions used in this analysis included, "Generally speaking, would you say you are very satisfied, somewhat satisfied, or not satisfied with the way the initiative process is working in California?" and "Do you think the citizens' initiative process in California is in need of major changes, or minor changes, or that it is it basically fine the way it is at this time?"

The August and November 2006 surveys included the following questions on trust in government, concerns about special interests, and partisanship: "How much of the time do you think you can trust the state government in Sacramento to do what is right—just about always, most of the time, some of the time, or never?" and "Would you say the state government is pretty much run by a few big interests looking out for themselves or that it is run for the benefit of all of the people?" and self-identified party registration (Independent, Republican, Democrat, other). The November survey also included the following question on populism: "How much trust and confidence do you have in California's voters when it comes to making public policy at the ballot box—a great deal, a fair amount, not too much, or none at all?"

Both the August and the November 2006 surveys measured opinions of the Governor by asking respondents "Overall, do you approve or disapprove of the way that Arnold Schwarzenegger is handling his job as Governor of California?"

All surveys included demographic questions on age of the respondent, annual household income, education, and gender, which were used in multivariate analyses. The November 2006 survey also included a series of questions on specific perceptions and experiences voting on the 13 measures on the state general election ballot, including perceptions of the quality and quantity of ballot initiatives, the amount of money spent by initiative campaigns, and support for proposals to change the initiative process, such as legislative reviews and funding source disclosures.

Public Opinion on the Initiative Process.

Attitude trends over time. Nearly three in four California voters expressed positive opinions of their state's direct democracy provision in the August 2006 survey, with 74 percent calling it a "good thing" that a majority of voters could make laws and change policies with initiatives, while 21 percent called it a "bad thing." (See Table 4.9.) The proportion giving a favorable assessment in 2006 represented a 6-point increase from October 2000 (68%), while the number making a negative assessment and those with no opinion both fell by 3 points.

The belief that public policy decisions made by voters through initiatives were probably better than those made by the state legislature also grew during the decade, with 60 percent in August 2006 holding this view, up from 53 percent in October 2000. While there was no change over time in the numbers saying that policies made by initiatives were probably worse than those made by the state's elected officials (25% in October 2000, 24% in August 2006) and the proportion rating both types of policy-making as equal (7% vs. 6%), those with no opinion fell five points between

Table 4.9   Trends in support for direct democracy

"In general, do you think it is a good thing or a bad thing that a majority of voters can make laws and change public policies by passing initiatives?"

	October 2000	August 2006
Good thing	68%	74%
Bad thing	24	21
Other/don't know	8	5
$N =$	1,096	1,011

"Overall, do you think public policy decisions made through the initiative process by California voters are probably better or probably worse than public policy decisions made by the governor and state legislature?"

	October 2000	August 2006
Better	53%	60%
Worse	25	24
Same (volunteered)	7	6
Don't know	15	10
$N =$	1,091	1,009

Sources: PPIC Statewide Surveys, October 2000 and August 2006. Likely voters.

2000 and 2006 (15% to 10%), indicating that increasing use of the initiative process had made voters more familiar with it and, at the same time, more positively inclined.

In the 2006 survey, the general perception that it is a good thing that voters can make laws and change public policies by passing initiatives was strongly correlated with the general belief that California voters make better public policy decisions than the Governor and legislature. Of those who said that the initiative process was a good thing, 73 percent believed policy decisions made by the voters were probably better than those made by [the] state's elected officials, while of those who said the initiative process was a bad thing, 62 percent said voters made worse policies than the state's elected officials. This relationship remained significant in a multiple regression equation that also included measures of voter distrust and special interests, approval of the Governor, partisanship (i.e., independent voters), and demographic factors (i.e., age, education, gender, income).

Voters' satisfaction in 2006.   After the 2006 general election, voters were asked about their experiences voting on the 13 initiatives that were on the statewide ballot and their perceptions of the need for initiative reform. Overall, voters were favorable about the way the initiative system was working in California, with 19 percent saying they were very satisfied and 50 percent somewhat satisfied. Only 27 percent were not satisfied. (See Table 4.10.)

Table 4.10   Satisfaction and perceived need for change, November 2006 election

"Generally speaking, would you say you are very satisfied, somewhat satisfied or not satisfied with the way the initiative process is working in California today?"

	November 2006
Very satisfied	19%
Somewhat satisfied	50
Not satisfied	27
Don't know	4
$N =$	1,995

"Do you think the citizens' initiative process in California is in need of major changes, minor changes or that it is basically fine the way it is."

	November 2006
Major changes	35%
Minor changes	32
Fine the way it is	26
Don't know	7
$N =$	1,993

Sources: PPIC Statewide Survey, November 2006. General election voters.

What are the effects of the four enabling factors on voters' satisfaction? In bivariate analyses, satisfaction with the initiative process was positively related to a measure of populism demonstrated by confidence in voters' ability to make public policy at the ballot box. Of those expressing a "great deal" of confidence in voters' decisions, 41 percent were very satisfied with the initiative process, compared to 21 percent of those who said they had a "fair amount" of confidence and only 12 percent of those who had little or no confidence. Trust in government also showed a positive correlation with initiative satisfaction. Twenty-six percent of voters who said they trusted state government to do what is right always or most of the time were very satisfied with the way the initiative process was working at the time, compared to 16 percent of those who trusted state leaders only some or none of the time. Similarly, the belief that state government was being run for the benefit of all the people increased satisfaction with the initiative system. Among those holding this view, 27 percent said they were very satisfied with the system's current performance, compared to 17 percent of those who believed state government was being run by a few big interests. As for the role of partisanship, 22 percent of Republicans were very satisfied with the initiative process today, compared with 18 percent of independents and 17 percent of Democrats. Approval of the Governor was also related to satisfaction with the initiative process. Of those approving of Schwarzenegger's job performance, 22 percent were very satisfied with the way the initiative system was working, compared to 14 percent of those who disapproved.

Although satisfaction with California's initiative provision was generally high in the wake of the November 2006 election, sizeable numbers were calling for reforms of the citizens' initiative process. Two in three voters said changes to the system were needed, including 35 percent saying "major changes" and 32 percent wanting "minor changes." Only 26 percent said the system was "fine the way it is."

How are the four enabling factors related to support for changing the initiative process? Confidence in California voters' performance at the ballot box significantly affected assessments of the need for initiative reforms in bivariate analyses. More than four in 10 (42%) of those with little or no confidence in voters' ability to make good policy decisions said that major reforms were needed. Among those with a great deal and those with a fair amount of confidence in voters' decisions, 28 percent in each group called for major reforms to the initiative system. Trust in government showed a similar relationship with the perceived need for reforms. Of those who said they trusted state leaders always or most of the time, 26 percent believed that the initiative process needed major changes. That position rose to 38 percent among voters with little or no confidence in their state government. Concerns about the influence of special interests also spurred perceptions of the need for initiative reform, with 39 percent of voters who believed state government was being run by big interests looking out for themselves saying that major changes were needed. Among those who thought government was run for the benefit of all of the people, only 24 percent wanted major changes to the initiative system. As for partisan differences, 40 percent of Democrats and 38 percent of independents believed major reforms were needed, compared to only 27 percent of Republicans. Opinions of the Governor were also significantly related to perceptions of the need for change, with 44 percent of those who disapproved of Schwarzenegger's performance in office saying major initiative reforms were needed, compared to 30 percent of those who approved of the way he handled his job.

To further investigate the relationship between the four enabling factors of populism, voter distrust, special interests and partisanship on attitudes about initiative use, we conducted multivariate linear regressions on data from the November 2006 survey of general voters. Table 4.11 presents the determinants of satisfaction with the initiative process, using survey items on trust in state government (voter distrust), concern about government being run for the benefit of a few large interests (special interests), confidence in voters' ability to make good policy decisions (populism), and creating a variable contrasting independent voters against party members (partisanship). We also included approval of the Governor as an independent variable. Gender, age and education were used as demographic control variables.

The results indicate that satisfaction with the way the initiative process was working in the context of the November 2006 election was positively correlated with populism, meaning that those with greater confidence in the voters were more satisfied with the initiative process. Concerns about the influence of special interests on state government, meanwhile, were negatively associated with satisfaction with the initiative process, with those who perceived state government as being run for the benefit of a few large interests expressing less satisfaction with the initiative provision. Voter trust and the measure of partisanship yielded no significant correlations.

Table 4.11   Determinants of voter satisfaction with the initiative process in the November 2006 general election

November 2006:	B	S.E.	Significance
Confidence in Voters	.188	.021	.001
Approve of Schwarzenegger	.183	.037	.001
Concern about Special Interests	−.098	.044	.03
Trust in Government	.026	.033	NS
Independent	−.026	.049	NS
Age	.091	.026	.001
Education	.050	.023	.03
Gender	.004	.034	NS
Income	−.005	.023	NS

R Square $= .100$

Note: NS refers to not significant
Source: PPIC Statewide Survey, November 2006 ($N = 2,000$).

Approval for Schwarzenegger's performance as Governor led to greater satisfaction with initiatives. Of the demographic variables, greater satisfaction was correlated with younger age and lower education levels, while gender and income were not significant factors.

Table 4.12 examines the factors determining perceptions that the initiative process is in need of reform. We used the same independent variables as in Table 4.11 for our measures of populism, distrust, special interests, and partisanship, as well as opinions of the Governor. The results of this analysis show that support for initiative reform was negatively correlated with populism; that is, lower confidence in voters' ability to make good public policies at the ballot box created the desire for greater reforms to the process and higher confidence in voters' reduced support for changing the initiative process. Support for initiative reform was positively correlated with the belief that state government is being run by special interests. As for partisanship, support for major reforms to California's initiative process was positively correlated with being an independent voter. There was no significant relationship with trust in government. However, disapproval of Schwarzenegger's performance was associated with calls for greater initiative reforms, while approval of the Governor reduced support for changing the initiative process. Of the demographic variables, the perception that the initiative system is in need of reform was greater among women and older voters, while there were no significant correlations with education or income.

Conclusions. Ballot box legislation is a dominant force in California governance today. Public support for the state's initiative provision is high across all segments of society. And, as shown by increases in approval during a decade with the highest ini-

Table 4.12   Determinants of perceived need for changes to initiative process in the November 2006 general election

November 2006:	B	S.E.	Significance
Confidence in Voters	−.158	.026	.001
Approve of Schwarzenegger	−.192	.044	.001
Concern about Special Interests	.154	.053	.01
Trust in Government	−.007	.039	NS
Independent	.125	.059	.03
Gender	−.114	.041	.01
Age	−.063	.031	.04
Education	.036	.028	NS
Income	.028	.027	NS

R Square $= .072$

Note: NS refers to not significant
Source: PPIC Statewide Survey, November 2006 ($N = 2,000$).

tiative use ever, it appears that the more Californians use this tool of direct democracy, the more they like it.

A main reason for growing support of the initiative process is the widespread perception that public policy decisions made by voters at the ballot box are better than those made by the Governor and state legislature. This populist view among California voters has also grown as initiatives gained increasing prominence in state politics. Californians' confidence in their ability to make public policy decisions seems to have been bolstered by the state's voters being increasingly asked to do so.

In the context of the November 2006 election, voters reported an overall positive experience, with nearly seven in 10 satisfied with the way the initiative process was working. This appears to negate fears that voters would become "burned out" on direct democracy after being called to the polls six times in four years (including twice for special elections). Yet a similar proportion call for changes in the initiative system, reflecting underlying concerns about the process and a perceived need for some reforms.

The multivariate analyses found that the four enabling factors in evidence during the growing use of initiatives had mixed effects in the 2006 election context. The measure of populism—confidence in voters' ability to make good public policy at the ballot box—consistently contributes to support of and satisfaction with the process. Trust in government, meanwhile, yields significant correlations with support for the initiative process in bivariate analyses, but not in the multiple regressions. The influence of partisanship, as measured by contrasting the responses of independent voters against those belonging to political parties, contributes to support for initiative reforms, but is not significantly correlated with satisfaction with the use of initiatives.

Future research should use attitudinal measures of partisanship, such as satisfaction with the major parties, which were not available in our 2006 post-election survey. Meanwhile, voters' concerns about the role of special interests—generally viewed as bolstering support for the use of initiatives in making public policy—actually had the effect of lowering satisfaction with the way the process is currently working, and raising calls for reforms.

An important implication of the latter finding is that voters' concerns about special interests having undue influence in state politics not only diminish their confidence in elected representatives (Baldassare et al., 2004; Baldassare, 2002) but also lower their satisfaction with the initiative process today. These results reflect the powerful role that special interest groups are playing in financing California's initiative campaigns. Efforts to enact legislation through direct democracy today generally involve the services of professional political consultants, requiring sizeable outlays for expenses, such as signature-gathering, campaign management, and advertising. The huge amounts of funding and professional know-how now required to mount an initiative campaign increase the role of wealthy individuals and special interest groups (see Broder, 2000; Gerber, 1998, 1999; Magleby, 1984; Magleby and Patterson, 2000; McCuan et al., 1998; McCuan and Stambough, 2005).

Spending on initiatives has skyrocketed in recent elections in California, with the various campaign committees laying out a total of about $317 million on the eight ballot measures in the 2005 special election and about $330 million going to 13 propositions in November 2006. Much of this money comes from special interest groups that are spending millions of dollars to fight initiative campaigns, such as the $84 million spent by pharmaceutical companies to quash a November 2005 initiative seeking to lower prescription drug costs and the $95 million spent by oil companies to defeat a November 2006 California proposition taxing oil producers to fund research on alternative energy (Baldassare and Katz, 2007). In this context, it is not surprising that perceptions of the role of special interests are related to dissatisfaction with the initiative process.

What specific complaints are emerging as the initiative process takes on a greater and more frequent role in policymaking? Once again, concerns about special interests have considerable influence on voter attitudes. In the November 2006 post-election survey, many voters felt that there were too many measures on the ballot (60%) and that the wording of these measures was too complicated and confusing (63%). But the biggest complaint was that there was too much money spent by the initiative campaigns (78%). These perceptions of ballot initiatives were widely held across all political and demographic groups.

Building on these specific areas of dissatisfaction, many voters express support for changes in the initiative process. In line with concerns about the number of ballot items and confusion about their content, 80 percent of November election voters favored including a period of time for legislative review to see if compromises could be reached before initiative proponents placed their measure on the ballot, while 72 percent of likely voters in the October pre-election survey favored a system of review and revision to avoid legal issues and drafting errors. Once again, special interest influence was the chief concern, with 84 percent of November election voters favoring an increase in public disclosure of funding sources for initiative campaigns. All three of these proposals for initiative reform enjoyed broad support across political and demographic groups.

Another interesting finding is the powerful role of approval of the state's leading elected official, Governor Schwarzenegger, in shaping current opinions of the initiative process. Those who approve of Schwarzenegger's performance in office are more likely to express satisfaction with the way that the initiative process is working today and less likely to favor changes in the system of direct democracy. One might expect that positive opinions of an elected representative would be related to negative opinions of direct democracy, but the opposite is true in the current political context. Clearly, this reflects the extent to which Schwarzenegger has linked himself to governing through the initiative system since his election in the 2003 recall. Moreover, during his campaigns and time in office, he has continually stressed the themes of populism, distrust in the political establishment and disdain for the power of special interests; he has positioned himself as an independent, "post-partisan" politician. The correlation between approval for Schwarzenegger and satisfaction with the initiative process in 2006 is all the more impressive in light of voters' opposition to his 2005 initiative-only special election and their rejection of all the measures he endorsed on that ballot.

Californians can look forward to the latest sequel in the growing use and influence of initiatives in 2008—a year in which voters will face initiatives in a February presidential primary, June primary, and November general election. As of this writing, three measures have qualified; 11 are in circulation for signatures; and five are pending approval, including measures to reform legislative term limits and redistricting. It would not be surprising for the decade's total to exceed 100 state propositions. The 2008 election cycle offers another opportunity to test the popularity of direct democracy, the effects of the enabling factors of populism, distrust, partisanship, and special interests, and voter opinions on possible ways to improve the process.

Acknowledgments.   We would like to thank the Public Policy Institute of California and the James Irvine Foundation for their support of the PPIC Statewide Survey and acknowledge the research support by Dean Bonner, Jennifer Paluch, and Sonja Petek. The opinions and conclusions expressed are those of the authors, and do not represent the views of the Public Policy Institute of California.

References

Abrams, Elliott (ed.) (2002). *Democracy: How Direct? Views from the Founding Era and the Polling Era,* Lanham, MD.: Rowman & Littlefield.

Allswang, John M. (2000). *The Initiative and Referendum in California, 1898–1998,* Stanford, CA: Stanford University Press.

Baldassare, Mark (2000). *California in the New Millennium: The Changing Social and Political Landscape,* Berkeley, CA: University of California Press.

——— (2002). *A California State of Mind: The Conflicted Voter in a Changing World,* Berkeley, CA: University of California Press.

——— (2006). *California's Exclusive Electorate,* San Francisco, CA: Public Policy Institute of California.

Baldassare, Mark; Cain, Bruce E. ; Apollonio, D.E.; and Cohen, Jonathan (2004). *The Season of Our Discontent: Voters' Views on California Elections,* San Francisco, CA: Public Policy Institute of California.

Baldassare, Mark and Katz, Cheryl (2007). *The Coming Age of Direct Democracy: California's Recall and Beyond,* Lanham, MD: Rowman and Littlefield.

Bowler, Shaun and Donovan, Todd (2000). *Demanding Choices: Opinion, Voting and Direct Democracy,* Ann Arbor, MI: The University of Michigan Press.

Bowler, Shaun; Donovan, Todd; and Tolbert, Caroline J. (eds.) (1998). *Citizens as Legislators: Direct Democracy in the United States,* Columbus, OH: Ohio State University Press.

Broder, David (2000). *Democracy Derailed,* San Diego, CA: James H. Silberman/Harvest/Harcourt.

Cain, Bruce E., and Miller, Kenneth P. (2001). The Populist Legacy: Initiatives and the Undermining of Representative Government, *Dangerous Democracy? The Battle over Ballot Initiatives in America,* Sabato, Larry J.; Ernst, Howard R.; and Larson, Bruce A. (eds.), Lanham, MD: Rowman and Littlefield, 33-62.

California Commission on Campaign Financing (1992). Democracy by Initiative: Shaping California's Fourth Branch of Government, Los Angeles, CA: Center for Responsive Government.

Citrin, Jack (1996). Who's the Boss? Direct Democracy and Popular Control of Government, *Broken Contract? Changing Relationships between Americans and their Government,* Craig, Stephen C. (ed.), Boulder, CO: Westview Press, 268–293.

Cronin, Thomas E. (1999). *Direct Democracy: The Politics of Initiative, Referendum and Recall,* Cambridge, MA: Harvard University Press.

Gerber, Elisabeth (1998). *Interest Group Influence in the California Initiative Process,* San Francisco, CA: Public Policy Institute of California.

———— (1999). *The Populist Paradox: Interest Group Influence and the Promise of Direct Legislation,* Princeton, NJ: Princeton University Press.

Gerston, Larry N., and Christensen, Terry (2004). *Recall: California's Political Earthquake,* Armonk, N.Y.: M E. Sharpe.

Hofstadter, Richard (1955). *The Age of Reform,* New York, NY: Random House.

Kazin, Michael (1995). *The Populist Persuasion: An American History,* New York, NY: Basic Books.

Lubenow, Gerald C. (ed.) (2003). *California Votes: The 2002 Governor's Race and the Recall that Made History,* Berkeley, CA: Berkeley Public Policy Press (University of California, Berkeley).

Magleby, David B. (1984). *Direct Legislation: Voting on Ballot Propositions in the United States,* Baltimore, MD: Johns Hopkins University Press.

Magleby, David B., and Patterson, Kelly D. (2000). Campaign Consultants and Direct Democracy: The Politics of Citizen Control, *Campaign Warriors: The Role of Political Consultants in Elections,* Thurber, James A. and Nelson, Candace J. (eds.), Washington, DC: The Brookings Institution, 133–152.

McCuan, David; Bowler, Shaun; Donovan, Todd; and Fernandez, Ken (1998). California's Political Warriors: Campaign Professionals and the Initiative Process, *Citizens as Legislators: Direct Democracy in the United States,* Bowler, Shaun; Donovan, Todd; and Tolbert, Caroline (eds.), Columbus, OH: Ohio State University Press, 55-79.

McCuan, David, and Stambough, Steve (eds.) (2005). *Initiative-Centered Politics: The New Politics of Direct Democracy,* Durham, NC: Carolina University Press.

National Conference of State Legislatures (2006). Ballot Measures Database, available on the Internet at: *www.ncsl.org/programs/legismgt/elect/dbintro.htm,* December 1, 2006.

Smith, Daniel A., and Tolbert, Caroline J. (2004). *Educated by Initiative: The Effects of Direct Democracy on Citizens and Political Organizations in the American States,* Ann Arbor, MI: University of Michigan Press.

Waters, M. Dane (2003). *Initiative and Referendum Almanac,* Durham, NC: Carolina Academic Press.

4.4 SUMMARY OBSERVATIONS

The readings in this chapter focus on the 2006 mid-term elections, but have much broader application. While their research efforts vary considerably, the authors apply methodological innovations to multiple regression models to make predictions from pre- and post-election polls in their areas of interest. Whether the topic is polling accuracy, identifying likely voters, predicting the impact of mid-term elections on the make-up of the Senate and House of Representatives, or determining support for ballot initiatives as a means of governance, the authors all provide information on different techniques that will be useful in upcoming elections, as well. Their approaches certainly suggest means for improving predictions in 2008.

The 2006 mid-term elections were also a time when methods were tried that—if successful—could be of use in the next presidential election. (See, for example, Pedraza and Barreto, 2007; Keeter, 2007; Chang and Holland, 2007; and Mulrow and Scheuren, 2007.) Because of their close connection to the 2008 elections, we decided to cover these activities in Chapter 6 of this volume.

Editors' Additional References

Bafumi, Joseph; Erikson, Robert; and Wlezien, Christopher (2006). Forecasting House Seats from Generic Congressional Polls, available on the Internet at: *http://www.temple.edu/ipa/workingPapers/*.

Bafumi, Joseph; Erikson, Robert; and Wlezien, Christopher (2007). Ideological Balancing: Generic Polls, and Mid-term Congressional Elections, paper presented at the meeting of the American Association for Public Opinion Research, Anaheim, CA, available on the Internet at *http://www.temple.edu/ipa/workingPapers/*.

Baldassare, Mark and Katz, Cheryl (2007). The Role of Direct Democracy in California Governance: Public Opinion on Making Policy at the Ballot Box, paper presented at the meeting of the American Association for Public Opinion Research, Anaheim, CA.

Chang, LinChiat and Holland, Keating (2007). Evaluating Follow-up Probes to *Don't Know* Responses in Political Polls, paper presented at the meeting of the American Association for Public Opinion Research in Anaheim, CA.

Harrison, Chase (2007). Poll and Races: The Impact of Methods and Conditions on the Accuracy of 2006 Pre-Election Polls, paper presented at the meeting of the American Association for Public Opinion Research, Orlando, FL, May 18, 2007.

Keeter, Scott (2007). How Serious Is Polling's Cell-Only Problem? The Landline-less Are Different and Their Numbers Are Growing Fast, paper presented at the U.S. State Department Conference on Survey Methodology, July 16, 2007.

Mulrow, Edward and Scheuren, Fritz (2007). Producing a 2008 National Election Voter Scorecard, paper prepared for this volume.

Murray, Gregg R.; Riley, Chris; and Scime, Anthony (2007). A New Age Solution for an Age-Old Problem: Mining Data for Likely Voters, paper presented at the meeting of the American Association for Public Opinion Research, Anaheim, CA.

Newport, Frank (2006). Frank Newport on Likely Voters, Guest Pollster's Corner, Pollster.com, Oct. 1, 2006, available on the Internet at: *http://www.pollster.com/blogs/frank_newport_on_likely_voters.php*.

Pedraza, Francisco and Barreto, Matt (2007). Exit Polls and Ethnic Diversity: How to Improve Estimates and Reduce Bias Among Minority Voters, paper presented at the meeting of the American Association for Public Opinion Research, Anaheim, CA.

CHAPTER 5

GLOBE-TROTTING CONSULTANT

5.1 INTRODUCTION

In recent years, Warren Mitofsky did a lot of international consulting on exit polling. In this chapter, he relates some of his experiences. Then, we provide some research papers that describe polling efforts in other countries. This begins with a historical account of election polling in Mexico. Next, a small selection of charts present some information from political polls in Muslim countries. These are followed by a case study that discusses several methodological approaches introduced to address statistical issues in Mexico. Two subsequent readings look at use of the Internet for election polling—one in Japan and the other in Sweden. The chapter concludes with a reprint of a paper that examines voters in a Québec, Canada referendum on independence.

We begin with Mitofsky's recollections of international technical assistance. As he points out, he often faced challenges that were very different from those he had dealt with in America.

Editors' Note: Text in gray shading is provided by the editors of this volume. Boxed-in text represents excerpts from interviews with Warren Mitofsky. The balance of the chapter is made up of papers prepared by contributing authors. [Text in square brackets was added by the editors.]

"So was the 1994 Russian Duma Election in Moscow the first overseas election?"

"No, I did one in March or April of 1993 for the Pool. It was in Russia. The Russians had four referenda on the ballot for that election. I did the first national exit poll in a number of countries. I did the first in the U.S., in Russia and in Mexico. There are some state exit polls earlier, but I did the first national one. I can't call them exit polls, but I did the first national projections for the Marcos/Aquino election in the Philippines in 1986. I did the first exit poll for the 2004 Taiwan presidential election. Those were [some of] the countries where I have worked."

An Eastern European Country

"...I did [an Eastern European country] too. That was a parliamentary election; we did 124 parliamentary districts. The 125th was a little difficult. It was made up of the refugees all over the country. The precincts were scattered, but they all counted for the same district. We didn't do that one...I got emails from people...which I answered. I answered all the email about methods and what we were doing...'You're working for the government,' one email said. I said, 'If we're working for the government, I certainly haven't been told about it...'"

"...I managed to find people to work with; one company through this NGO [non-governmental organization], and two others through my Russian partner, Vladimir Andreenkov. Vladimir did the ...election with Joe Lenski and me. Nobody told us how to do the exit poll or anything else about what to do. We had a contract that said we make the decisions about what we are doing..."

"The partisans did not like our results and neither did the pro-government people. I may be off by 1 or 2 districts, but in a 110 or 112 of the districts, there was nothing unusual about the result. They were about what the election commission had reported. In the others, we had vast differences with the election commission about the final result. We didn't say they committed fraud, but we said that we could not substantiate their results and, furthermore, for those districts that were also exit polled by the group working for the U.S. State Department, we had the same results as their exit polls."

"To finish the ...story, everything was fine there until election night. We hired three survey organizations..., two to do the interviewing at the precincts—one from the NGO we worked for, one through Vladimir. The third organization did none of the interviewing. We hired them to monitor the work of the other two, to go around to the polling places and observe everything and keep records and call us when there were differences with the procedures we wanted. We weren't able to observe all of the polling places. We had close to 1,500 of them. This exit poll was as big as a U.S. election. But they monitored about a third of them, all over the country. We did all the precincts...where the problems would occur. There were only about 20 districts in the country where there was expected to be a close contest. The rest of them were about 80-to-20 races. Each district had 40 to 45 precincts each. In the closer ones, we selected 20 polling places...and in the others we just sampled 10 precincts each."

"We used the same election computer system that we used for the 2004 U.S. presidential election. The computer system was here in the U.S.; the programmers were in an office just north of New York. Some were at the computer site, and we were in [the country where the election was taking place]. We had two phone rooms . . . taking vote returns in. We had the vote recorded on paper and others put the data into the computer."

"On election night, our clients didn't want us to release any results. Our whole relationship with them changed. It started when a limousine came to pick me up at our election headquarters. Two people I knew were in the car, and the three of us and the driver drove through the back streets . . . at 7 o'clock at night just as the polls closed. We eventually turned down a dark alley. Then, we turned down another alley, and the cars stopped. We all got out. There was a steel door on what was a wall enclosing a courtyard. We went into this darkened courtyard, up some stairs, and there was this long conference table. I must tell you, I was getting a little nervous. No one from our group came with me. . . . There were people that I recognized, including a guy . . . who was scary. We discussed the results of the exit poll. I had two sets of results: the individual district-by-district results and a summary of seats won by each party. . . Everyone at the meeting knew the party line-up was meaningless. They said they would get back to me and let me know what I could distribute."

"Once back at our office, I finally got a call telling me I could give out the meaningless party summary and that I should not give out the winners of individual seats. Well, we had scheduled a press conference for about 9:00 p.m. on election night to announce our results, except no one notified the press. I got to the press conference and I was the only one there!"

"Meanwhile, we discovered that the polling organization that worked for the NGO had given us phony results for [one set of] districts. They had the supervisors call-in the phony results to us. Except what they forgot to do was to tell the interviewers not to call us! So, before the supervisors called, the interviewers had already called. And the interviewers gave us the right data. Vladimir's daughters, Anna and Alena, were supervising the operators and spotted the duplicate reports— the fraudulent ones—and held them out of the computer. . . . What the supervisors did was to rearrange the numbers so their preferred candidate was ahead. They were putting the big vote with the pro-government candidate and the small numbers with the opposition candidate. Late that night, we managed to get real vote tallies from a few hundred precincts in [that area], and they confirmed the first call from the interviewers."

"We knew how the whole thing happened by the next morning—who arranged it and how it was carried out. Monday, we were told we could release our projection for some of the districts, but not all of them. We released all but 30. Tuesday we got on a plane and went home. Then, we published all remaining results. Our sponsors were furious! . . . What the dissidents did not like is that we said that in 112 districts, there were no problems. They were mad because of that. The other side, our NGO, was mad because we said in 12 districts we disagreed with the election commission."

"A newspaper in [the country] caught [the full results that we posted on the Web in the U.S.] and they . . . published our article in their paper after they translated it. . . And if you think the people that hired us were mad when we put up the numbers on our Web site, they saw the article and they were perplexed!'"

Another Eastern European Country

"I finished with the networks on a Saturday night and [on] Sunday . . . I was in the capital of another country organizing an exit poll. They said I should give them the results of my work for no charge, as I worked for them. I said I did through Saturday, 'but on Sunday, when the election takes place, I don't work for you.'"

"When I did the first . . . exit poll, the biggest problem was how to get the data in from the precincts. The sample was 120 precincts spread over 11 time zones. So, how do we get the data in? My first election working with Vladimir was in 1993. He had only three telephones and one of them was connected to a fax. This was in the days when it was almost impossible to get telephones. . . . His offices were at the typical . . . institute, where they're all academics that are working in the commercial world, trying to make a few bucks doing survey research, or, I should say, market research. So Vladimir got some of his neighbors to drop phone lines out the windows. We set up some other neighbors in the area near the institute to receive some of the calls and, then, we developed a scheme to speed the whole thing up. Interviewers did not get paid much, so at every polling place we hired two interviewers, and we also had one supervisor for every two polling places. We had the interviewers transcribe the answers to each exit poll questionnaire onto answer sheets. There were only 12 questions on the exit poll. We gave a row on the answer sheet to each respondent and a column to each answer, and they transcribed maybe 20–25 respondents per page; then, the supervisor faxed it to us. Every post office had a fax machine. The election was on a Sunday, so we gave the Postmaster a few bucks to let us use the fax machine. We were able to fax in a lot of the data. That took much less telephone time than reading each answer to each question over the phone. That's how we got all the data into a computer."

". . . I said that when we started we were estimating the number of votes, not percentages. . . In '93, we tried the same kind of a model, estimating the number of votes and I had a terrible time. The reason I had a terrible time was that I still couldn't get decent measures of size for the precincts. And finally, I think I gave up and sampled them with equal probability. . . "

A Country in Southeast Asia

"The same problem collecting the data happened in Southeast Asia, for the 1986 . . . election, but for a different reason. . . [The incumbent] controlled the country. All the phone systems were bugged, and we were convinced none of them would work on Election Day. We did not believe we could collect the results from our interviewers by telephone. So, what we did to get the data was to set up a ham radio network. We built a tower . . . 500 feet above sea level."

"What we did to get the data in was to transmit everything in code. We gave the interviewer a precinct ID that was a number. We gave them a script and they filled it in. They just read numbers that identified coded answers. We gave them a whole bunch of questions about activities taking place at the polling places. Was there something going on at the polling places that prevented people from voting? Were there goons trying to influence the vote? Did the election appear to be operating normally? Things like that."

"We asked the interviewer a whole bunch of questions and all the answers had codes, and, of course, the votes were numbers. The people receiving the calls were working very near our office on top of a hotel. They would get a call and the interviewer would say '3, 1, 2, 4, 9, 6, 8.' All they would hear is a string of numbers, and they would write it all down on the forms we gave them and then key them into the computer."

"So maybe the people writing this down didn't know what it meant either?"

"Oh they knew. We had been working with them; I'm sure we told them. We had to tell them to make sure they didn't put out numbers for each category that were out of range."

"You bought the computers in Hong Kong?"

"Remember, this was in '86 and computers weren't as available as they are now. I remember I sent someone to Hong Kong to buy the computers. If I was to bring two computers on the plane from New York..., first off they would be heavy...they [the Hong Kong computers] were knock-offs. I left the computers in [the country when I left]. I gave them to somebody. I wasn't going to haul this stuff back to New York. They were 220 volts and heavy."

A Latin American Country

"[The election] was '94. I had actually been there in '93, but I didn't do any work until '94. In July '93, I went . . . to observe how they were doing exit polling. This was the polling organization attached to the President's office. There was a fellow there from Spain, Bernardo—something Bernardo—a well known. . . Bayesian. . . The presidential election was held in the summer of '94. Originally, I was just going to be an advisor. They had brought together 11 local survey companies to do the work. They wanted it to be as broad and ethical as possible. They were going to do it for all the media, and it was going to be paid for by the equivalent of their national association of broadcasting, but that's not what they call it. All of the organizations, one after the other, kept falling by the wayside. We ended up with two organizations, one of which we. . . gave the interviewing to do and the other one did all of the field work. I didn't much like them either. Well, I worked with them that year but not after that. And we did the presidential election in '94."

"Well [the election in] '88 had been widely believed to be fraudulent. I don't believe it, but *The New York Times* does, because they kept mentioning it in every story. So [the President] wanted exit polls to help the credibility of the election."

"Anyway, [Bernardo] came [from Spain] with software. There was someone there from France. There were a couple of us from the U.S... So, on Election Day, they said 'Aren't you going to make an estimate? Bernardo is making an estimate.' He brought his software and his Bayesian model; I hadn't brought anything. I said, 'Would you like an estimate?' and they said, 'Yes.' So I had a little credit card calculator. I said 'Can you give me the precinct returns as percentages?'... because they had selected them PPS [probability proportionate to size]. I said 'Give me the precincts as percentages' and I told them how to organize them. And I used the little calculator and I made a post-stratified estimate and mine was better than Bernardo's!"

"So, the next day, the fellow who ran the President's office said '[The President] would like to meet with you.'... I've never met a U.S. President. So, I went... with this fellow who invited me, Ulises Beltrán... The chief of staff is there, the President is there. We talked for a while about exit polls and he said, 'I want you to have exit polls done in the presidential election. I want the election to be more credible and exit polls are independent of the vote count.' So he says 'What do I have to do?'"

"I said, 'There are a few things you can do: one is you can't do them here in your office. They have to be done independently by a neutral source or a source perceived to be neutral. The second thing is they have to be reported promptly after poll closing by the media.' He said, 'Is there anything else?' I looked at him and said, 'Your party could lose an election every once in a while.' The other two laughed, but he didn't. He didn't think much of that."

"... [the President] ... was to come make a speech at the U.N. [United Nations] a few days before he left office. He stayed at The Plaza [Hotel] and they had a big dinner for him at the Metropolitan Museum of Art on a Sunday, when the museum was closed. I was invited to come over to The Plaza to meet with [him] and he thanked me for doing the exit poll that confirmed the results. Then, he made this speech and in the middle of his speech he publicly thanked me."

"... They lost in 2000. The likely predictor was losing;... they'd been in power since 1929, the people tired of them, but that's not why they lost. The base for the [ruling party] was the poor, the uneducated, and the rural people. But... by 2000, their base had eroded... because the President's, and his successor's, programs were so successful. What they did with their programs was managed to raise the level of education. Many people left the farming areas, and ... [the ruling party] managed to reduce its base by about 10 percent... The support for the opposition parties was ... in the cities, better off, the middle class and up. So, you have a growing middle class, a declining rural population, and programs that were successful..."

The excerpt above highlights some of the difficulties that Warren Mitofsky and his colleagues encountered in conducting election polling in other countries. While hardly representative of polling efforts worldwide, the rest of this chapter provides a few recent papers that discuss historical and methodological issues encountered in polling in countries other than the United States. This is but a brief smattering of reports, but serves to demonstrate that election polling has been found useful throughout other democracies and that some solutions may bear application in other areas to address similar challenges.

5.2 A HISTORICAL LOOK AT EXIT POLLS FROM A MEXICAN PERSPECTIVE

We began this chapter with Mitofsky's reminiscences about his experiences in introducing exit polls in other countries. In what follows, Bautista et al. (2007) briefly recount the history of exit polling in Mexican elections.

5.2.1 Excerpt from: Exit Polls as Valuable Tools to Understand Voting Behavior: Using an Advanced Design in Mexico—History
René Bautista, Marco A. Morales, Francisco Abundis, and Mario Callegaro

Exit polls appeared in Mexico in the late 1980s as a tool for strategic decision-making for politicians who needed early information to assess the cost of tampering with electoral results. Yet the media became interested in them by the mid-1990s, only after they proved their usefulness to politicians.

The Milestone Elections. The history of exit polling in Mexico can be better understood through milestone elections, where the behavior of significant actors highlights the political relevance of exit polls. In that regard, five federal and local election years can be examined: the 1989 gubernatorial election in Baja California and the Institutional Revolutionary Party's (PRI) primary election in Acapulco in the State of Guerrero; the 1994 presidential election; the 1997 election of Mexico City's Mayor; the 2000 presidential election; and the 2006 presidential election. We look at them as specific events to understand their particular context, but also as part of a larger political process to underscore their relevance and explain the state of exit polls in Mexico today.

Before 1989, attempts to conduct exit polls publicly had been frustrated. The clearest example of this was the 1988 presidential election, when the Gallup organization and its Mexican partner—the Instituto Mexicano de Opinión Pública (IMOP)—tried to carry out an exit poll on July 6. On July 1, however, the executive secretary of the Federal Electoral Commission sent a letter prohibiting such an exit poll, arguing that it would violate Article 41 of the Mexican Constitution, which guarantees the secrecy of voting (Basañez, 1995; Domínguez and McCann, 1996; Mitofsky, 1994). As a matter of fact, the few public opinion polls carried out before 1989 were conducted privately by the then-ruling PRI or individually by party members (Camp, 1996:5).

For all practical purposes, the history of exit polls begins in the administration of President Carlos Salinas (1988–1994), who appointed full-time pollsters to his staff, although reserving polling information to a few decision-makers. To the best of our knowledge, the first exit poll was conducted on July 1989, to gauge the outcome of the gubernatorial election in the State of Baja California. As the early results from the exit poll arrived, it became clear that the PRI candidate had been defeated. PRI had lost its first gubernatorial election since its founding in the late 1920s, and an exit poll had informed the party's decision to acknowledge its defeat.

The events of 1989 did not end there. In October, José Francisco Ruíz Massieu, Governor of the State of Guerrero, hired a polling firm—Centro de Estudios de Opinión Pública (CEOP)—to carry out an exit poll during the PRI's primary election, to select the mayoral candidate for Acapulco. Historically, party leaders were in charge of selecting PRI's candidates without consulting its base. Thus, a primary election and an exit poll to select candidates were previously unseen events. It seems that by using the information on the exit poll, Governor Ruíz Massieu tried to anticipate the winner of the Acapulco primary election to build strategic alliances before the actual vote count was delivered (Moreno, 1996).

During the early 1990s, exit polls had only been used privately by PRI politicians, but no public use of them had been made. All this changed with the federal electoral reform of 1994, when polls and quick counts were first publicly discussed.[68] Politically, the inclusion of exit polls and quick counts in the 1994 reform was necessary to vouch for the confidence that opposition parties and the general public were required to give to the recently created Federal Electoral Institute.

With the explicit acknowledgment of exit polls and their methodology in the law, the media's interest turned to them and crystallized on the first media-sponsored exit poll in the 1994 presidential election. However, doubts about the neutrality of the electronic media and the effectiveness of exit poll methodology were such that the backing of the entire broadcasting industry was necessary. The National Chamber for the Radio and Television Broadcasting hired Mitofsky International—Warren Mitofsky's firm—along with two other polling firms—Buró de Investigación de Mercados (BIMSA) and Indemerc-Louis Harris—to conduct the exit polling project (Beltrán, 2006, 2007; Mitofsky, 1994). The print media also ventured their own exit poll, *El Nacional* newspaper sponsored another exercise to be carried out by the polling agency Gabinete de Estudios de Opinión (GEO). PRI managed to win the election by more than 20% of the vote, so exit polls were not used to "validate" electoral results. Importantly, once the media took upon themselves the responsibility of conducting exit polls for Mexican elections, there was no turning back. Furthermore, as confidence in

---

[68] Without regulations, exit polls were vulnerable to the "secrecy of voting" argument; hence, the importance of establishing regulations in the secondary law. On the other hand, the 1994 reform to the Electoral Code obliged polling firms to submit their polling methodologies to the Federal Electoral Institute, but also forbade them to publish pre-election polling results a week prior to the Election Day. As per exit polls, it obliged polling agencies to keep their results secret until the final polling place had closed on Election Day.

the predictive power of exit polls increased, along with the perception that they could contribute to foster fair elections, the pressure to use them in subsequent elections increased.

It would take three more years for exit polls to show their effects in the political arena. The Mayor of Mexico City was to be elected for the first time in 1997, after decades of being a presidential appointment post. By the end of the campaign, it was clear that the PRI-controlled government of Mexico City was bound to be transferred to a candidate from the left: Cuauhtémoc Cárdenas, the most powerful opposition figure to PRI at that time. The once-unthinkable outcome became *fait accompli* once the two largest TV networks revealed that their exit polls projected the leftist candidate as the winner of the election. The effects were twofold: on the one hand, exit polls showed once again that they could be accurate instruments to measure electoral outcomes, but on the other hand, they proved that the media could be relied on to reveal election night projections, even when they did not favor PRI. On this occasion, Warren Mitofsky was hired by Televisa—one of the two major television networks—and partnered with Consulta, an emerging Mexican polling firm (Beltrán, 2006). Since then, Mexico's most important newspapers, along with the two major TV networks, have sponsored, reported, and analyzed exit polls as a matter of course (Camp, 2003).

In the 2000 presidential election, exit polls consolidated their role as a fixture of democratic normality in Mexico. The 2000 election was a momentous event, not only because the Mexican President, PRI-affiliated Ernesto Zedillo, refused to select his successor, as per tradition, but because PRI would finally lose the presidency after almost seventy years in power. During the last week of the presidential campaign, when the law forbids pre-election polling results to be made public, rumors about the possibility of a PRI defeat increased, and a remarkable uncertainty surrounded Election Day. Under an expected tight race, exit polls and quick counts would be crucial to give credence to the outcome of the election. Several exit polls were conducted on Election Day. Some of them had political intentions, like the one promoted by Democracy Watch, which hired the polling firm Penn, Schoen & Berland to support Vicente Fox's victory. But the majority contributed to improve the perception of independent exit polls. That was certainly the case of Televisa's exit poll—which hired the Mexican polling agency Consulta, now associated with Mitofsky International (Mitofsky, 2000)—and the *Reforma* newspaper exit poll, both contributing to increase the perception of genuine media-sponsored Election Day surveys. They all accurately projected the official results, which implicitly substantiated Vicente Fox's victory speech and President Zedillo's public message acknowledging PRI's defeat. This proved that exit polls could work as an instrument of certainty in the atmosphere of uncertainty that the Mexican democracy implied.

In spite of the credibility amassed by exit polls in Mexico as of 2000, a boisterous setback occurred regarding the 2006 presidential election. In the weeks preceding the election, polls had shown a tight race between the conservative candidate Felipe Calderón from the National Action Party (PAN) and the leftist candidate Andrés Manuel López Obrador from the Party of the Democratic Revolution (PRD), while PRI's candidate—Roberto Madrazo—was trailing in a relatively distant third place.

Furthermore, López Obrador had publicly denounced a "fraud" committed against his candidacy. With these conditions, exit polls seemed bound to face their ultimate litmus test: call accurately the winner of an election that had been tied for too long.

On the Election Day, eight exit polls and quick counts were carried out by a dozen different polling firms. Their results were published by nine different media outlets on election night (Beltrán, 2007). Exit poll results commissioned and publicized by political parties and candidates became, themselves, news that night, given the vacuum generated by media-contracted pollsters, who, because of the narrow difference, were not able to call the winner of the election. Despite the fact that the media presented a united front to explain that the election was too close to call, the parties and their candidates released exit poll data which presumably supported their victory without giving consideration to the constraints of the method that revealed a tie. As a matter of fact, it took weeks until the official result was delivered, and PAN's Felipe Calderón turned out to be the winner by a razor-thin margin of less than a half percentage point. Needless to say, there were negative consequences on the credibility of exit polls resulting from both leading presidential candidates going public on election night presenting only numbers favorable to themselves; it was an election simply *too* close to call.

Unquestionably, exit polls responded to the needs of those purposely looking for information in 2006. Both media and political parties were interested in knowing the winner of the election. However, during the Election Day, the media waited for as much information as possible, using not only exit polling data but quick count data, to project a winner, whereas parties required a constant updating of the information over the course of the day, only exit polling data, to fine-tune the operations that would increase their chances of victory and, ultimately, to declare themselves winner of the election. But what happened on July 2, 2006, simply reflected what had been happening in local elections during previous years: the two leading candidates proclaimed themselves winners on election night using their own exit polling numbers. In the end, 2006 became the year when exit polls were reduced to partisan instruments and were practically stripped away from the media.

Thus, exit polls have played—and still play—a political function in Mexico: first, as instruments of politicians to assess political costs, later to "prevent" electoral frauds and, today, as means to support claims of victory in tied elections.

References

Basañez, M. (1995). Problems in Interpreting Electoral Polls in Authoritarian Countries: Lessons from the 1994 Mexican Election, *International Social Science Journal*, 47(4), 643-650.

Beltrán, U. (2006). Warren in Mexico, paper presented at the American Association for Public Opinion Research, New York, NY.

Beltrán, U. (2007). Warren in Mexico: Elections Become Citizens' Events, *Public Opinion Pros,* January.

Camp, R. A. (1996). Introduction, *Polling for Democracy: Public Opinion and Political Liberalization*, Camp, R.A. (ed.), Wilmington, DE: Scholarly Resources Inc.

Camp, R. A. (ed.) (2003). *Politics in Mexico: The Democratic Transformation*, New York, NY: Oxford University Press.

Domínguez, J. I. and McCann, J. A. (1996). *Democratizing Mexico: Public Opinion and Electoral Choices*, Baltimore, MD: Johns Hopkins University Press.

Mitofsky, W. J. (1994). The 1994 Mexican Elections: Electoral Credibility Was the Issue, *Public Perspective,* November/December, 17–18.

Mitofsky, W. J. (2000). At the Polls: Mexican Democracy Turns the Corner, *Public Perspective,* September/October, 37-40.

Moreno, A. (1996). The Political Use of Public Opinion Polls: Building Popular Support in Mexico under Salindas, *Polling for Democracy: Public Opinion and Political Liberalization,* Camp, R.A. (ed.), Wilmington, DE: Scholarly Resources Inc.

5.3   MUSLIM OPINIONS ON POLITICAL REFORM

Next, we provide some results from public opinion data on politics and democracy in Muslim countries. In a recent speech, Fares Braizat (2007) drew from several sources to present information on Islamic sentiments regarding political reform. Unfortunately, we were unable to obtain his written text, but felt it was useful to include several of Braizat's charts, which demonstrate that political polls are, indeed, being conducted globally, including in the Arab world.

5.3.1   Excerpt From: Muslims and Democracy: Stability and Change in Public Opinions[69]
Fares Braizat

Figure 5.1 compares support for radical reform, gradual change, and maintenance of the status quo. It demonstrates that most countries polled indicated a strong sentiment for gradually introducing political reform. At the time of the surveys, the two countries most interested in maintaining the status quo were Bangladesh and Jordan—both countries with majority Muslim populations.

Figure 5.2 compares support for democracy versus authoritarian rule. All the countries polled expressed preferences for a democratic form of government, although Turkey, in particular, shows strong support for authoritarian leadership. There was an election coming up in Turkey at the time of the polling and, apparently, the need for a continuation of the strong secular traditions arising out of Turkey's creation as a modern nation state was very much on everyone's minds.

---

[69]Muslims and Democracy: Stability and Change in Public Opinions, by Fares Braizat, prepared for the U.S. State Department, 2007.

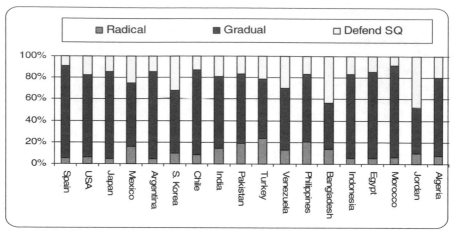

Note: SQ is Status Quo

Figure 5.1   Demand for reform.

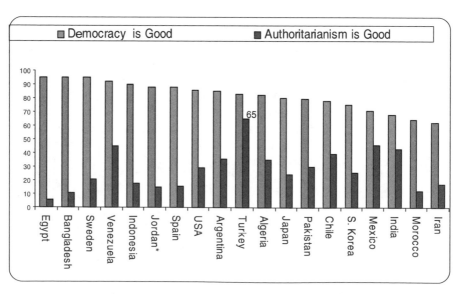

Figure 5.2   Support for democracy and authoritarianism.

		Country		
		Algeria 2004	**Palestine 2003**	**Iraq 2004**
Political	**Democracy with Religion**	39.0	45.1	42.7
	Secular Democracy	45.0	37.2	43.3
System	**Religious System without Democracy**	10.0	11.2	6.8
Preference	**Secular Non-Democracy**	6.0	6.5	7.2

Figure 5.3   Proportion of respondents preferring various models of government.

Figure 5.3 compares results of a poll where respondents from selected Arab countries were asked about different political systems. All expressed preferences for a form of democracy, although respondents from Algeria leaned more heavily towards a secular government, while those from Palestine and Iraq expressed more interest in a form of government tied to Islam.

Another point made in the polling, but not fully drawn out in the charts shown here, is that, from an Arab public opinion perspective, political reform is needed, demanded, and supported, while political violence is increasingly rejected. These data provide a glimpse at the variety of opinions and the importance of tracking those sentiments as they change over time.

5.4   A MEXICAN CASE STUDY

In another look at Mexican elections, we examine three different issues related to exit polls—one concerning sample design and the other two involving nonresponse and measurement error. First, René Bautista et al. (2005) consider a mixed mode approach for exit polls of limited literacy populations and describe a solution for collecting voting information in a confidential manner. This approach is particularly important in the developing countries, where—especially among older and rural people—literacy rates may be low.

5.4.1   Excerpt from: Nonresponse in Exit Poll Methodology: A Case Study in Mexico—Mixed Modes
René Bautista, Mario Callegaro, Jose Alberto Vera, and Francisco Abundis

Since the beginning of the first exit poll–Kentucky's gubernatorial election of November 1967 (Mitofsky,1991)—exit polling data in the U.S. have exclusively been collected using self-administered methods, i.e., interviewers hand ballots to every kth voter leaving the polling place. During the self-administered process, interviewers approach potential respondents and, after a brief introduction, provide a ballot on a pad with a pen so the respondent can fill it out. After completion, respondents drop it in a ballot box next to the interviewer. In the early exit polls, only ballots were given out; however, after Mitofsky's suggestion of lengthening the ballot, demographic questions, as well as relevant political questions concerning the electoral race, were added (Mitofsky, 1991). This data collection method has two assumptions:

1. Voters can read and understand questions well enough to give a reasonable answer; and

2. Self-administered ballots minimize socially desirable responses.

While the first hypothesis is taken for granted and it has not been explicitly mentioned in the exit polling papers, the second assumption has been discussed by several authors, such as Frankovic (1992), Levy (1983), and Mitofsky (1991). As Bishop and Fisher (1995) pointed out, the idea of self-administered questionnaires as an attenuating condition of socially desirable responses was not empirically tested in the early days of exit polls; nevertheless, pollsters extended some public opinion findings to the exit poll methodology field. A formal experiment on face-to-face vs. self-administered ballots was done in 1992 (Bishop and Fisher, 1995). The results show that respondents in the face-to-face condition refused to reveal the candidate whom they voted for in the presidential election, at a higher rate than those in the self-administered condition. An important finding is that the secret ballot technique was more accurate in estimating the final outcome on the most socially sensitive issue of the ballot: a vote against a tax levy for elderly services. Similar results were found by Traugott and Price (1992) concerning the 1989 State of Virginia gubernatorial race. They found that the failure of correctly predicting the election outcome was more related to the face-to-face data collection method employed by the survey organization, than to the sampling design. All these findings suggest that a self-administered ballot design is a better methodology to collect data in exit polling than an interviewer-administered methodology.

Not all aspects of data collection are under the researcher's control (Couper and Groves, 1996; Groves and Couper, 1998). Consider, for example, social environment and respondent characteristics. The researcher can only adapt the best research methodology in order to maximize the quality of the data collection. One of these characteristics is the literacy level of the target population. Since a self-administered questionnaire requires that the respondent (voter) can read and comprehend questions, low literacy levels preclude the use of a self-administered instrument. If such instruments are used anyway, they can create biased results.

This is an important issue in other contexts different from the U.S., because in those contexts it is not possible to assume that potential respondents have adequate reading skills to answer questions without interviewer assistance. Therefore, exit polling in countries outside the U.S. is dramatically different (e.g., Mexico).[70] To provide an example of this difference, when Mitofsky did the first tests of exit polls in 1976 in the U.S., the national median school years completed was 11.8 for persons 15 years old and over (U.S. Census Bureau, 2000) and nowadays it is 12.1; in Mexico, for 2004, the national estimated average for persons aged 15 or older was 8.04 years of education (SEP, 2004). Thus, unlike the U.S., an interviewer-administration mode is critical in the Mexican context. Due to voters' weak abilities to read and write, a mixed mode data collection method may be best. It can help minimize possible biases and increase exit poll data quality.

During 2004 and 2005, four state-level exit polls were conducted in Mexico by Parametria SA de CV, an independent Mexico City-based polling firm. The exit polls were conducted as follows:

1. The interviewer approached the kth voter after exiting the voting place and introduced him/herself explaining the purpose of the interview.

2. A face-to-face interview took place collecting demographic data, government approval, highest grade in school, self-perception of social stratum, and political awareness.

3. At this point, a simulated ballot was given to the respondent who filled it out in a self-administered mode and then placed it in a portable ballot box carried on the interviewer's shoulder. This ballot was a black-and-white reproduction of the official ballot, containing candidates' names and political parties' logotypes.

4. The interview is then concluded with some more demographic questions in a face-to-face mode.

The entire procedure tries to minimize social desirability even if it cannot be eliminated completely (DeMaio, 1984), allowing, at the same time, collection of additional information. The average time to complete the interview was about 12 minutes.[71] Mixed mode data collection methods (interviewer-administered and self-administered) work quite well for collecting data in contexts where voters have weak

---

[70] A low level of education can be an issue for some minorities in the U.S., as well. For example, in an exit poll of Mexican-American voters in Chicago (Michelson and Pallares, 2001) the research team gave the voter the option to complete the exit poll either face-to-face or in self-administered mode (Michelson, 2005).

[71] A data center facility located in Mexico City was prepared to receive the telephone calls from the field team. Interviewers were calling, through a free toll number, to report questionnaire by questionnaire, including ballots. Given that voting places were open from 8:00 a.m. to 6:00 p.m., there were three waves of data transmissions. The first was around 10:30 a.m., the second at about 1:30 p.m.; and the third one about 5:00 p.m. Data entry personnel at the facility data center received telephone calls and the information was entered by means of ad hoc computer software.

reading and writing skills. Exit poll results gathered using these mixed modes are pretty close to the population parameter.

References

Bishop, G. F. and Fisher, B. S. (1995). Secret Ballots and Self-reports in an Exit-Poll Experiment, *Public Opinion Quarterly, 59*(4), 568–588.

Couper, M. P. and Groves, R. M. (1996). Household-level Determinants of Survey Nonresponse, *New Directions for Evaluation,* Braverman, M. T. and Slater, J. K. (eds.), 70, San Francisco, CA: Jossey Bass, 63–79.

De Maio, T. J. (1984). Social Desirability in Survey Measurement: A Review, *Surveying Subjective Phenomen,* Turner, C. F. and Martin, E. (eds.), 2, New York, NY: Russell Sage, 257–282.

Frankovic, K. A. (1992). Technology and the Changing Landscape of Media Polls, *Media Polls in American Politics,* Mann, T. E. and Orren, G. R. (eds.), Washington, DC: The Brookings Institution, 32–54.

Groves, R. M., and Couper, M. P. (1998). *Nonresponse in Household Interview Surveys,* New York, NY: John Wiley & Sons, Inc.

Levy, M. R. (1983). The Methodology and Performance of Election Day Polls, *Public Opinion Quarterly, 47*(1), 54–67.

Michelson, M.R. (2005). Personal communication, April 3, 2005.

Michelson, M.R., and Pallares, A. (2001). The Politicization of Chicago Mexican Voters: Naturalization, the Vote, and Perceptions of Discrimination, *Aztlán, 26*(2), 63–85.

Mitofsky, W. J. (1991). A Short History of Exit Polls, *Polling and Presidential Election Coverage,* Lavrakas, P. J. and Holley, K. K. (eds.), Newbury Park, CA: Sage, 83–99.

SEP (2004). *Informe de Labores 2003–2004,* Mexico City, Mexico: Secretaria de Education Publica.

Traugott, M. W., and Price, V. (1992). A Review: Exit Polls in the 1989 Virginia Gubernatorial Race: Where Did they Go Wrong? *Public Opinion Quarterly,* 56(2), 245–253.

U.S. Census Bureau. (2000). *Years of School Completed by Persons 14 Years Old and Over, by Age, Race, and Sex, for the United States: March 1967,* Washington, DC: U.S. Census Bureau, Population Division.

Speed Bump reprinted with permission.
Copyright ©2004: Dave Coverly.

Bautista et al. (2005) also examined nonignorable nonresponse.[72] This very common problem is also addressed in earlier chapters.

5.4.2 Excerpt from: Nonresponse in Exit Poll Methodology: A Case Study in Mexico—Nonignorable Response
René Bautista, Mario Callegaro, Jose Alberto Vera, and Francisco Abundis

... Nonignorable nonresponse occurs when the probability of not answering depends on the variable of interest that cannot be explained completely by means of information in the sample and adjustable by modeling. The main concern about nonignorable nonresponse is that inferences can have serious faults if the average of the nonrespondent segment differs significantly from the average of respondents in the same segment regarding the parameter of interest. Unfortunately, in most cases it is highly expensive to take a sample from nonrespondents in exit polls in order to verify if true differences between respondents and nonrespondents exist. As an initial step toward the study of nonresponse in exit polls, it is important to determine the type of nonresponse. In exit polls there are two kinds of nonrespondents: refusals and misses (Merkle and Edelman, 2002). A *refusal* occurs when a sampled unit (kth voter) does not respond to the request to be surveyed. A *miss* happens when the interviewer is too busy to approach the selected voter or when the voter does not pass the interviewer. According to Merkle and Edelman (2002), about three-fourths of nonresponse consists of refusals and about a quarter are misses. Thus, more time and effort should be devoted to the analysis of the former than the latter, because it can be argued that misses are more likely to occur at random than refusals, which clearly imply an explicit reluctance from the kth voter. Setting aside the "misses" part in nonresponse (assuming that it is "missing at random"), two concerns about refusals can be considered:

1. Those who refuse to participate may have different voting behavior from those who were actually interviewed. Also, typically,

2. Refusals during the exit poll (meaning a low response rate) will decrease precision in exit poll estimates, because they can lower the final n.

In order to test the first statement, it is necessary to take a sample of those who refused to participate during the exit poll. This kind of sample can be expensive and somewhat inconclusive (i.e., if a sample of nonrespondents were taken, it could still have nonrespondents, still leaving uncertainty about the hypothetical behavior of true nonrespondents). On the other hand, the statement about the relationship between refusals and an exit poll's precision can be better approached for studying.

---

[72] For a discussion of nonignorable nonresponse, see Rubin (1976).

Table 5.1   Percentages of refusals in exit poll by state

Group	Veracruz	Tlaxcala	Puebla	Guerrero
Male ($\leq$ 40 years)	26.0	21.1	23.6	22.8
Male ($>$ 40 years)	30.3	27.5	25.9	29.7
Female ($\leq$ 40 years)	22.6	25.0	24.8	22.4
Female ($>$ 40 years)	21.1	26.4	25.7	25.1
X^2	38.05**	18.68**	2.35	19.90**
Sample size	1,908	1,918	1,688	1,497

** $p < 0.05$

In order to keep track of those kth voters who refuse to participate, each interviewer was given a worksheet, which was attached to the back of the interview pads. To the best of their ability, the interviewers recorded age (coding either younger or older than 40) and gender of those who refused to participate, based on observation alone. Moreover, a survey among interviewers concerning problems during the exit poll was conducted at the end of the Election Day. Interviewers were asked about problems with electoral officials and political parties' representatives, conflicts in general, number of polling place's exits, distance from the polling place, troubles keeping up with the interviewing rate, problems understanding the question's wording, if they received proper training, experience as interviewer, and demographic questions.

One of the most important questions in such a follow-up survey is who the non-respondents are and how they behave. The first question can be better answered than the second one. In our Mexican case study, evidence suggests the most likely group to refuse at exit polls is men age 40 or older, in particular those living in rural or mixed areas.[73] As can be seen in Table 5.1, men 40 years old or more are more likely than women of the same age to refuse to be interviewed. These findings are congruent with previous findings in the U.S., which illustrate that women's response rates are slightly higher than men's response rates (Groves and Couper, 1998).

It is not possible to verify if [nonrespondents] truly have a different voting behavior from the respondents belonging to the same segment (a representative nonrespondent sample would be needed). However, it can be hypothesized that refusals are missing data at random, given covariates. Such a hypothesis implies that it is enough to determine their voting behavior by considering preferences from the respondent group holding the same demographic characteristics. [This, at best, is only a rough approximation. Evidence from the U.S. suggests that there is a polling bias which, even after adjustment, favors Democratic candidates.]

[73] Mixed areas are a combination of urban and rural areas.

References

Groves, R. M., and Couper, M. P. (1998). *Nonresponse in Household Interview Surveys,* New York, NY: John Wiley & Sons, Inc.

Merkle, D. M., and Edelman, M. (2002). Nonresponse in Exit Polls. A Comprehensive Analysis, *Survey Nonresponse,* Groves, R. M.; Dillman, D. A.; Eltinge, J. L.; and Little, R. J. A. (eds.), New York, NY: John Wiley & Sons, Inc., 243–257.

In their full paper, Bautista et al. (2005) go on to explore whether there is a relationship between response rates and exit poll errors, as well as analyzing other factors that could impact the data quality.[74] The paper provides insights that may have bearing on U.S. quality concerns, as well. As already noted, the nonresponse issue experienced in the recent U.S. elections was a source of considerable controversy—especially in the 2004 presidential election. (See Chapter 3 of this volume.)

Bautista et al. (2007) examine ways to reduce measurement error in exit polls by using variations in questionnaire design to address missing data by applying multiple imputation with planned missingness. The excerpt below, from their 2007 paper, describes this approach. In a comparable discussion relating to U.S. exit polls, Edison Media Research and Mitofsky International (2005) examine factors contributing to inter-precinct variability. See Chapter 3 (3.2.1) for an excerpt of this effort.

5.4.3 Excerpt from: Exit Polls as Valuable Tools to Understand Voting Behavior: Using an Advanced Design in Mexico—Measurement Error

René Bautista, Marco A. Morales, Francisco Abundis, and Mario Callegaro

The Scholarly Use of Exit Polls in Mexico: A Rising Tide Lifts All Boats. One remarkable advantage of using exit polls for voting behavior research is that they can help reduce measurement errors prevalent in pre- and post-election surveys. Those measurement errors typically are a function of three factors:

1. Closeness to the time when vote choice is made (i.e., volatile vote intention in pre-election polls)

2. Faulty vote reporting in post-election surveys (i.e, social desirability bias as long as the identity of the winner is known and misremembering when some time has elapsed between the election and the survey interview)

3. Screening for likely voters (i.e., pre- and post-election polls inevitably interview respondents that may not go to vote).

---

[74] The complete version of Bautista et al. (2005) is available on on request through *www.votingsystems.us*.

As exit polls collect information *minutes after* the vote has been cast, while the winner of the election is still *unknown*, these particular problems can be minimized. Yet, one of the most relevant reasons why exit polls are not commonly used by scholars analyzing elections relates to the need to limit interview time, otherwise risking measurement error in vote choice, which is the primary aim of exit polls.

In other words, the main challenge to enable exit polls to become an essential tool in voting behavior research is to have a design that *can* collect more information *without* jeopardizing the collection of vote choice. The design we advance here—*planned missingness* with *multiple imputation* (PM-MI) *exit polls*—enables the collection of a plethora of information by using variations in the questionnaires, to later use all available information *as if* it had been collected simultaneously. *Planned missingness* (PM)—collecting several questionnaires from the same target population—produces various small *n* data sets. If these groups of observations were treated as independent when performing an analysis, they might produce biased conclusions, as relevant information is excluded. So the necessary additional step is to use *multiple imputation* (MI) techniques to "fill in" the missing values and generate plausible values for the missing information to produce a "completed" data set (Rubin, 1987).[75] The imputation is not deterministic in the sense that only one value is attributed to each respondent that can be replicated later on, but stochastic in the sense that it produces a set of $m > 1$ plausible values that are generated using the available information and taking into account the covariation among variables. In the PM-MI case, the imputation is "clean," since all survey responses share sampling methodology, are collected on the same date by the same polling organization, all respondents answer a series of questions that are presumably adequate predictors of the missing values, and "almost all of the missingness is due to unasked questions" (Gelman et al., 1998: 847).[76] By virtue of this procedure, all the available data from the exit poll can be used *as if* all questions had been asked to all respondents (Graham et al., 1996; Gelman et al., 1998), producing consistent and efficient estimates of the parameters of interest (Rubin, 1987; King et al., 2001).[77]

The idea of having different short questionnaires in the same exit poll was pioneered in Mexico by Warren Mitofsky in the 1994 presidential election exit poll and tested again by Mitofsky International and Consulta in the PRI presidential primary exit

[75] Readers interested in methods devised to deal with missing data are referred to Weisberg (2005). Readers interested in further details on multiple imputation and the algorithms developed to implement it are referred to Rubin (1987); King et al. (2001); Gelman et al. (2004); and Weisberg (2005). Readers interested in available software for multiple imputation are referred to Horton and Kleinman (2007). Readers interested in MI applications to political science are referred to King et al. (2001) and Honaker and King (2006).

[76] A good imputation for survey data should account for the sample design (Rubin, 1996; King et al., 2001), otherwise one risks producing inconsistent estimates. Researchers might want imputations that include "cluster effect" dummy variables for every sampling unit and the survey-design weights (Reiter et al., 2006) and eliminate imputed *y*'s from the analysis, especially in data sets with high missingness or a low *m* (Von Hippel, 2007).

[77] To account for the uncertainty associated with MI, the estimation involves three steps (Rubin, 1987): 1) impute $m > 1$ values for each missing observation so to produce m data sets; 2) perform the usual data analysis on each of the *m* data sets; and 3) use these m estimates to compute point estimates and variances for the parameters of interest.

poll in 1999 (Mitofsky, 2000).[78] Mitofsky's contribution is taken a step further in Parametría's 2006 exit poll to enhance scholarly research. Given the high demand for information on election night 2006—which resulted from a tight presidential race—and its inelastic supply, given the fixed number of exit polling agencies—it would not be unusual to have syndicated exit polls—i.e., various stakeholders sharing the cost of the exercise–to reduce costs while providing information to multiple clients. Parametría—one of Mexico's largest polling firms—was commissioned to conduct a nationwide exit poll by three different clients with particular needs for information: *Excélsior* newspaper, the Party of the Democratic Revolution (PRD) and the Federal Electoral Institute (IFE). All sponsors, except IFE, agreed to allow Parametría to use their respective portion of information for academic research.

Parametría's exit poll collected information from 7,764 voters on Election Day, which yielded an approximate sampling error of ± 1.1% with 95% of statistical confidence. It was the result of a multistage sampling design, having precincts as primary sampling units, which were selected with probability proportionate to size. A total of 200 precincts was drawn as a nationwide sample, where a range of [7, 70] voters were interviewed.[79] Because of literacy problems in the country, a mixed mode data collection method was implemented (Bautista et al., 2007).[80] First, the interviewer approached the selected respondent in order to ask—in a face-to-face mode—the respondent's information on gender, age, education, income, social class, presidential approval, and vote choice for President and the Chamber of Deputies. Then, a black-and-white facsimile of the official ballot was handed to the respondent who would deposit it in a portable ballot box.[81] Lastly, a final set of questions—those that varied across respondents—was administered to the respondent face-to-face. The questionnaires fielded maintained a set of common questions to produce a database with information common to *all* respondents, with certain pieces of information that were common to only *subsets* of respondents.

Four different versions of the last battery of questions were rotationally administered. Hence, Version "A" asked respondents their recollection of having watched President Fox's administration's campaign ads and whether respondents were beneficiaries of several social policy programs. Version "B" asked respondents to assess which candidates and parties had produced the most negative or positive tone during the presidential campaign. Version "C" asked respondents to place candidates, parties, and themselves on a 7-point ideological scale, their party identification, and motivations for vote choice. Version "D" gathered information to document voters' perceptions of the voting experience. While being able to provide its clients with the required information, Parametría was also able to produce accurate electoral projections. As shown in Figure 5.4, exit poll estimates were highly accurate and well

---

[78][See Section 5.2.1 of this volume for a historical account of recent Mexican elections.]

[79]The number of voters varied since respondent availability is a function of the turnout rate and response rate in each particular precinct.

[80][See Section 5.4.1 in this volume for information on the mixed mode approach.]

[81]Each ballot facsimile contained a control number that would allow matching reported vote choice with the information collected in the questionnaires.

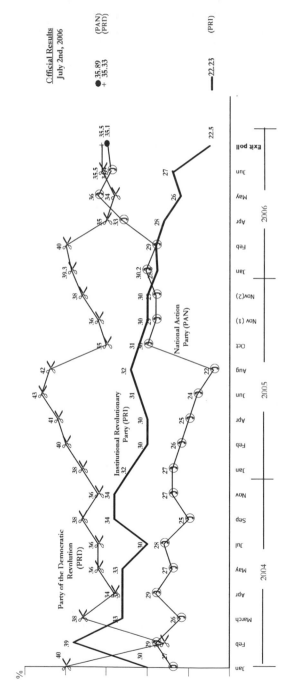

Figure 5.4 Parametría's pre-election series on presidential vote intention: 2004–2006.

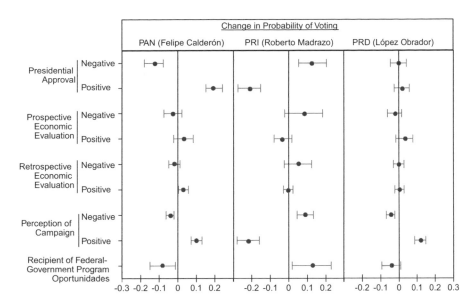

Figure 5.5   First-difference estimates and 95% confidence intervals on voting in the 2006 Mexican presidential election.

within the margin of error, suggesting that the variations in the questionnaires did not represent a source of error in the vote choice estimation.

Analyzing Data from a Planned Missingness-Multiple Imputation Exit Poll.   The most interesting feature of *planned missingness* with *multiple imputation* in the exit polling framework is that it provides the means to conduct more profound vote choice analyses. Remarkable accounts were advanced by political pundits and journalists to explain the outcome of the 2006 presidential election in Mexico. Among the most important were the intensive negative campaigns in which the National Action Party (PAN) and PRD camp engaged; the controversial "continuity campaign" promoted by Fox administration; and the federal cash-transfer and social spending programs, the state of the economy, and the "coat-tail" effects of the high approval of President Fox that presumably helped the PAN candidate. At the time we write, only two analyses of the outcome of the 2006 Mexican election using exit polling data have been published (Moreno, 2007; Estrada and Poiré, 2007). Although these academic pieces shed light on the 2006 result, their scope was limited by data availability. Interestingly, the PM-MI design of Parametría's exit poll allowed accumulating more information to address all competing arguments advanced to explain the outcome of the election in the same econometric model.

Figure 5.5 shows a graphical summary of the main findings from a multinomial probit analysis of vote choice performed on the PM-MI exit poll data. The expected values (dots) of *first differences* (King, 1998) and their associated 95% confidence intervals correspond to the simulated changes for an average voter in the probability of voting for a given party in the 2006 election—when variables change from the reference to the indicated category. Briefly, the analysis (detailed elsewhere, Morales et al., 2007) shows that the so-called "continuity campaign" had no effects on vote choice; the tone–negative or positive—of the campaigns *did* matter; economic conditions had a differentiated effect on vote choice; presidential approval was the single most important predictor of voting for Calderón (PAN), but had a negligible effect on voting for López Obrador (PRD); and President Fox's important poverty-alleviation federal program, *Oportunidades,* which provided cash subsidies to people in need, seems to have had a strong impact on vote choice, but favoring the Institutional Revolutionary Party's (PRI) Roberto Madrazo instead of the PAN's Felipe Calderón.[82] In a nutshell, by virtue of the PM-MI design, we were able to have accurate projections of vote choice on election night and perform comprehensive analysis of vote choice taking advantage of measures with a minimal dose of certain types of measurement errors.

The Importance of Exit Polls in Mexico: Concluding Remarks.   Exit polls are undoubtedly the most influential contribution of the media industry to the study of public opinion and voting behavior. Not only have exit polls become a powerful tool for networks to project the winner of elections as the last vote is cast; they are also a powerful instrument to understand why voters cast their ballots the way they do. However, even when exit polls have helped reduce the prevalent uncertainty on Election night, they also seem to have lagged behind in their potential to understanding voting behavior.

This [excerpt] illustrated that a particular exit polling design combined with the appropriate statistical techniques can put exit polls at the center of voting behavior research. That is, *planned missingness-multiple imputation* (PM-MI) *exit polls* can help overcome some otherwise insurmountable problems in data collection and reduce some sources of measurement error, at minimal cost for both exit pollsters and voting behavior researchers, especially since such a design seems unlikely to affect the accuracy of vote choice estimates.[83] As exit polls will continue to be used as a critical source of information for a variety of political actors—ranging from political parties to electoral authorities—syndicated exit polls are likely to be a more demanded product in elections to come. Taking advantage of this particular context, we were able to empirically explore additional explanations for the outcome of the 2006 election, which otherwise would have been impossible using traditional exit poll designs.

[82] This last finding is controversial, at best, and different results have been advanced elsewhere by applying techniques to correct unbalances in the data (Díaz-Cayeros et al., 2006) But our purpose here is simply to illustrate the use of the data from the exit poll to perform the econometric analysis, not produce an accurate estimate of the effect of *Oportunidades*.

[83] Evidently, this is still a matter that would require further empirical exploration, especially as there might be some unforeseen sources of non-sampling error (Weisberg, 2005).

References

Bautista, R.; Callegaro, M.; Vera, J.A.; and Abundis, F. (2007). Studying Nonresponse in Mexican Exit Polls, *International Journal of Public Opinion Research,* 19 (4), 492–503.

Díaz-Cayeros, A.; Estévez, F. et al. (2006). Buying Off the Poor: Effects of Targeted Benefits in the 2006 Presidential Race, Conference on the Mexico 2006 Panel Study, Boston, MA.

Estrada, L. and Poiré, A. (2007). The Mexican Standoff: Taught to Protest, Learning to Lose, *Journal of Democracy,* 18(1), 73–87.

Gelman, A.; Carlin, J. B. et al. (2004). *Bayesian Data Analysis*, New York, NY: Chapman & Hall/CRC.

Gelman, A.; King, G. et al. (1998). Not Asked and Not Answered: Multiple Imputation for Multiple Surveys, *Journal of the American Statistical Association,* 93(443), 846–857.

Graham, J. W.; Hofer, S. M. et al. (1996). Maximizing the Usefulness of Data Obtained with Planned Missing Value Patterns: An Application of Maximum Likelihood Procedures, *Multivariate Behavioral Research,* 31(2), 197–218.

Honacker, J., and King, G. (2006). What to Do about Missing Values in Time Series Cross-Section Data, Los Angeles, CA: University of California, Los Angeles-Harvard University.

Horton, N. J. and Kleinman, K.P. (2007). Much Ado About Nothing: A Comparison of Missing Data Methods and Software to Fit Incomplete Data Regression Models, *The American Statistician,* 61(1), 79–90.

King, G. (1998). Unifying Political Methodology, *The Likelihood Theory of Statistical Inference*, Ann Arbor, MI: University of Michigan Press.

King, G.; Honaker, J. et al. (2001). Analyzing Incomplete Political Science Data: An Alternative Algorithm for Multiple Imputation, *American Political Science Review*, 95(1), 49–69.

Mitofsky, W. J. (2000). A Week in the Life of an Exit Pollster, *Public Perspective,* March/April, 38–42.

Morales, M.; Bautista, R. et al. (2007). The 2006 Mexican Election: Overcoming Limitations in Analyzing a Complex Outcome, New York University-University of Nebraska-Lincoln, unpublished manuscript.

Moreno, A. (2007). The 2006 Mexican Presidential Election: The Economy, Oil Revenues, and Ideology, *Political Science and Politics,* January, 15–19.

Reiter, J. P.; Raghunathan, T. E. et al. (2006). The Importance of Modeling the Sampling Design in Multiple Imputation for Missing Data, *Survey Methodology,* 32(2), 143–149.

Rubin, D. (1987). *Multiple Imputation for Nonresponse in Surveys*, New York, NY: John Wiley & Sons, Inc.

Rubin, D. (1996). Multiple Imputation After 18+ Years, *Journal of the American Statistical Association,* 91 (434), 473–489.

Von Hippel, P.T. (2007). Regression with Missing *Y*'s: An Improved Strategy for Analyzing Multiply Imputed Data, *Sociological Methodology,* 37.

Weisberg, H.F. (2005). *The Total Survey Error Approach: A Guide to the New Science of Survey Research*, Chicago, IL: University of Chicago Press.

5.5 PRE-ELECTION POLLING IN NAGANO, JAPAN

Nicolaos Synodinos and Eiji Matsuda also examine a mixed mode approach. In this case, however, they compare results of mail, Web, and telephone pre-election polls for a local election in Japan.

5.5.1   Differently Administered Questionnaires: Findings from Surveys of *Asahi Shimbun* for the 2006 Nagano Gubernatorial Election
Nicolaos Synodinos and Eiji Matsuda

Introduction.   This article compares the findings of some similar surveys of *Asahi Shimbun* (Newspaper) conducted for the 2006 Nagano prefecture gubernatorial election. Nagano is a rural Japanese prefecture best known to foreigners for hosting the 1998 Winter Olympics. In 2000, Nagano politics received national attention when Yasuo Tanaka (a famous novelist) contested the governorship as an independent with a reform, anti-construction industry, and pro-environmental stance. This was the first time Tanaka ran for public office and he was elected governor on October 15, 2000, for a 4-year term. For that election, there were four candidates and Tanaka received 49.1% of the vote (Nagano-Ken, 2007).

Tanaka was re-elected on September 1, 2002, following his stepping down after a no confidence vote from the prefecture's assembly in July 2002. For that election, there were six candidates and Tanaka won convincingly with 64.3% of the vote, a little more than double the percentage of his nearest opponent (Nagano-Ken, 2007).

The contest of August 6, 2006, was a pivotal scheduled gubernatorial election and generated national interest. There were only two candidates: the incumbent Tanaka and Jin Murai, a former Liberal Democratic Party (LDP) member of the House of Representatives.

Description of the Surveys.   *Asahi Shimbun* is the second largest nationwide Japanese newspaper (Foreign Press Center Japan, 2007).[84] Along with its regularly scheduled public opinion surveys, it conducts investigations to keep abreast of new developments that potentially may result in methodological improvements in research operations. For the 2006 Nagano gubernatorial election, *Asahi* carried out a mail questionnaire and two comparable Web surveys. In addition, a telephone survey with some somewhat similar questions was conducted. These four surveys mostly inquired about the Nagano gubernatorial election and political issues, in general.

Mail survey.   About 1,500 persons were randomly selected from the Nagano Prefecture Voter Registration Lists. These individuals were mailed a notification postcard on July 14 and the questionnaire on July 22. A reminder postcard was sent on July 29 to those from whom replies were not received by that day.

The 4-page questionnaire consisted of 46 questions plus a space at the end for additional comments, if any. There was no separate cover letter; the request to participate and the instructions were printed on the top of the questionnaire's first page. The questionnaire was mailed in a large business envelope with *Asahi's* name and return address prominently shown on its backside, where such information is customarily printed. The envelope contained the questionnaire, an addressed return envelope stamped with an actual stamp, and an incentive consisting of a ballpoint pen (that

---

[84]According to Foreign Press Center Japan, *Asahi* had a combined daily circulation (of morning and evening editions) of 11.8 million copies in 2006.

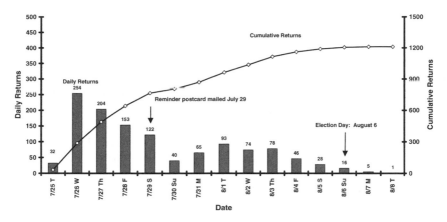

Figure 5.6 Daily and cumulative mail questionnaires received.

retails for about 500 yen) embossed with the name "*Asahi Shimbun.*" In addition, respondents were promised a book card of 500 yen upon returning the questionnaire.

Figure 5.6 shows the daily and cumulative returns of the mail survey. The effect of the reminder is easily discernible. By August 8, 1,211 replies were received consisting of 1,204 acceptably completed questionnaires and seven that were rejected as incomplete. For election prediction reporting, only questionnaires returned by August 4 (the Friday before the Sunday election) could be used. By that date, 1,161 questionnaires were returned and 1,155 were deemed usable (RR1 = 77.0%). For the mail survey, the analyses are based on these 1,155 questionnaires.

Importantly, the August 4 cutoff date of the mail survey coincided with that used for the Web questionnaires. However, for the Web surveys the completion and received date coincided, whereas for the mail questionnaire the completion date was at least one day prior to August 4.

Web surveys. *Asahi Shimbun* also commissioned two Web surveys using two different panels of Nagano prefecture residents. Both of these panels were composed of opt-in volunteer members. Currently in Japan, there are no *nationwide* probabilistically created Web panels.

Web1 was administered to NetMile panel members. NetMile is a mileage administration company and persons join to use this service. Survey participation is one possible way that panel members can earn mileage points. When joining, persons complete a short Web form with some basic information about themselves, including address, gender, and birthday. At the time of the Nagano survey, there were approximately 3 million members nationwide (Matsuda, 2006b).

For this survey, eligible panel members were those that resided in Nagano Prefecture and were '20 and over.' In Japan, the Voter Registration Lists are created from the Basic Resident Registers and a person reaching age 20 is automatically registered and eligible to vote.

A selection button was presented to all eligible panel members when they accessed the Webpage to check their NetMile mileage account balance. The button was titled "Nagano Survey" and indicated the number of points that could be earned by completing this survey. By selecting this button, the person was transferred to the survey window. The button and the survey were posted July 26 and remained available up to and including August 4. In addition, approximately two days after posting the message, NetMile notified nonrespondents by email and invited them to participate in the survey.

Web2 was administered to panel members of Intage Interactive. Intage Interactive specializes in Web surveys and is a joint venture of Intage (a marketing research company) and Yahoo Japan. When joining, persons answer approximately 50 questions about themselves and enroll in E-bank. Panel members earn points by completing surveys and these points are exchangeable for yen. There were approximately 500,000 panel members at the time of the survey (Matsuda, 2006b).

Of the approximately 6,000 panel members eligible to vote in Nagano Prefecture, 4,503 were selected for this survey. Specifically, all those eligible were chosen, except for males 20–49 and females 20–39, who were randomly selected among eligible panel members. An invitation email containing a URL link to the survey was sent on July 27 to all those selected. The survey was available from that date up to and including August 4.

The wording of the Web surveys was similar to that of the mail questionnaire. Web1 and Web2 were almost identical and each consisted of 23 actual questions. Their differences were mostly due to the different software authoring systems. The opening screen noted that the survey was commissioned by *Asahi Shimbun* and had brief instructions about completing the survey. Web1 had a space for any additional comments as its last question; Web2 did not have such a space.

In all, 1,882 persons completed Web1 during the 10 days the questionnaire was available and 2,495 persons completed Web2 during its 9-day availability. Figure 5.7 shows the daily and cumulative completed Web questionnaires: within the first three days of posting the surveys, over three-quarters (78% for Web1 and 87% for Web2) of the total completions were received.

RDD telephone survey. An RDD [Random-Digit Dialing] telephone survey was conducted among registered voters of Nagano Prefecture during the same time period. A total [list] of 6,554 numbers was created (list-assisted, 100-banks). Based on Auto Call Check, 3,805 of the numbers were determined to be working. Of these, 2,836 comprised the list used for this survey.

The fieldwork was on July 29 and 30 (Saturday and Sunday). The 2-day fieldwork period is typical of *Asahi* telephone surveys. Calling was from 9:00 a.m. to 10:00 p.m. on the first day and until 9:30 p.m. on the second day, with the provision that specific appointments could be arranged for at hours outside these times. Each number was called up to six times at different hours and days and for every attempt the phone was allowed to ring up to 15 times.

The called numbers were classified into four categories: households with at least one voter (1,738); businesses (177); households without a voter (62); and of undeter-

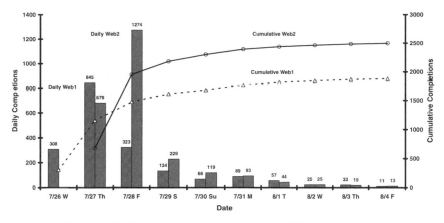

Figure 5.7   Daily and cumulative completed Web questionnaires.

mined disposition, such as ring no-answer, hang-up, and other (859). Voter households were asked the number of voters and via a computer algorithm the nth oldest was selected for the interview.

A total of 1,218 interviews were completed. The 520 non-completions were coded as 'refusals' (221), 'absences of the selected person during the survey period' (186), 'unable to do the interview because of illness or very old age' (76), 'not at home or busy' (24), and 'other reasons' (13).

Results.   The phrasing and choices of several questions were essentially identical in the mail and Web surveys. Some of their differences stemmed from either mode specific idiosyncrasies or variation between the two Web authoring software. The exact Japanese phrasing of the mail survey questions can be found in Matsuda (2006b, 208–215). There were differences in the comparable questions of the telephone survey as these were designed for aural interviewer-administration rather than visual self-administration.

The following three sections summarize some of the major findings. The first section provides a general comparison of the composition of those responding, including their reported frequency of Internet use and political party preferences. The second section summarizes the respondents' gubernatorial candidate preferences and compares them with the election outcome. The last section presents findings from questions about the respondents' likelihood of participating in surveys by different modes.

Composition of responding samples.   Table 5.2 presents the gender and age composition of the respondents in the four surveys and includes population figures for comparison. The gender composition of Web respondents differed from the population with the differences being especially large for Web1. For age, the Web surveys substantially over-represented respondents '20 to 49' and under-represented the rest. The under-representation of persons '60 and over' was especially acute in the Web

Table 5.2 Gender and age distribution (%) of respondents in the surveys and in the population of Nagano prefecture

	Population	Mail (n_{gender} = 1,148) (n_{age} =1,149)	Web1 (n = 1,882)	Web2 (n = 2,495)	RDD (n = 1,218)
Gender[1]					
Male	48.2	47.8	59.1	52.3	46.1
Female	51.8	52.2	40.9	47.7	53.9
Age[2]					
20–29	12.4	9.5	18.1	25.5	6.2
30–39	17.0	14.4	43.2	32.7	17.7
40–49	14.7	15.1	27.7	26.9	17.6
50–59	18.0	18.7	9.1	12.1	18.8
60–69	15.2	17.9	1.3	2.3	17.9
≥ 70	22.7	24.4	0.5	0.6	21.9

Totals may not add exactly to 100% due to rounding.
[1] Population percentages for *gender* are of eligible voters in Nagano for the August 6, 2006 election (Nagano-Ken Senkyo Iinkai, 2007).
[2] Population percentages for *age* are of persons in Nagano as of October 1, 2006 (Statistics Bureau, 2007b). These figures include some persons (e.g., foreigners) who are not eligible to vote.

surveys. Also, there was a large under-representation of persons in their 20s' in the RDD survey.

Mail survey respondents were asked whether they used the Internet and if they replied affirmatively the frequency of using it. Respondents of the two Web panels were asked only about their frequency of using it, employing exactly the same response categories as those of the mail survey. The RDD survey asked whether respondents used the Internet, but did not inquire about frequency.

There were substantial differences between mail and Web panel respondents (see Figure 5.8). The results of mail and RDD were almost identical: 37.8% of mail respondents said they used the Internet compared to 39.1% in RDD. The relatively low use of the Internet among mail and RDD respondents was not unexpected, as Nagano is a rural prefecture.

The party affiliation question inquired about the political party supported now. Mail and Web survey respondents had to select one of ten choices. The first eight categories consisted of political parties presented in order of approximate magnitude: *Jimintō* (Liberal Democratic Party [LDP]), *Minshutō* (Democratic Party of Japan [DPJ]), *Kōmeitō* (New Komei Party [NKP]), *Kyōsantō* (Japanese Communist Party [JCP]), *Shamintō* (Social Democratic Party [SDP]), *Kokumin Shintō* (People's New Party), *Shintō Nihon* (New Party of Japan), and *Jiyū Rengo* (Liberal League). The ninth choice was 'do not support any party' and the tenth was 'other, please specify.' Blank responses (only possible in the mail survey) were coded in the 'other' category.

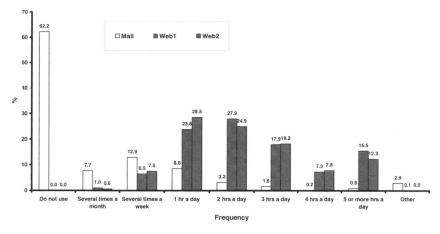

Figure 5.8 Frequency of Internet use.

A somewhat similar party affiliation question was asked in the RDD telephone survey but in an *open-ended* format. Answers for the first nine choices were coded using the same categories as in the mail and Web surveys. The tenth category of the telephone survey was 'no answer/don't know.'

Table 5.3 summarizes the results of the party affiliation question for mail, Web1, Web2, and RDD[85]. Four points can be gleaned from this table. First, results of the two Web surveys closely resembled each other. Second, for all four surveys there was a very sizable portion indicating that they do not support any party. Third, although about a third of the respondents in the mail survey indicated no party support, it was much smaller than in the RDD and Web surveys. Fourth, the proportion of persons indicating that they supported the ruling LDP was substantially larger in mail than in the other three modes.

Preferred candidate for the Nagano prefecture gubernatorial election. Respondents of the mail and Web surveys were asked to indicate which gubernatorial candidate they would choose if they were to vote 'today.' The question's three choices consisted of Jin Murai, Yasuo Tanaka (this listing sequence corresponded to the order they filed their official candidacy), and 'don't know.' In the telephone survey, respondents were asked if they had already decided for whom they would vote. Those who replied affirmatively were asked to name the candidate they would select if they were to vote today. Those who said they have not decided or replied 'don't know' were given the candidates' names and asked which one they would support if they were to vote today.

---

[85] The RDD results presented here are not weighted. The telephone survey results were weighted (RDDw) based on eligible voters in the geographic area, number of landlines and eligible voters in the household, and census demographics for gender and age. RDDw results for party affiliation were similar to the RDD results presented in Table 5.3. That is, the RDDw percentages were: LDP 21.7%; DPJ 13.1%; NKP 2.9%; JCP 2.2%; SDP 2.6%; Other Party 0.8%; No Party 49.8%; and NA/DK 6.8%.

Table 5.3   Political party preferences in four surveys

Political party supported now (Q14[1])	Mail (%)	Web1 (%)	Web2 (%)	RDD (%)
Liberal Democratic Party (LDP)	31.9	17.5	18.4	21.9
Democratic Party of Japan (DPJ)	23.5	18.0	17.1	13.8
New Komei Party (NKP)	2.3	2.7	2.6	2.4
Japanese Communist Party (JCP)	2.9	2.6	3.0	2.5
Social Democratic Party (SDP)	2.2	1.1	1.6	2.3
Other Party/Multiple Parties[2]	1.6	0.7	0.8	0.9
No Party	32.5	56.9	55.6	48.9
Other/NA/DK	3.3	0.6	1.0	7.3
N	1,155	1,882	2,495	1,218

Totals may not add exactly to 100% due to rounding.

[1] Indicates the question number in the Mail questionnaire.

[2] Multiple parties: Only possible in mail surveys where a respondent can check more than one category in a single answer question. There were only 9 such cases for this item.

Jin Murai won with 53.4% to Tanaka's 46.6% (Nagano-Ken, 2007). Of the approximately 1.75 million eligible voters, 66.0% voted in this election (Nagano-Ken, 2007). This voting ratio was lower than the two previous (69.6% in 2000 and 73.8% in 2002) Nagano gubernatorial elections (Nagano-Ken, 2007). However, the 2006 Nagano election had a relatively high voting ratio compared to other recent gubernatorial elections. Among 47 Japanese gubernatorial elections during 2002–2006, the average voting ratio was 50.9%, ranging from 72.2% in Fukui Prefecture (April 2003) to 27.1% in Hiroshima Prefecture (November 2005), with the current Nagano election having the 7th highest turnout (Statistics Bureau, 2007a).

Table 5.4 summarizes the findings of the preferred gubernatorial candidate for mail, Web1, Web2, and RDD[86]. Several important points can be deduced from this table. First, DK [don't know] responses were approximately equal in all methods accounting for about a fifth of the responses. Second, the Web surveys were similar to each other and both showed Tanaka (the loser) as having approximately double the support of Murai. Third, mail and RDD results were very similar and both indicated close races.

In addition, Table 5.4 shows the percentages for the same question but excluding DK responses. These percentages make the candidate preferences more readily comparable with the election result. The underlying assumption is that most persons responding DK will distribute their votes equally between the candidates or they will

[86] The RDD results presented here are not weighted. RDDw results for preferred gubernatorial candidate differed somewhat from the RDD results presented in Table 5.4. The RDDw percentages were: Murai 40.1%; Tanaka 37.1%; and DK 22.8%. Excluding DK responses, the RDDw percentages were 52.0% for Murai and 48.0% for Tanaka.

Table 5.4 Preferred gubernatorial candidate in four surveys

Preferred candidate (Q7[1])	Mail		Web1		Web2		RDD	
	%	%	%	%	%	%	%	%
MURAI, Jin	41.4	51.2	26.6	34.4	26.9	33.9	38.6	49.9
TANAKA, Yasuo	39.5	48.8	50.7	65.6	52.3	66.1	38.8	50.1
DK	19.1	—	22.7	—	20.8	—	22.7	—
N	1,155	934	1,882	1,454	2,495	1,976	1,218	942

Totals may not add exactly to 100% due to rounding.
[1] Indicates the question number in the Mail questionnaire.

not vote. The conclusions (presented in the preceding paragraph) remained the same because the DK portions were approximately equal in the four surveys.

Respondents were asked if they were likely to vote in the Nagano gubernatorial election. For the mail and Web surveys, this question consisted of seven categories of voting likelihood (ranging from 'definitely will vote' to 'definitely will not vote') plus 'other.' The RDD question provided three options ('will vote for sure,' 'will vote if possible,' and 'will not vote') plus 'other.' For the mail and Web surveys, *very likely* voters were defined as those who replied either 'definitely will vote' or 'almost certain will vote.' For the RDD survey, *very likely* voters were defined as those who indicated that they 'will vote for sure.' Most respondents of these four surveys presented themselves as *very likely* voters (77.7% for mail; 67.6% for Web1; 69.1% for Web2; and 82.5% for RDD). Compared to the actual voting ratio of 66.0%, the expressed likelihood of voting was higher by 1.6 points (Web1) to 16.5 points (RDD).

The responses for preferred candidate were examined for *very likely* voters only. Among *very likely* voters, the portion of DK responses decreased in all four surveys. This decrease ranged from 5.4 (in RDD) to 9.0 points (in Web1).

Figure 5.9 shows the election outcome and the four surveys' predictions, based on the preferred candidate among *very likely* voters answering substantively. As the portion of such voters was high, the preferred candidate results of *very likely* voters were similar to those obtained by using all respondents (compare Figure 5.9 with Table 5.4).

Likely to participate in surveys by various modes. The mail and Web surveys asked respondents to indicate the likelihood of participating in four types of surveys (personal interviews, telephone interviews, mail surveys, and Web surveys). The question's main stem was essentially identical in mail, Web1, and Web2. The parts of this question consisted of four survey modes presented in a matrix format. Each mode was described briefly and respondents had to select one of three options.

In the mail questionnaire, respondents were asked to use one of three symbols to indicate if their participation for that mode was 'likely' (O), 'not likely' (X), or

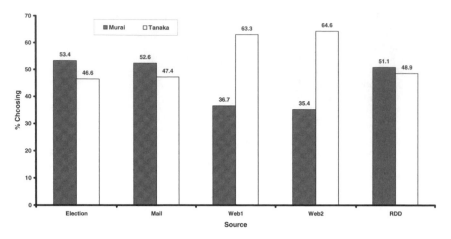

Figure 5.9   Election outcome compared with likely voters' preferences in four surveys (based on substantive answers only).

'it depends' (Δ). Further, mail respondents could leave an item blank ('no answer'). The Web versions presented the same three substantive choices as radio buttons, and respondents had to indicate their likelihood of participating in each of the four survey administration modes.

- **Personal interviews.** The most striking finding was that the vast majority (70% in mail, 84% in Web1, and 85% in Web2) indicated that they were not likely to participate in personal interviews (see Table 5.5). The proportion obtained in the mail survey was of similar magnitude to that of the Web surveys, *if* it is assumed that many persons that left this item unanswered (16%) were probably politely indicating 'not likely.'

- **Telephone interviews.** Very few respondents indicated that it was likely that they would participate in telephone interviews, but the likelihood of participation was higher than in personal interviews (see Table 5.5). The 'not likely' portion in the mail survey was of similar magnitude as in the Web surveys, *assuming* that the 15.7% 'no answers' mostly reflected negative predisposition towards participation.

    The proportion of Web respondents indicating that they were 'likely' to participate in telephone interviews was higher than that of mail respondents. Probably, this was due to the fact that the Web respondents were part of a panel eager to participate in various types of surveys to earn points.

- **Mail surveys.** As was to be expected, substantially more mail survey respondents expressed willingness to participate in mail surveys. Interestingly, about two thirds of the Web respondents indicated that they were likely to participate in mail surveys and about a sixth gave conditional responses (see Table 5.5).

Table 5.5   Mail and Web survey respondents' likelihood of participation in four modes of surveys

	Mail (%)	Web1 (%)	Web2 (%)
Likely to participate in:			
Personal Interviews (Q17a[1])			
Likely	4.8	4.7	4.4
Not likely	70.0	83.8	85.1
It depends	9.1	11.4	10.5
No answer	16.0	—	—
Telephone Interviews (Q17b[1])			
Likely	7.5	13.4	14.3
Not likely	53.2	68.4	67.1
It depends	23.5	18.1	18.6
No answer	15.7	—	—
Mail Surveys (Q17c[1])			
Likely	86.3	67.7	68.5
Not likely	2.3	16.3	14.9
It depends	6.6	16.0	16.6
No answer	4.8	—	—
Web Surveys (Q17d[1])			
Likely	26.4	92.2	94.5
Not likely	31.8	2.7	1.5
It depends	22.4	5.1	4.0
No answer	19.4	—	—
N	1,155	1,882	2,495

Totals may not add exactly to 100% due to rounding.

[1] Indicates the question number in the Mail questionnaire.

Presumably, respondents to a self-administered survey (e.g., Web) are relatively willing to participate in another self-administered mode (e.g., mail).

Few mail respondents (4.8%) left unanswered the item about their likelihood of participating in mail surveys. In comparison (see Table 5.5), the *same respondents* gave substantially more 'no answers' in personal interviews (16.0%), telephone interviews (15.7%), and Web surveys (19.4%). This is consistent with the previously proposed interpretation that, for many mail respondents, leaving an item of likely participation unanswered indicated a polite 'not likely' response.

- **Web surveys.** Almost all Web respondents indicated willingness to participate in Web surveys; after all, these were volunteer members of Web panels. Approximately a quarter of the mail respondents (26.4%) indicated willingness to participate in Web surveys (see Table 5.5). This portion was substantially smaller than Web respondents willing to participate in mail surveys (67.7% for Web1 and 68.5% for Web2). However, willingness to participate in Web surveys was 52.2% among mail respondents who were Internet users. Thus, for a sizeable number of mail respondents the unwillingness to participate in Web surveys stemmed from not being Internet users.

  Among mail respondents, there was 19.4% 'no answer' regarding their likelihood of participating in Web surveys and probably this reflected a polite 'not likely.' This was the largest proportion of 'no answer' given in these four questions by the *same* respondents (see Table 5.5).

Discussion.   The mail survey had a high return rate, even with a short fieldwork period and only a single reminder postcard. A previous *Asahi* survey indicated that mail questionnaires tend to have similar response rates in rural and urban areas and tend to have higher returns from persons in their twenties, compared to RDD and personal interviews. Thus, mail surveys may be a useful alternative in cases where the expected return rate is high, when the opinions are relatively stable, and when there is adequate time for preparation and fieldwork (i.e., in an expected election). For mail surveys to specified individuals, it is necessary to have access to an accurate and comprehensive list for sampling (e.g., Basic Resident Register, Voter Registration List).

Web surveys can be tempting because of their fast turnaround (see Figure 5.7) and relatively lower costs. However, there were no *nationwide* probabilistically created panels in Japan at the time of this survey and this is still the case to date. Also, there is no reservoir of experience for weighting the Web results to reflect the population of Japanese voters.

Opt-in panels rely on volunteers to join and some persons participate in a given survey motivated solely by the presence and/or type of reward. By their nature, such panels are neither representative of the general nor the Internet population. In this particular case, the demographic composition of Web respondents differed from the population substantially (see Table 5.2). The vast differences between Web and mail respondents were clearly shown in the question about their use and frequency of use of the Internet (see Figure 5.8).

In Japanese, unaffiliated voters are called *mutōha* that literally means 'no party group.' Some of the *mutōha* are undoubtedly truly unaffiliated voters, whereas others are persons unwilling to report their preferences. Thus, an important advantage of this mail survey was that the *mutōha* portion was smaller than in the other modes (see Table 5.3). The proportion of *mutōha* among mail respondents was similar to those reported in some other similar recent mail surveys (Synodinos and Matsuda, 2007).

The election outcome provided a yardstick against which survey results can be compared. Two important points stem from the comparison of the four surveys with the election results. First, responses about the supported candidate showed that the mail and telephone surveys were close to each other and to the election outcome (see Figure 5.9). Importantly, the results of the mail and telephone surveys were similar despite the differences in the structure of the question inquiring about the supported candidate. Second, Web1 and Web2 responses were similar—and differed substantially from the election outcome, indicating a lead of approximately two-to-one for the loser (see Figure 5.9).

Decreases in Japanese response rates have been noted for a number of years now (Matsuda, 2006a; Onodera, 1995; Synodinos and Yamada, 2000; Ujiie and Takeshita, 1995). The mail and Web surveys reported here indicated some very alarming views towards participation in personal and telephone interviews. Respondents of the three surveys were negatively predisposed towards participating in telephone interviews but this predisposition was somewhat more positive than for personal interviews (see Table 5.5). Respondents of Web1 and Web2 expressed very similar opinions about participating in different modes of surveys (see Table 5.5). By and large, these opinions were fairly similar to those expressed by mail respondents.

In recent years, accessing and getting the cooperation of some demographic groups has been difficult. Typically, few Japanese persons in their 20s participate in personal (Matsuda, 2006a; Synodinos and Yamada, 2000) and telephone interviews (Matsuda, 2006a). The ability of Web surveys to reach young adults (albeit self-selected) is an important advantage. Thus, the potential of using Web questionnaires for mixed-mode surveys for persons in under-represented groups should be explored, with the important provision that such persons be selected probabilistically. Indeed, mixed-mode surveys seem to have evolved from a curiosity to a necessity (Dillman, 2007). However, the use of such surveys involves another set of problems revolving around comparability of data gathered by different modes.

Technological and societal changes increased the difficulties of obtaining representative samples by any method. Even in cases where an appropriate sampling frame is available, the representativeness of those that participate can be questionable given the falling response rates. Thus, Web questionnaires may have a place in a mixed-mode framework. To this effect, substantial further research on Web surveys within the Japanese context and with elaborate controls is urgently needed.

Acknowledgment. We are grateful to the *Institute of Statistical Mathematics* (Tokyo), where portions of this manuscript were completed while the first author had visiting appointments.

References

Dillman, Don A. (2007). *Mail and Internet Surveys: The Tailored Design Method 2007 Update with New Internet, Visual, and Mixed-Mode Guide* (2nd ed.), Hoboken, NJ: John Wiley & Sons, Inc.

Foreign Press Center Japan (2007). Mass Media, *Facts and Figures of Japan 2007*, Chapter 17, 155–160, on the Internet at: *http://fpcj.jp/old/e/mres/publication/ff/pdf_07/17_mass%20media.pdf/* (last accessed November 23, 2007).

Matsuda, Eiji (2006a). Chōsa ō meguru konnichiteki kadai to tembō: Denwa (RDD)-ho no genkai, mensetsu-ho no sanjyō, yūso-ho no saihyoka, soshite intānetto wa [Current Issues and the Future of Research: Limitations of the Telephone (RDD) Method, Miserable Condition of the Personal Interview Method,

Reevaluation of Mail Surveys, and Internet], *Shin Joho* [*New Information*], 94, October 20, 2007, 8–17.

Matsuda, Eiji (2006b). Yoron chōsa hōkoku Nagano-ken chiji senkyo (2006, 7–8 gatsu): Yūso, intānetto hikaku chōsa de yoron chōsa no kanōsei o saguru [Public Opinion Survey Report of Nagano Prefecture Governor Election (July–August 2006): Mail, Internet Comparative Study Exploring the Possibilities of Public Opinion Surveys]. *Asahi Sōken Reporto* [*Asahi Comprehensive Report*], 198, November, 190–219.

Nagano-Ken Senkyo Iinkai (2007). Kako no Senkyo Kekka [Election Commission of Nagano Prefecture: Past Election Results], April 9, 2007, available on the Internet at: *http://www.pref.nagano.jp/senkyo/past/past.html/*, (last accessed June 26, 2007).

Onodera, Noriko (1995). Yoron chōsa ni okeru chōsa funō to seido [Nonresponse and Accuracy in Public Opinion Surveys], *Hōsō Kenkyū to Chōsa* [*The NHK Monthly Report on Broadcast Research*], 45 (9), 54–57.

Statistics Bureau (2007a). Elections for Prefectural Governors and Prefectural Assemblymen (2002-06), *Japan Statistical Yearbook 2007,* Table 24-10 of the electronic version, available on the Internet at: *http://www.stat.go.jp/data/nenkan/zuhyou/y2410000.xls* (last accessed on May 7, 2007).

Statistics Bureau (2007b). Table 10. Population by Age (5-Year Group) and Sex for Prefectures–Total Population, October 1, 2006, available on the Internet at: *http://www.stat.go.jp/english/data/jinsui/2006np/zuhyou/15k18-10.xls/* (last accessed on June 26, 2007).

Synodinos, Nicolaos E. and Mastuda, Eiji (2007). Searching for Alternative Modes of Questionnaire Administration: Recent Findings from Surveys by the *Asahi* Newspaper, paper presented at the meeting of the American Association for Public Opinion Research, Anaheim, CA.

Synodinos, Nicolaos E. and Yamada, Shigeru (2000). Response Rate Trends in Japanese Surveys, *International Journal of Public Opinion Research*, 12 (1), 48–72.

Ujiie, Yutaka and Takeshita, Yoshiaki (1995). Yoron chōsa no chōsa funō ni tsuite [Nonresponse in Public Opinion Surveys), *Nihon Kodō Keiryō Gakkai* [*Behaviormetric Society of Japan*] (Ed.). Dai 23 Kai Taikai Happyō Ronbun Sairokushū [Collection of papers presented at the 23rd Conference], Kansai University, Suita City, Osaka: Nihon Kodō Keiryō Gakkai, 272–275.

5.6 SWEDISH PUBLIC OPINION POLLS

Isaksson and Forsman (2005) discuss weighting adjustment methods for Web and telephone surveys of Swedish public opinion polls. In so doing, they attempt to address the fact that most Web surveys in Sweden are dependent on nonprobability samples—a problem affecting most Web surveys.[87] Many researchers, including Warren Mitofsky, have expressed concern about the use of nonprobability samples as the starting point for Internet surveys. Isaksson and Forsman present a potential solution for ameliorating at least some of the bias inherent in such studies.

[87] Knowledge Networks uses an alternative approach by developing a national probability sample of households recruited by Random-Digit Dialing, who either have been provided Internet access through their own computer or are given a WebTV device. See, for example, Chatt, C; Dennis, M. et al. (2005).

5.6.1   A Comparison Between Using the Web and Using Telephone to Survey Political Opinions
Annica Isaksson and Gösta Forsman

Introduction.   In Sweden, as in many other countries, the telephone is the predominant mode for surveying opinions among the general population. In recent years, however, the World Wide Web has become a viable alternative for the survey industry. Among the appealing features of Web surveys are low data collection costs and a large flexibility in the design of the questionnaire. Web surveys, however, also suffer from several methodological problems, including dubious representativity due to low Internet penetration in the general population, lack of sampling frames, and high nonresponse rates. Data are often collected from easily recruited nonprobability samples, or convenience samples, of Web users. Inference from such samples relies heavily on appropriate weighting methods to reduce selection bias.

Weighting adjustment methods are often based on the assumption that a population can be divided into groups (or *weighting classes*) within which the respondents characteristics are "representative" for the entire group. If the assumption holds, the selection bias may be completely eliminated. This is, however, an optimistic—or even naive—assumption, when inference is made from nonprobability samples. A reasonable motivation for using weighting class adjustment for nonprobability samples is that the selection bias may be somewhat reduced, rather than completely eliminated.

An important key to a successful weighting adjustment is the identification of appropriate weighting classes. We compare two procedures that rely on different sources of information for weighting class construction: *poststratification* and *propensity score adjustment*. In poststratification, the weighting classes are defined and limited by available external data, e.g., the distribution of the population over one or more variables, such as population distribution by age, sex and race available from standard population estimates. In propensity score adjustment, on the other hand, the weighting classes are defined by a number of covariates, observed on both the nonprobability (Web) sample and a parallel probability (telephone) sample.

In this paper, we present results from a study conducted by the Swedish commercial survey institute TEMO in a political opinion poll before the Swedish September 2002 election.

Propensity Score Adjustment.   Propensity score adjustment was originally developed to reduce selection bias in observational studies (quasi-random experiments), see, e.g., Rosenbaum and Rubin (1983, 1984); D'Agostino and Rubin (2000). Typical for such studies is the comparison of treatment effects between two subpopulations, such as smokers and non-smokers. Propensity score weighting for survey nonresponse adjustment has been proposed by Little (1986) and Little and Rubin (1987). The application of the method on Web surveys utilizes a "control" survey based on a traditional probability sampling method (e.g., a telephone survey based on Random-Digit Dialing (RDD)) that runs parallel to a Web survey, based on a nonprobability sample. In the original applications, participation in the telephone and Web surveys then corresponds to "treatments." The United States company Harris Interactive has promoted this method, see Terhanian et al. (2001). The use of propensity score weighting in the survey field, including nonresponse adjustment as well as nonprobability sampling, is further discussed in Danielsson (2002).

The control survey. Propensity score weighting relies on the existence of a control survey, i.e., a separate survey based on a representative (probability) sample and conducted successfully in a way that population parameters and distributions can be estimated unbiasedly (at least approximately). The control survey is used to adjust the Web survey data to the population level. Or, more precisely, the control survey is used to adjust the Web survey data to the control survey level, which is assumed close to the population level. Therefore, the quality of the control survey is critical for the method.

In practice, the control survey is often conducted by telephone for an RDD sample, see Terhanian et al., (2001). In the United States, RDD telephone surveys are the dominant type of probability (or approximately probability) sample surveys in market research. Any mode that permits approximately unbiased estimates of population parameters and distributions may, however, be used for the control survey.

The theory assumes that the control survey and the Web survey are conducted at the same time. For cost reasons, however, the control survey may be conducted at certain time intervals, e.g., each month, while the Web survey is conducted daily. When the data collection is conducted, the Web sample and the control sample are merged and the propensity score is defined on the merged sample.

The covariates. The covariates (here assumed to be categorical) are used to partition the merged sample into groups. Within these groups, the Web sample respondents and telephone (control) sample respondents are—ideally—assumed to have identical distributions of the target variable y. Thus, the choice of covariates (sometimes called "webographics") is critical for propensity score weighting. In particular, the procedure is sensitive to any kind of mode effects when the covariates are measured on the Web and the telephone samples. Typically, the "webographic" questions touch issues such as lifestyle, attitudes, and self-perception.

The propensity score and the forming of weighting classes. In each of the groups defined by the covariates, the propensity score is defined as the proportion of Web sample respondents of all respondents in the group. Formally, for a given individual k, let Z_k take the value 1 if he participates in the Web survey, 0 if he participates in the telephone survey. Further, let X be a vector of covariates, possibly associated with Z, available for both the control and the Web survey. The vector of covariates for individual k is X_k. The main problem is to estimate the expected value π_k of Z_k: the propensity score. The prevalent strategy is to formulate a logistic regression model (see, e.g., Neter et al., 1996, equation 14.37) for π_k as function of X_k. Under such a model, it is straightforward to estimate π_k. Finally, groups with similar estimated scores are collapsed, thus forming a few (usually five) weighting classes.

Let n_{Th} denote the number of responses by telephone within weighting class h. The total telephone sample size is denoted by n_T. In the estimation, within class h, the mean of the Web responses is weighted (multiplied) by n_{Th}/n_T.

Poststratification. A variation of stratification, poststratification is treated in most standard textbooks on survey sampling, see, for instance, Särndal et al. (1992, chapter 7) or Cochran (1977, section 5A.9). As the name suggests, the idea is simply to stratify the sample *after* it has been selected, instead of before. Originally, poststratification was used simply as an alternative to stratification and for the same reasons (such as improved precision in the estimates)—not to adjust for missing data. Common reasons for not stratifying beforehand, then, include that practical considerations favor some simpler design or that the stratum identity only can be established for sampled individuals.

The case when poststratification is used as a means to adjust for nonresponse bias is treated, e.g., by Särndal et al. (1992, section 15.6). The poststrata are usually formed to agree with response homogeneity groups. To separate the different purposes of poststratification, Kalton and Kasprzyk (1986) suggest the name "population weighting" for this application. Technically, poststratifiction, or population weighting, can also be applied to nonprobability samples to adjust for selection bias.

Research Questions, Data Set, and Study Design. In our study, two modes of data collection, telephone and Web, were compared, as well as the weighting procedures poststratification and propensity score adjustment. The telephone sample was the TEMO omnibus sample, an approximate probability sample of phone numbers (households) combined with a probability sample of one individual within each household. The phone number sample is a list-based sample of phone numbers according to the "Plus one" sampling technique (Lepkowski, 1988; Forsman and Danielsson, 1997). Phone number noncontacts are replaced by noncontacts from earlier waves of the omnibus survey: a technique similar to one described by Kish and Hess (1959). The selection of an individual within household is conducted according to the "last-birthday" method, which involves identifying the person in the household who had the last birthday among all eligible household members. (For details on the "lastbirthday" method, and comparisons with other procedures for selection within household, see, e.g., Binson et al., 2000; Forsman,1993; or Oldendick et al., 1988).

For the Web data collection, a stratified sample of the TEMO Web panel is used. Although this panel originally was recruited from earlier waves of the TEMO omnibus, since the dropout rate is very high, we regard it as a nonprobability sample of Web users. More precisely, we treat it as a quota sample rather than a stratified probability sample with very low response rate.

The data set was collected in a political opinion poll before the Swedish September 2002 election. The telephone sample included responses from 1,001 individuals; the Web sample was composed of nearly thrice as many (2,921 individuals). We consider the problem of estimating the election outcome from these data. This implies estimating party sympathies within the 'voter population': the group of Swedes who, on Election Day, are entitled to vote (that is, are at least 18 years of age and Swedish nationals), use this right, and return valid ballot-papers. More precisely, we

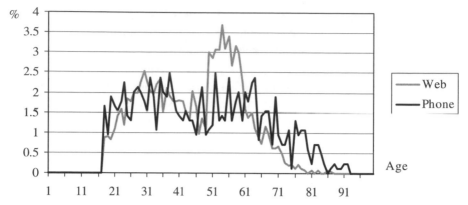

Figure 5.10 Sample distributions over ages (respondents belonging to the voter population only).

want to estimate the proportion of this group that will vote for each political party. Our poststrata are formed the way TEMO usually does it; that is, by sex, age class, and dwelling (big-citydweller or not). The propensity scores, on the other hand, are estimated from a set of "webographic" questions. (See 7.4 in this volume) We also try using both the lifestyle questions *and* the poststratification variables when estimating the propensity scores.

The estimation task is made slightly more difficult by the fact that the exact number of voters is unknown until the election has taken place. We handle this lack of information by estimating the size of the voter population from the sample data. Our approach, which corresponds to treating a population proportion as the ratio of two unknown population totals, is well established in the statistical literature (see, e.g., Särndal et al., 1992, result 5.8.1).

Before we present our results, for reference, let us take a brief look at how the samples are composed. We restrict our attention to respondents who belong to the voter population. The proportion of male respondents is 54.0 percent in the Web sample, 50.0 percent in the telephone sample. The sample distributions over ages are presented in Figure 5.10; the sample distributions over educational levels are shown in Figure 5.11. From Figure 5.10, the Web sample has a peak in its age distribution in the age span 50–60 years that is not present in the telephone sample. From Figure 5.11, the average Web respondent seems to be better educated than the average telephone respondent is.

Results. Our estimates based on the telephone data are presented in Figure 5.12; our estimates based on the Web data are shown in Figure 5.13. For the Web data, one possible approach is to use only the lifestyle questions to estimate the propensity scores.

In Figure 5.13, the columns representing the resulting estimates of party sympathies are labeled *Propensity score 1*. When both lifestyle questions and poststratification variables are used to estimate the propensities, we label the columns *Propensity score*

Table 5.6   Numbering of the Swedish political parties

No.	Party name
1	Conservative
2	Liberal
3	Center
4	Christian Democrat
5	Social Democrat
6	Left Party
7	Green
8	Other

2. In both Figures 5.12 and 5.13, the numbers on the horizontal axes represent the Swedish political parties (a translation of the numbers is provided in Table 5.6).

For comparison, the election outcome, as well as the unweighted sample proportions, are included in both figures. An individual's party sympathy is not necessarily stable over time. Even though the sample data were collected close to Election Day, the time lag is still expected to create some unavoidable discrepancies between the sample- based estimates of the election outcome and the actual outcome.

It is interesting to compare the estimates based on the Web data not only with the election outcome, but also with the poststratified estimates based on the telephone data. To facilitate this, the poststratified telephone estimates from Table 5.7 reappear in Figure 5.13.

In Figure 5.12, the poststratified column heights differ only marginally from the unweighted counterparts. Figure 5.13 shows some promising results for the propensity score weighting technique: the propensity score estimates typically come closer to the election outcome (as well as to the poststratified telephone estimates) than do the

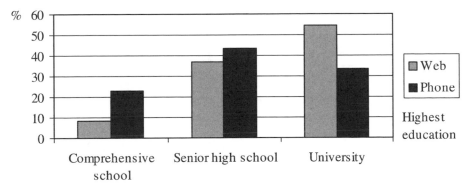

Figure 5.11   Sample distributions over educational levels (respondents belonging to the voter population only).

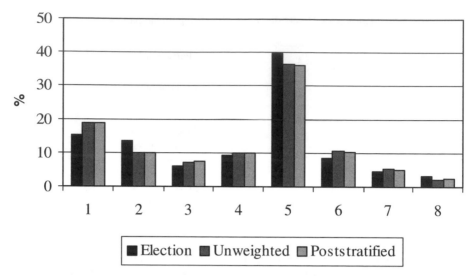

Figure 5.12   Poststratified estimates of election outcome, based on the telephone data (by political party).

sample proportions and the poststratified estimates. The differences between the two variations of propensity score estimates are quite small.

As an aid in judging the accuracy of the Web estimates, we present a number of sums in Table 5.7. In the table, separately for each type of estimate, we give:

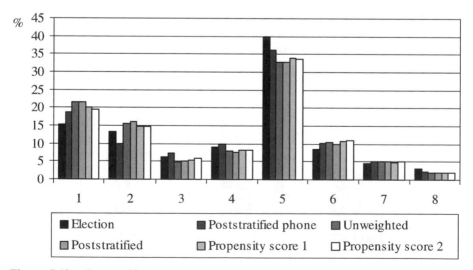

Figure 5.13   Poststratified and propensity score weighted estimates of election outcome, based on the Web data (by political party).

1. The sum of absolute differences between the Web estimates and the election outcome, and

2. The corresponding sum of absolute differences.

Discussion. The findings from our study show, not surprisingly, that unweighted estimates, based on the TEMO Web panel, differ from those based on the telephone sample. The poststratification does not improve the Web estimates appreciably, whereas the propensity score adjustment at least seems to make some difference to the better. This result holds no matter if the Web estimates are compared with the poststratified estimates based on the telephone data or with the election outcome. More research is needed, however, before we understand the usefulness of propensity score weighting. Among the research issues yet to be investigated, we identify the choice of covariates and the choice of size of the Web panel sample.

The *choice of covariates* is critical for propensity score weighting. In our study, we used covariates not particularly tested for Swedish conditions. There is a strong need for a systematic evaluation of different kinds of covariates. According to the theory of propensity score weighting, the covariates are assumed to partition the sample into groups within which Web respondents and telephone respondents have identical distributions of the target variable. We agree to the suspicion, discussed by Stenbjerre (2002), that such covariates, if they exist, may be specific. For example, in the Scandinavian countries, with homogeneous populations and generally high Internet penetrations, it may be harder to find discriminating covariates than in countries where a considerable part of the population has a low Internet penetration.

In our study, the *size of the Web panel sample* was 2,921. In studies reported from the United States, sample sizes are often much higher (typically, well over 10,000). Since the number of cells defined by the covariates is large, and large sample sizes in the cells produce better estimates of the propensity scores, the weighting should be more efficient with larger samples. Experiments with larger samples should be possible as the TEMO Web panel increases.

Finally, there are, of course, other weighting adjustment methods than those used in our study. For example, the propensity score weighting procedure applied by Harris

Table 5.7 Sums of absolute differences

	Election outcome	Poststratified estimates from telephone data
Unweighted	21.51	16.95
Poststratified	21.60	17.87
Propensity score 1	17.45	13.27
Propensity score 2	17.42	12.79

Interactive differs somewhat from the procedure used in this study. Harris Interactive uses a raking procedure (see, e.g., Oh and Scheuren, 1983) with the telephone sample distribution of the weighting classes and population distributions of background variables (such as age and gender) as marginal auxiliary information. We have not used this procedure in this study, but we find it likely that the resulting estimates resemble those based on our procedure *Propensity score 2*, since the same information is used.

Acknowledgment. The financial support of this work by the Bank of Sweden Tercentenary Foundation (Grant no. 2000- 5063) is gratefully acknowledged.

References

Binson, D.; Canchola, J.A.; and Catania, J.A. (2000). Random Selection in a National Telephone Survey: A Comparison of the Kish, Next-Birthday, and Last-Birthday Methods, *Journal of Official Statistics*, 16, 53–59.

Cochran, W.G. (1977). *Sampling Techniques*, 3rd edition. New York, NY: John Wiley & Sons, Inc.

Couper, M.P. (2000). Web Surveys–A Review of Issues and Approaches, *Public Opinion Quarterly*, 64, 464–494.

D'Agostino, R.B., Jr. and Rubin, D.B. (2000). Estimating and Using Propensity Scores with Partially Missing Data, *Journal of the American Statistical Association*, 95, 749–759.

Danielsson, S. (2002). The Propensity Score and Estimation in Nonrandom Surveys—an Overview, *Research Report,* Department of Statistics, Linköping University.

Forsman, G. (1993). Sampling Individuals Within Households in Telephone Surveys, *Proceedings of the Section on Survey Research Methods*, American Statistical Association.

Forsman, G. and Danielsson, S. (1997). Can Plus Digit Sampling Generate a Probability Sample? *Proceedings of the Section on Survey Research Methods*, American Statistical Association.

Kalton, G. and Kasprzyk, D. (1986). The Treatment of Missing Survey Data, *Survey Methodology*, 12 (1), 1–16.

Kish, L. and Hess, I. (1959). A Replacement Procedure for Reducing the Bias of Nonresponse, *The American Statistician*, 13 (4), 17-19.

Lepkowski, J.M. (1988). Telephone Sampling Methods in the United States, *Telephone Survey Methodology*, Groves, R.; Biemer, P.; Lyberg, L.; Massey, J.; Nicholls, W.; and Waksberg, J. (eds.), New York, NY: John Wiley & Sons, Inc., 73–99.

Little, R. (1986). Survey Nonresponse Adjustments for Estimation of Means, *International Statistical Review*, 54, 139–157.

Little, R. and Rubin, D. (1987). *Statistical Analysis with Missing Data,* New York, NY: John Wiley & Sons, Inc.

Neter, J.; Kutner, M.H.; Nachtsheim, C.J.; and Wasserman, W. (1996). *Applied Linear Statistical Models,* 4th ed., Chicago, IL: Irwin.

Oh, H.L. and Scheuren, F. (1983). Weighting Adjustment for Unit Nonresponse, *Incomplete Data in Sample Surveys, Volume 2, Theory and Bibliographies,* Madow, W.G.; Olkin, I.; and Rubin, D.B. (eds.), New York, NY: Academic Press.

Oldendick, R.W.; Bishop, G.G.; Sorenson, S.B.; and Tuchfarber, A.J. (1988). A Comparison of the Kish and Last Birthday Methods of Respondent Selection in Telephone Surveys, *Journal of Official Statistics*, 4, 307–318.

Rosenbaum, P.R., and Rubin, D.B. (1983). The Central Role of the Propensity Score in Observational Studies for Causal Effects, *Biometrika*, 70, 41–55.

Rosenbaum, P.R. and Rubin, D.B. (1984). Reducing Bias in Observational Studies Using Subclassification on the Propensity Score, *Journal of the American Statistical Association*, 79, 516–524.

Särndal, C.-E.; Swensson, B.; and Wretman, J. (1992). *Model Assisted Survey Sampling,* New York, NY: Springer-Verlag.

Stenbjerre, M. (2002). Online Live Exit Poll During Danish General Elections, paper presented at the meeting of the American Association for Public Opinion Research, St. Pete Beach, FL.

Terhanian, G.; Smith, R.; Bremer, J.; and Thomas, R.K. (2001). Exploiting Analytical Advances: Minimizing the Biases Associated with Internet-Based Surveys of Non-Random Samples, ARF/ESOMAR: Worldwide Online Measurement, ESOMAR Publication Services, 248, 247–272.

5.7 ELECTION POLLING IN CANADA

Jason Cawley and Paul Sommers examine irregularities in a 1995 referendum on Québec sovereignty. Using regression techniques, they explore the impact of rejected ballots on the special election to determine whether the predominantly French-speaking province should secede from Canada. Researchers who still question the validity of recent U.S. presidential elections may find the approach here to be of interest.

5.7.1 Voting Irregularities in the 1995 Referendum on Québec Sovereignty [88]
Jason Cawley and Paul Sommers

On October 30, 1995, the citizens of Québec went to the polls to decide the future of their province and ultimately that of Canada. At issue, should the predominantly French-speaking province secede from Canada and become a sovereign country? The choice for voters was a simple "Oui" or "Non," although the wording of the ballot question was a little confusing.[89] On referendum night, once the returns were in, the "No" side was declared the winner by the narrowest of margins, 52,448 votes or barely more than one percentage point.[90] The victory was secured in large part by the nearly 95% of nonfrancophone voters who voted overwhelmingly against independence. Immediately following the vote, however, allegations began to surface that the result was closer than it should have been.[91]

Poll scrutineers, who are appointed by the governing political party (in this case, the separatist Parti Québecois) and whose job is to decide which ballots are rejected, were accused by "No" campaign observers of orchestrating a campaign of "systematic electoral fraud"[92] by voiding thousands of "No" votes in the predominantly allophone[93] and anglophone (electoral division) ridings of Montréal.

The purpose of this brief article is to examine the relationship between the proportion of rejected votes and the proportion of allophones and anglophones in each

---

[88] Voting Irregularities in the 1995 Referendum on Québec Sovereignty, by Jason Cawley and Paul Sommers, reprinted with permission from *Chance*. Copyright ©2007: American Statistical Association. All rights reserved

[89] The exact wording of the question was: "Do you agree that Québec should become sovereign after having made a formal offer to Canada for a new Economic and Political Partnership, within the scope of the Bill respecting the future of Québec and of the agrement signed on June 12, 1995?"

[90] See *The Globe and Mail*, 1995.

[91] [Nonfrancophone refers to persons for whom French is not their first language.]

[92] See Mulcair (1995).

[93] An "allophone refers here to someone whose first language is neither English nor French.

riding. This will be done by means of a regression fitted to the cross-section of 125 electoral divisions in the October 1995 referendum.

The Regression Results. Equation (5.1) relates the number of rejected ballots expressed as a percentage of the total votes cast in each riding to a number of demographic variables. ALLOPHONE is the percentage of residents in each riding whose first language is neither French nor English. ANGLOPHONE is the percentage of residents in each electoral division whose first language is English. The percentage of rejected ballots in the most recent general election, REJECT94, is included in the regression to determine whether or not electoral divisions experienced roughly the same rejection rates in the 1994 general election as they did in the 1995 referendum vote. LAVAL is equal to 1 if the riding is Chomedey, Fabre, Laval-des-Rapides, Mille-Iles, or Vimont. Finally, the specification takes into account in an important positive interaction between ALLOPHONE and LAVAL. (Failure to do so would imply that a given increment to the percentage of allophones has the same effect on the rejection rate in all regions of the province.) Therefore, the coefficient on ALLOPHONE in Equation (5.1) was allowed to change for ridings in Laval.[94]

The 1995 referendum results and the breakdown of each riding's population by first language were obtained from sources on the Internet. Their URL addresses are as follows: REJECT (*http://www.dgeq.qc.ca/dge/anglais/resultat/res95-01.htm*) and ALLOPHONE, ANGLOPHONE (*http://www.quebec-oui.org/en/infos/bottin/ bottin.htm*). The statistics for each electoral division on rejected ballots from the 1994 general election are from *Rapport des résultats officiels du scrutin. Élections générales du 12 septembre 1994 et élection du 24 octobre 1994 dans la circonscription electorate de Saint-lean* (53–93).

The regression results are (*t*-statistics in parentheses):

$$REJECT = \underset{(5.68)}{1.112} + \underset{(4.34)}{.020} \ ALLOPHONE$$

$$+ \underset{(0.12)}{.001} \ ANGLOPHONE + \underset{(2.64)}{.223} \ REJECT94$$

$$\underset{(-8.61)}{-3.773} \ LAVAL \tag{5.1}$$

$$+ \underset{(15.62)}{.387} \ (LAVAL \times ALLOPHONE)$$

$$R^2 = .759 : DF = 119$$

[94] Apart from Laval, 15 other regions were defined for the province of Québec: (1) Bas-Saint-Laurent, (2) Saguenay/Lac-Saint-Jean, (3) Québec, (4) Mauricie-Bois-Francs. (5) Estrie, (6) Montréal, (7) Outaouais, (8) Abitibi-Témiscamingue, (9) Côte-Nord, (10) Nord-du-Québec. (11) Gaspésieiles-de-la-Madeleine, (12) Chaudiére-Appataches, (13) Lanaudiére, (14) Laurentides, and (15) Montérégie (see *http://www.bsq. gouv.qc.ca/bsq/ang/regions.htm*). Regressions with dummy variables defined for each region revealed no geographical bias, with the notable exception of Laval.

The percentage of rejected votes was discemibly higher ($p < .001$) in ridings with heavy concentrations of allophones. For every 10-percentage-point increase in the percentage of allophones, the percentage of rejected ballots rises by two-tenths of 1%. The impact of the ALLOPHONE variable is abnormally larger (over 20 times larger) in Laval than in other regions of Québec.[95] Contrary to popular wisdom, ridings with high proportions of English-speaking Canadians were not similarly targeted ($p = .901$). The positive and statistically significant coefficient on REJECT94 ($p = .009$) suggests that ridings with a high (low) proportion of rejected ballots in the 1994 general election tended to exhibit a high (low) proportion of rejected ballots in the October 1995 referendum, after allowing for differences in ethnicity.[96]

Concluding Remarks. The evidence presented here suggests that there were voting irregularities in the October 1995 Québec referendum on quasi-independence from Canada, with rejection rates running substantially higher in ridings with high concentrations of allophones, especially in Laval.

This article represents a modest attempt to assess the significance and relative weight of a small number of demographic variables on the pattern of rejection rates. The statistical model employed allows one to isolate the "effect" of ethnicity after allowing for regional differences and differences in the voting patterns of the most recent general election. A study of electoral bias or "malfeasance" that ignores this particular disaggregation of the data is questioned, largely because one would otherwise confuse ridings with, say, a high percentage of anglophones (and, hence, strong opposition to the referendum) with a high rejection rate.

Further disaggregation of the data to, say, the polling station level[97] would make it practically impossible (without detailed exit survey polls) to explicitly account for differences among voters in the number of anglophones and allophones and, hence, what effect ethnicity had on rejection rates.

Although Québecers very nearly voted to secede from Canada, the curious pattern of high rejection rates across ridings where the "No" side dominated suggests that the vote was not as close as it was. Although the election results do not permit one to know which votes ("Oui" or "Non") were rejected, it is clear that the rejection rates were highest in ridings with large ethnic populations.

---

[95] A test of whether this change is significant is provided by a test of the null hypothesis that the coefficient of (LAVAL × ALLOPHONE) is 0. In this case, the null hypothesis can be rejected at better than the .0001 level.

[96] Of the 38 ridings in which rejection rates rose from the 1994 general election, 25 voted against the referendum. Of the 87 ridings in which rejection rates dropped, 67 voted for the referendum. The strong relationship between higher (lower) rejection rates and "No" ("Yes") ridings is statistically discernible at better than customary levels of significance ($\chi^2 = 21.029$, $p < .0001$).

[97] See, for example, Baker (1996).

Two other papers recently presented at the 2007 meetings of the American Association for Public Opinion Research in Anaheim, CA also focus on Canadian polling efforts—Francois Petry spoke about the role of polling in the health care field[98] and Andrea Rounce discussed the impact of publishing polling results on public opinion research.[99] For space reasons we have not excerpted these papers. Nonetheless, we found their work has considerable merit and a value beyond the Canadian context.

5.8   SUMMARY OBSERVATIONS

Public opinion researchers have much to learn from quality improvement efforts in other countries. As we have pointed out, the progress on nonresponse adjustment described in this chapter has broad applications worldwide.

The experiences other countries have had with mixed mode surveys and the use of the Web—as opposed to on-location, self-administered exit polls, typical of the United States—are useful, as well. It is important to remember, though, that rural areas like those covered, in part, by the Mexican and Japanese polls, differ substantially from the challenges posed in a large heterogeneous industrial country like the U.S. On the other hand, Sweden and many very homogeneous European countries offer a further contrast.

Editors' Additional References

Bautista, René; Callegaro, Mario; Vera, Jose Alberto; and Abundis, Francisco (2005). Nonresponse in Exit Poll Methodology: A Case Study in Mexico, paper presented at the meeting of the Association for Public Opinion Research, *Proceedings of the Section on Survey Research Methods*, American Statistical Association, 3802–3809.

Bautista, René; Morales, Marco A.; Abundis, Francisco; and Callegaro, Mario (2007). Exit Polls as Valuable Tools to Understand Voting Behavior: Using an Advanced Design in Mexico, working paper of the Survey Research and Methodology program of the University of Nebraska, Lincoln.

Braizat, Fares (2007). Muslims and Democracy: Stability and Change in Public Opinions, presentation at the U.S. State Department Conference on Survey Methodology, July 16, 2007.

Cawley, Jason and Paul Sommers (1996). Voting Irregularities in the 1995 Referendum on Québec Sovereignty, *Chance*, 9 (4), 26–27.

Chatt, Cindy; Dennis, Michael; Li, Rick; and Pulliam, Paul (2005). Data Collection Mode Effects Controlling for Sample Origins in a Panel Survey: Telephone Versus Internet, paper presented at the meeting of the American Association for Public Opinion Research, Miami Beach, FL.

Edison Media Research and Mitofsky International (2005). Evaluation of Edison/Mitofsky Election System 2004, available on the Internet at: *http://www/exit-poll.net/election:night/EvaluationJan192005.pdf*, January 19, 2005.

Isaksson, Annica and Forsman, Gösta (2003). A Comparison Between Using the Web and Using Telephone to Survey Political Opinions, paper presented at the meeting of the American Association for Public Opinion Research, *Proceedings of the Section on Survey Research Methods*, American Statistical Association, 100–106.

---

[98] Petry, Francoise (2007).
[99] Rounce, Andrea (2007).

Petry, Francoise (2007). The Role of Polling in Health Care Policy Making in Canada, presentation at the meeting of the American Association for Public Opinion Research, Anaheim, CA.

Rounce, Andrea (2007). Public Opinion Research in the Saskatchewan Government: The Impact of Publishing Polling Results, presentation at the meeting of the American Association for Public Opinion Research, Anaheim, CA.

Rubin, Donald (1976). Inference and Missing Data, *Biometrika*, 63 (3), 581–592.

Synodinos, Nicolaos and Eiji Matsuda (2007). Differently Administered Questionnaires: Findings from Surveys of *Asahi Shimbun* for the 2006 Nagano Gubernatorial Election, unpublished.

Frank and Ernest ©Thaves/Dist. by Newspaper Enterprise Association, Inc.

CHAPTER 6

LOOKING AHEAD: RECOMMENDATIONS FOR 2008 AND BEYOND

6.1 INTRODUCTION

The controversies of the early part of the decade led to numerous efforts not only to prove—or disprove—the existence of fraudulent voting procedures, but also to extensive reevaluation of existing exit polling methods, in order to improve procedures for the future. This chapter presents some recommendations for improvements and some examples of research that have been conducted recently. For the readings, we begin with an excerpt from the Edison-Mitofsky Evaluation Report (2005) that suggests some future steps. This is followed by three papers that describe current data quality improvement studies—on ethnic diversity, cell phone users, and questionnaire probes. Two more write-ups provide professional association standards for polling and the chapter concludes with two papers that suggest improvements to the election process to ensure better quality official vote counts. First, though, Warren Mitofsky talks about one pre-test he planned.

Editors' Note: Text in gray shading is provided by the editors of this volume. Boxed-in text represents excerpts from interviews with Warren Mitofsky. The balance of the chapter is made up of papers prepared by contributing authors. [Text in square brackets was added by the editors.]

Elections and Exit Polling. Edited by Fritz Scheuren and Wendy Alvey
Copyright © 2008 John Wiley & Sons, Inc.

"We're doing a little pre-test in Alabama for the 2008 election. The pre-test has the logo back on the questionnaires. The only reason we're doing the pre-test is they [the National Election Pool consortium] want to know if anything about this system will prevent us from doing both the Democratic and Republican primary on the same day. In 2004 I sampled proportionate to Democratic vote. So that's why the estimates were so good for the primary. Now, I'll get to sample in proportion to the total vote and do two primaries."

"They said, 'Well, we're not sure if that's going to work.' I said, 'With the exception of 2004, every primary I ever did was that way... I've never had the luxury of sampling on the one party alone. What do you mean, if it's going to work? Of course, it's going to work. I've already tested it and it's going to work just fine.' ''

"What I did was I selected the samples initially in 2004 proportional to the total vote total. Then I subsampled them is such a way that it would have been the equivalent of having initially selected them proportionate to the Democratic vote. So all I did to test it was to undo the weighting. It did just what I said—it made pretty good estimates. The variances were a little bit bigger, but there was no problem. The thing that surprised me the first time I discovered all of this was that I figured you'd get poor estimates for one party or the other, and you don't. And the reason that you don't is that a big vote is a big vote. So, when you sample on total vote, you're doing fine for both parties. Even if the precinct is heavily for one party or the other, even the small one is a bigger number for a big precinct.''

6.2 RECOMMENDATIONS FOR THE FUTURE

Much of the research conducted during the decade yielded suggestions for further study and recommended enhancements for elections and exit polling. One paper that received a lot of attention was the 2005 Edison-Mitofsky evaluation report. Below are the formal recommendations that came from that report.

6.2.1 Excerpt from: Evaluation of Edison/Mitofsky Election System 2004 (NEP) [100]
Edison Media Research and Mitofsky International

Our goal is to have smoother and more accurate election nights in the future. After two months of study [of the results from the exit polls of the 2004 presidential election], we recommend the following changes in what we do and how we function. We believe these recommendations will result in improvements for NEP [the National Election Pool] and the subscribers.

---

[100]Excerpt from: Evaluation of Edison/Mitofsky Election System 2004, by Edison Media Research and Mitofsky International, reprinted with permission. Copyright ©2005: National Election Pool. All rights reserved.

Improving Our Estimates

Exit poll errors. One way to reduce error is to take additional steps to keep the interviewers focused on strictly following their interviewing rates in order to properly sample voters within each polling location. This will be made an even greater priority in the future. We will develop additional steps in the recruiting and training process to make certain that the interviewers are following the detailed instructions that we give them. Another way to potentially reduce the overall error in the exit poll is to improve the completion rates. We plan to examine more closely how the size, design and layout of the questionnaires may affect the percentage of sampled voters who choose to complete the exit poll questionnaire.

While we have identified factors that we can control in order to lessen Within Precinct Error, we cannot eliminate the possibility of any statistical bias based upon differential nonresponse by Democratic and Republican voters. We plan to identify indicators in the exit poll data that will give those who are using the exit poll data an early warning that there may be the possibility of Within Precinct Error in the exit poll estimates.

Recruiting exit poll interviewers. We plan to make enhancements to recruiting exit poll interviewers.

- We will use augmented recruiting methods to reduce the proportion of students and young adults as interviewers.

- We will add a standardized training script for all individual training phone conversations that occur prior to our main training/rehearsal call.

- We will evaluate other training techniques, such as the video training guide and interviewer tests.

- We will use the Internet more effectively as an interviewer training tool.

Distance issues. We need to be more proactive in gaining cooperation from state and local election officials who try to impose distance restrictions of 50 feet or more on exit poll interviewers. Compared to the 2000 data collected by VNS [Voter News Service[101]], more than twice as many of our exit poll interviewers in 2004 reported that they were forced to stand more than 50 feet away from the polling location. There is convincing evidence that both the response rates and the accuracy of the exit poll data decrease once an interviewer is forced to stand more than 50 feet away from the polling location. The priority states in our efforts should be Arizona, Florida, Minnesota, Nevada, New Mexico, Ohio and South Dakota, which all tried to impose distance requirements greater than 50 feet on exit poll interviewers during this election.

Improving our computer system. We plan to make the following changes in the operations and technical features of the computer system:

[101] [VNS was the pre-cursor to NEP, which was established in 2003.]

We will have one dedicated person on both the decision and technical conference bridge when there are any problems with the system, so that there is no confusion in communication. That person will have no other system responsibilities.

Technical changes will include testing the system with at least twice the number of users and a compressed time during simulations. Testing will also include twice the expected data as on Election Day. This may have exposed the database server problem we experienced with the screens "freezing up" starting at 10:35 p.m. [on election evening in 2004.]

We plan to use more sophisticated monitoring tools for the database on election night that will pinpoint hot spots for us prior to any problems occurring. The only circumstance where untested code will be executed on election night is to correct a problem causing the system not to function. Otherwise, no untested code will be executed as long as the system is up and running.

During failover we will isolate each data center, so that all primary servers point to the primary database server and all backup servers point to the backup database server. This eliminates the need for running scripts for this purpose.

Improving the Analysis

National Exit Poll and Cross Survey. We will look for ways to better achieve consistency between the Cross Survey results for a few selected characteristics and the National Exit Poll.

If we want to improve the National Exit Poll estimates for minority groups or other characteristics that are highly clustered, we need to increase the number of polling locations in the national sample or oversample polling locations with the characteristics of interest.

Absentee/early voter telephone surveys. Ideally, the state absentee/early voter telephone survey sample sizes should be increased in the states where [the] absentee vote is a large proportion of the total vote. With only 500 respondents in 2004 representing nearly one-fifth of all voters nationally who voted early or by absentee, the results of the national and regional breakouts for some of the smaller demographics are based upon sample sizes that are too small.

Other Improvements

Dealing with the leaking of NEP exit poll data. The decision by the NEP members to withhold the distribution of exit poll information within their organizations until 6:00 p.m. ET [Eastern Time] on Election Day will help prevent—or at least delay—the use of exit poll data before poll closing by those who have not purchased the data. We will work closely with the NEP members to develop security measures deemed appropriate to implement this policy.

Subscribers. In the general election, several subscribers felt that they were not given the same guidance about possible inaccuracies in the exit poll estimates that we had given the NEP members. On Election Day, at 4:30 p.m. ET, we convened a conference call with the Decision Teams of the NEP members and cautioned them that we expected sizeable errors in the exit polls in nine states; in seven states (Connecticut, Delaware, Minnesota, New Hampshire, Pennsylvania, Vermont and Virginia) we suspected that the exit poll estimates were [an] overstatement of the vote for Kerry; in two states (South Dakota and West Virginia) we suspected an overstatement of the Bush vote. We made these warnings based upon the discrepancies between the exit polls and our prior estimates in these nine states. We made a mistake in not sharing with the subscribers our concerns about the accuracy of the exit poll estimates in those nine states. In the future, we will need to make sure that whatever guidance we share with the NEP members is also communicated to the subscribers, so that they can feel comfortable using the data.

In the rest of the report, Edison-Mitofsky state that their intention was to "detail areas that we intend to improve for the next election cycle,"[102] identifying specific areas for additional research.

The next sections of this chapter present selected examples of some of the most current research efforts that are underway to improve data quality in election and exit polling.

6.3 SOME RECENT RESEARCH

One of the most critical issues with regard to data quality in survey research is whether the exit polling data collected are representative of the population— something especially important, given that for 2008 one of the presidential major candidates will most likely be either black or female, both a first for the U.S. In the following paper, Pedraza and Barreto (2007) explore this issue for the increasingly diverse U.S. population.

[102]Edison Media Research and Mitofsky International (2005: 18).

6.3.1   Excerpt from: Exit Polls and Ethnic Diversity: How to Improve Estimates and Reduce Bias Among Minority Voters
Francisco Pedraza and Matt Barreto

When political polling got its start, the median voter was White, male, middle class, and civically-minded. Today, the voting public is incredibly diverse and growing weary of pollsters. Despite considerable demographic changes in the American electorate, it is unclear whether survey research and exit polling have kept up with the changes. The most recent data from the 2007 Current Population Survey point out that 34% of Americans are not White and in many states Whites are the minority. [The population in] California is now just 43% White, New Mexico is 44% [White], and Texas is 47% White. These changing demographics have led some scholars (Leal et al., 2005) to question whether the data and results of today's exit polls are correct and reflective of our multicultural society, or do they tend to over-represent the opinions of White, middle-class, suburban voters? Accurate exit poll data are important to policy makers, the media and scholarly researchers, all of whom attempt to understand what is on the voter's mind.

During the 2000 presidential election, Voter News Service (VNS) exit polls stated that Al Gore had won Florida—in fact he lost. During the 2004 contest, the new national exit poll which replaced VNS showed that George Bush lost the states of Ohio and New Mexico—two pivotal states that he actually won.

In addition to many state level exit poll results being skewed in 2004, comparative vote results for minority groups, such as Latino voters, also appeared to be off. The National Exit Pool (NEP) reported on November 2, 2004 that Bush won 45 percent of the Latino vote, increasing by 10 points from his support in 2000. In contrast, an exit poll of *only* Latino voters implemented by the Willie C. Velasquez Institute found that Bush won just 32 percent of the Latino vote. Further, a pre-election survey of Latinos by *The Washington Post* showed Bush garnering just 30 percent of the vote in late October (Leal et al., 2005).

What explains these differences? One possibility is the methodology used to select the precincts where exit poll interviews are conducted is faulty. The goal is that respondents in the exit poll survey are accurate representatives of the entire city or state in which the election is being held. However, if the exit poll interviews respondents who are too conservative or too liberal, too young or too old, too poor or too rich, it could skew the overall results by a non-trivial margin. Given the increasing racial and ethnic diversity of America's largest cities, the question might now be "Are exit poll surveys accurately representing White and minority voters?. . . "

Racial Segregation in the United States.   The United States is an increasingly racial and ethnically diverse country. Somewhat paradoxically, however, the increase in diversity across the country over the past twenty-five years has been accompanied by an increasing level of racial and ethnic segregation. Nowhere is the state of segregation in our country more telling than in the racial distribution characterizing our school system. While it remains true that diversity has increased in the public K-12 school system as a whole, by the year 2000 much of the Black-White desegregation

Table 6.1 Index of racial segregation by city, 2000

City	White-Black	White-Latino
New York	77	64
Los Angeles	77	71
Chicago	88	64
Houston	78	66
Philadelphia	82	70
Phoenix	63	63
San Diego	67	64
San Antonio	60	55
Dallas	75	69
Miami	86	51
Detroit	68	65
Washington	84	65
Boston	78	65
Denver	71	63
Seattle	69	51

Note: Value of *0* reflects pure integration and value of *100*
reflects pure segregation
Source: Population Studies Center, University of Michigan,
2001

accomplished since *Brown v. Board of Education* (1954) had essentially been reversed to 1970 levels (Orfield and Lee, 2006: 14). Grimmer is the picture for Latinos who now hold the position of both the most racially and economically isolated group in our nation's public schools (Orfield and Lee, 2006: 10-11). The isolation of Black and Latino students in our schools can be traced to the racial and ethnic separation between neighborhoods.

Residential segregation is common and increasing throughout the United States. According to a study of racial segregation from the University of Michigan Population Studies Center, U.S. racial groups remain divided (Farley, 2001). The study provides a racial residential segregation index of dissimilarity, where a value of *0* is perfect integration and a value of *100* is extreme segregation. Table 6.1 lists for selected cities in 2000 the White-Black and White-Latino residential segregation scores. Seattle is included among some of the most segregated cities in our country, with a score of *69* for the degree of White-Black dissimilarity and a value of *51* in the comparison between Whites and Latinos. Moreover, patterns of racial group isolation show little sign of reversal. According to an analysis of Census data completed by the Harvard Civil Rights Project, racial residential segregation has increased from 1990 to 2000 in major metropolitan areas, including Boston, Chicago, and San Diego (Stuart, 2002; McArdle, 2003a; McArdle, 2003b). For each of these cities, comparisons of Census data in 1990 and 2000 reveal evidence of "White flight," suggesting that increasing levels of residential segregation are attributable to Whites moving out of central city communities and into the suburban neighborhoods of metropolitan areas, a pattern that geographers have documented in Seattle, as well (Guest, 2006)... [103]

... In the next section we discuss an exit poll experiment conducted by the University of Washington in the Seattle-King County metropolitan area that compares results from a random sample of precincts to those from a racially stratified selection. In each of the samples, interviews were available in English, Spanish, Chinese, Korean, and Vietnamese, in order to control for the importance of language of interview.

Methodology in the 2006 Seattle Study. Building on a pilot project conducted in Los Angeles during the 2005 mayoral election, we implemented a racially stratified homogenous precinct sample in an exit poll in King County, Washington during the 2006 mid-term election. In order to assess the racially stratified approach, we simultaneously implemented a second exit poll in King County in which precincts were selected at random. Thus, we are able to compare the results of two exit poll sampling strategies to determine if they yield different results.

... Of course, it is not realistic to recruit 2,000 volunteers to staff each of the 1,000 precincts in a state like Washington. Thus, the key is picking a select number of precincts that accurately represents the full universe of 1,000 precincts throughout the state. Because a small number of precincts is chosen to represent the universe, if the "wrong" precincts are selected, the results may be biased. Therefore, exit poll research teams take considerable care to select precincts. However, their selection criteria can sometimes be flawed—one possible problem during the 2000 and 2004 presidential elections. For example, in 2004, only five majority-Latino precincts were included in the NEP survey and, as a result, most Latinos interviewed came from majority White neighborhoods and tended to be more acculturated and conservative (Gomez, 2004; Tolson, 2004).

The reality is that the great majority of voters do not live in racially integrated neighborhoods. Instead, most voters reside—and vote—in precincts that are racially homogenous. According to an analysis of geographic segregation by the University of Michigan Population Studies Center, King County racial groups, like much of America, are still very residentially divided (Farley, 2001; see also Logan, 2002).

Do Asian Americans who vote at International Terrace (precinct # SEA37-1825) in the International District/Chinatown, and have a population that is 56 percent Asian, differ from Asians who voted at a suburban precinct? What happens to the overall results if Asian voters at the International Terrace are excluded from the exit poll because that specific precinct is not selected in the sample? Similarly, questions may arise about African Americans who vote at the Tabernacle Missionary Baptist Church, which is majority-Black. Thus, a more accurate representation of racial and ethnic voters—and, therefore, the city at large—might be to conduct exit poll interviews in high concentration racial precincts (homogenous) and a mix of racially diverse precincts.

---

[103][This is an excerpt of the original paper by Pedrazza and Barreto (2007). A more complete discussion of the issues is available upon request through *http://www.votingsystems.us*.]

Table 6.2   Types of precincts included in each sample method

	Distribution by Sample	
Precinct Type	Sample 1	Sample 2
Heavily White	6	20
Heavily Asian	8	5
Heavily Black	6	1
Heavily Latino	6	1
Racially Mixed	4	3
Total Precincts	30	30

Sample 1: Racially stratified homogenous precincts.   In order to select precincts for the sample, we first obtained information about the White, Black, Latino, and Asian American voting age population within each of the 510 voting precincts in King County. We then sorted the precincts into five categories: heavily White; heavily Black; heavily Latino; heavily Asian; and racially mixed. Within each category we randomly selected precincts from the top quartile, i.e. racially homogenous. Because of the diversity within the Asian population, and because Asians are the largest minority group in King County, we selected two extra Asian precincts. The stratified homogenous approach has two major advantages over any other approach, in particular a random selection. The first is that it improves the actual sample selection by focusing on precincts where most minority voters live. Instead of picking up minority voters in the suburbs, this approach ensures the inclusion of minority voters in majority-minority precincts. Second, it increases the sample size of minority voters, thereby decreasing the margin of error on the subgroup voting estimates.

Sample 2: randomly selected precincts.   The randomly selected precincts were picked without regard for the racial and ethnic composition within the precinct and, instead, were chosen purely at random throughout the County. Theoretically, this method should provide a fairly accurate picture of the overall county. However, it will likely yield a much smaller number of interviews with minority voters and is more likely to pick up minority voters outside majority-minority precincts. The 30 precincts in the racially stratified homogenous approach and the 30 precincts selected in the random sample are detailed in Table 6.2.

For both sampling approaches, student researchers were recruited from the University of Washington, and a two-person team was assigned to each precinct. Voters were recruited as they left the voting precinct, using a traditional skip pattern to randomize which voters were selected. Those selected completed a self-administered survey using pen and paper. The surveys were available to voters in multiple languages[104],

[104]Surveys were available in English, Spanish, Chinese, Korean, Vietnamese, Tagalog, Russian.

and the student researchers were appropriately assigned to precincts based on their language skills or race/ethnicity. The exit polls were conducted from 7:00 a.m. to 8:00 p.m., the entire time that the polls were open in King County. Prior to the November 7, 2006 election, student researchers attended two training sessions on the exit poll project.

The Results: Does Precinct Selection Matter?   In order to assess whether or not precinct selection impacts exit poll results, we present three levels of analysis. First, we compare the overall results for Sample 1 and Sample 2 for key questions to determine if the overall samples are yielding different frequency percentages. Second, we compare just the minority voters in Sample 1 to minority voters in Sample 2, to measure any potential differences in ideology, partisanship, and attitudes for non-Whites in majority-minority precincts and those in majority-White precincts. Since previous exit polls, such as during the 2004 presidential election, have not conducted significant over samples in majority-minority precincts, they have tended to over-represent Black and Latino voters living in suburban White neighborhoods. Here, we are able to directly compare minorities in both majority-White and majority-minority precincts to determine whether or not statistically significant differences exist. Third, we conduct a similar comparison for White voters. Comparing results among Whites in Sample 1 and Sample 2 is important, because Whites continue to make up a majority of the electorate, even in racially diverse cities and states. Thus, if there is significant variation in White responses by sample type, pollsters can make improvements in estimating White public opinion by including majority-minority precincts in their sample. Although such precincts are majority-minority and yield a large number of minority interviews, there are still a number of White voters in these areas whose opinion might be ignored if majority-minority precincts are not included in the sample.

Ideology and partisanship.   Arguably the most important individual character-istic that pollsters need to "get right" is the correct partisan and ideological balance of voters in the city or state they are sampling. While previous critiques of partisan imbalance have focused on interviewer effects (i.e., young people more successfully recruit liberals to take exit polls), we argue that precinct selection can dramatically alter the partisan and ideological affiliation of voters. Table 6.3 displays the results of self-reported political ideology and party affiliation among Whites and minorities in both samples. Overall, comparing all voters in both samples, respondents in the racially stratified homogenous precinct approach are more liberal and more Demo-cratic. While minority voters are consistently more liberal and more Democratic than Whites, White voters in the racially stratified sample appear to be considerably dif-ferent than White voters picked up in the random sample. In the random sample, 55 percent of Whites self-identify as Democrats, whereas nearly 70 percent of Whites in the stratified sample are Democrats. Among minorities, those in the random sample are more conservative (20%) than those in the stratified sample (14%) and also more likely to identify as Republican in the random sample. The data make clear that which precincts are selected in the exit poll sample can greatly impact the party and ideolog-

Table 6.3   Difference in voter ideology and partisanship by sample method

	All Voters		Minority		White	
Political ideology	Random	Race	Random	Race	Random	Race
Very liberal	22.8	36.3	22.3	22.0	23.5	42.6
Somewhat liberal	27.6	31.5	29.0	33.5	27.8	29.7
Moderate	25.5	20.3	28.6	30.2	25.2	16.6
Somewhat conservative	14.9	6.0	12.4	10.6	14.3	4.1
Very conservative	9.3	6.0	7.7	3.7	9.2	7.0
Party affiliation	Random	Race	Random	Race	Random	Race
Democrat	53.2	67.4	66.0	69.4	55.1	69.7
Independent/Other	21.8	21.1	15.6	18.9	17.6	17.5
Republican	25.0	11.5	18.4	11.8	27.3	12.8

ical balance of the voters. Not only is there a significant difference between the two polls, but we argue that the racially stratified sample is more accurate. According to records from the County Registrar of Voters, in the September 2006 election in King County, 68.3% of all voters were Democrats. In the racially stratified sample, 67.4% of all voters identified as Democrats, compared to 53.2% in the random sample—15 points lower than the known partisan balance of King County.

Most important issue.   The differences noted above are likely to have "trickle down" effects, given the importance of partisanship and ideology in explaining voter attitudes on issue and policy preference. Table 6.4 reports voter responses to the question, "What general issues were most important to you as you thought about how you would vote today," and respondents could select two issues. Once again, the two sample designs yield different results overall, which hold for both minority and White voters. Across the board, the war in Iraq was listed as the most important issue. However, the degree varied by sample and race. White voters were more likely to list the war in Iraq in the stratified sample compared to the random sample, while minorities were less likely to list Iraq in the stratified sample. The tradeoff appears to be with minorities being more likely to list education as a top concern in the stratified sample—perhaps because schools in majority-minority neighborhoods are under-performing and under-funded. Similarly, minorities in the racially stratified sample were more likely to list jobs or the economy as a top issue. This suggests, beyond partisan affiliation or political ideology, voters in the racially homogenous precincts are facing different issues than voters randomly selected throughout the county, which could impact the results of an exit poll on a school bond measure or statewide initiatives on the environment or taxes.

Table 6.4   Difference in issue importance by sample method

	All Voters		Minority		White	
Most Important Issue	Random	Race	Random	Race	Random	Race
War in Iraq	53.6	61.8	58.7	50.8	54.1	66.3
Education	25.7	31.2	32.7	41.7	24.5	26.0
Environment	20.5	24.2	13.8	12.9	21.6	29.4
Ethics in Government	22.7	21.0	19.8	12.9	23.5	24.1
Jobs/Economy	17.2	18.2	21.2	26.0	16.3	14.6
National Security	15.8	9.1	13.9	11.3	15.8	8.1
Health Care	8.7	11.5	13.5	11.7	8.6	12.4
Taxes	13.1	10.3	9.9	16.7	13.2	8.1
War on Terror	9.0	4.0	7.1	1.4	9.0	5.0
Illegal Immigration	7.5	3.2	5.2	3.1	7.6	2.6
Gas Prices	2.9	3.3	1.9	7.3	2.6	1.8

Immigration and discrimination.   Finally, we examined results for a very specific set of issues related to immigration policy and discrimination against immigrants. The racially stratified samples are not only higher in their minority population, but significantly higher in their immigrant population and proximity to immigrant communities. Given the national attention to immigration as an issue, and the numerous polls asking voters about their opinions of immigrants and immigration reform, it is important to examine the potential effects of sampling strategy on this issue. Table 6.5 displays results for two questions on our exit poll related to immigration. Voters were asked "Regarding immigration policy, do you favor deportation, temporary guest worker program, pathway to earned citizenship, or no change in policy?" and were later asked, "Do you think discrimination against immigrants is much of a problem in today's society?" Among Whites, those in the racially stratified sample were significantly more supportive of pathways to citizenship; 62% compared to 51% in the random sample. For minority voters the relationship is less clear. Among all minorities—Black, Latino, Asian—those in the racially stratified sample were actually somewhat less supportive of a pathway to citizenship. This is likely the result of anti-immigrant attitudes held by some African Americans and found to be particularly strong in heavily Black neighborhoods (Doherty, 2006). Thus, we provide results for only Latinos, the group often referenced during the immigration debate and the results are much more clear. Latino voters in the stratified sample had higher rates of support for a path to citizenship—66% compared to 55% in the random sample. Further, virtually no Latino voters in the stratified sample favored deportation, compared to 7% in the random sample. Finally, with respect to discrimination against immigrants, noticeable differences emerge in the expected direction. Both White and

Table 6.5 Differences in attitudes about immigration by sample method

Immigration policy	All Voters		Minority		Latino		White	
	Random	Race	Random	Race	Random	Race	Random	Race
Deportation	16.5	10.5	13.5	16.1	6.6	0.5	16.5	8.2
Guest worker	26.5	24.7	29.0	25.6	34.6	27.0	26.5	23.8
Path to citizenship	50.0	57.3	50.5	47.8	54.7	65.8	50.9	62.2
No change	3.0	4.3	3.6	7.0	4.1	3.0	2.9	3.5

Discrimination against immigrants?	All Voters		Minority		White	
	Random	Race	Random	Race	Random	Race
Yes very much	24.0	39.3	34.8	47.0	23.8	37.4
Yes somewhat	45.5	40.7	47.3	35.7	48.2	45.8
Not really	25.1	16.8	16.4	17.3	26.7	16.5

minority voters were more likely to state that discrimination against immigrants is "very much" a problem.

Conclusions. The data clearly demonstrate that exit poll methodology matters. In particular, as pollsters grapple with how to best survey the increasingly diverse American electorate, precinct selection is an important concern. In this paper, we have outlined one possible alternative approach to exit polling in racially diverse settings, to ensure the representation of majority-minority precincts where a high percentage of minority voters reside. In addition to the aim to reduce survey error, each of these considerations also has in common the potential to lessen the burden on voters of participating in an exit poll. The resources that go into making an exit poll a success aren't just drawn from a pollster's budget, but also from the time and energy of each participant. However, making it easier for voters to actually participate in an exit poll doesn't have to be costly for a pollster. As alternative methodology exit polls in Los Angeles and Seattle have demonstrated, it can be as easy as simply selecting heavily minority precincts or offering the exit poll in multiple languages. These kinds of quality touches are investments that promise returns, not only for any one particular exit poll, but will likely help in securing a more favorable public view of polling in general.

References

Doherty, Carroll (2006). Attitudes Toward Immigration: In Black and White, a report from the Pew Research Center for the People and the Press, available on the Internet at: *http://pewresearch.org/pubs/21/ attitudes-toward-immigration-in-black-and-white.*

Farley, Reynolds (2001). Racial Residential Segregation: Census 2000 Findings, a working paper, University of Michigan Population Studies Center, available on the Internet at: *enceladus.isr.umich.edu/race/racestart.asp*

Gomez, Lisa Marie (2004). Why Don't Hispanic Polls Seem to Add Up? *San Antonio Express-News,* November 7.

Guest, Pete (2006). Gentrification and Affordable Housing, presentation hosted by the University of Washington Institute for the Study of Ethnicity, Race and Sexuality, University of Washington, November 7, 2006.

Leal, David; Barreto, Matt; Lee, Jongho; and de la Garza, Rodolfo (2005). The Latino Vote in the 2004 Election, *P.S.: Political Science and Politics,* 38, January, 41–49.

Logan, John (2002). Hispanic Populations and Their Residential Patterns in the Metropolis, Working Paper, Lewis Mumford Center for Comparative Urban and Regional Research, May 8, available on the Internet at: *mumford1.dyndns.org/cen2000/HispanicPop/HspreportNew/MumfordReport.pdf.*

McArdle, Nancy (2003a). Race, Place, and Opportunity: Racial Change and Segregation in the Boston Metropolitan Area: 1990-2000, Cambridge, MA: The Civil Rights Project at Harvard University.

McArdle, Nancy (2003b). Race, Place, and Opportunity: Racial Change and Segregation in the San Diego Metropolitan Area: 1990-2000, Cambridge, MA: The Civil Rights Project at Harvard University.

Orfield, Gary and Lee, Chungmei (2006). Racial Transformation and the Changing Nature of Segregation, Cambridge, MA: The Civil Rights Project at Harvard University.

Stuart, Guy (2002). Integration or Resegregation: Metropolitan Chicago at the Turn of the New Century, Cambridge, MA: The Civil Rights Project at Harvard University.

Tolson, Mike (2004). Exit-Poll Math Doesn't Add Up, One Institute Says, *Houston Chronicle,* November 6, 2004.

Another fairly current paper by Barreto et al. (2006) looks at whether absentee voters differ from polling place voters. The authors conclude that their study did not show major demographic or political preference differences, but suggest that further research is needed. These findings may change as the tendency to cast absentee and early ballots increases.

Since much election polling is dependent on telephone survey techniques, another issue that is becoming an increasing concern to survey researchers and their users is that of respondents who only have cell phones and do not maintain regular "landline" telephones. Remember, with early voting becoming more and more common, the image of an exit pollster as a person standing outside a precinct is changing. The expected large increase in first-time younger voters (often in cell phone-only households) makes this research even more pressing for 2008. In the next paper, Keeter (2007) discusses the potential impact of this problem on data quality.

6.3.2 How Serious Is Polling's Cell-Only Problem? The Landline-less Are Different and Their Numbers Are Growing Fast
Scott Keeter

Twenty years ago the survey research profession—having grown comfortable with telephone interviewing as an alternative to personal interviewing for conducting surveys—worried mostly about the roughly 7% of U.S. households that could not be interviewed because they had no telephone. Today our concern is somewhat different, and potentially more serious. According to government statistics released last month, nearly 13% of U.S. households (12.8%) cannot now be reached by the typical telephone survey because they have only a cell phone and no landline telephone.[105]

If people who can only be reached by cell phone were just like those with landlines, their absence from surveys would not create a problem for polling. But cell-only adults are very different. The National Health Interview Survey found them to be much younger, more likely to be African American or Hispanic, less likely to be married, and less likely to be a homeowner than adults with landline telephones. These demographic characteristics are correlated with a wide range of social and political behaviors.

Polling's cell phone problem is a new one. In early 2003, just 3.2% of households were cell-only. By the fall of 2004, pollsters and journalists were openly worrying about the potential bias that cell-only households might create for political surveys. The National Election Pool's exit poll found that 7.1% of those who voted on Election Day had only a cell phone, and these cell-only voters were somewhat more Democratic and liberal than those who said they had a landline telephone. But pre-election telephone polls in that election were generally accurate, and pollsters felt that they had dodged the proverbial bullet. This fortunate outcome was a result of the fact that the statistical weighting employed by most telephone polls helped to correct for the missing respondents. The fact that the cell-only group in 2004 was still a relatively small part of the overall population also helped mitigate the impact of the problem.

But given the speed with which the number of cell-only households has increased, there is growing concern within the polling business about how long the landline telephone survey will remain a viable data collection tool, at least by itself. At [the 2007] annual meeting of the American Association for Public Opinion Research (AAPOR), survey research's top professional organization, an entire series of research panels focused on the cell phone issue. At one of the panels, a government researcher told the audience that the size of the cell-only group could approach 25% by the end of 2008 if the current rate of increase is sustained.

To monitor the impact of the cell-only phenomenon, the Pew Research Center conducted four studies in 2006 that included samples of cell phone numbers as well

---

[105] Adding in the 2.2% of households with no phone service whatsoever, a total of 15.0% cannot be reached by landline surveys. See Blumberg and Luke (2007).

Table 6.6   Summary of comparisons between landline samples and cell-only samples

Number of survey questions compared	46
Average (mean) difference between landline and cell-only samples across all 46 questions	7.8%
Range of differences (absolute value)	0%–29%
Maximum change in final survey estimate when cell-only sample is blended in	2%
Average (mean) change in final survey estimate when cell-only sample is blended in	0.7%

as a full sample of landline numbers.[106] The four surveys covered a very wide range of topics, including use of technology, media consumption, political and social attitudes, and electoral engagement. Comparing the cell-only respondents with those reached on landlines allowed us to assess the degree to which our traditional surveys are biased by the absence of the cell-only respondents.

We compared the cell-only and landline respondents on 46 different survey questions. [See Table 6.6.] Across these questions, the average difference between cell-only and landline respondents was approximately 8 percentage points (7.8%), with the range of differences running from 0% (for a question about whether the respondent is "bored" by what goes on in Washington, DC) to 29% (being registered to vote). But the good news is that none of the measures would change by more than 2 percentage points when the cell-only respondents were blended into the landline sample. Thus, although cell-only respondents are different from landline respondents in important ways, they were neither numerous enough nor different enough on the questions we examined to produce a significant change in overall general population survey estimates when included with the landline samples and weighted according to U.S. Census parameters on basic demographic characteristics.

The picture is not entirely positive, however. While the cell-only problem is currently not biasing polls based on the entire population, it may very well be damaging estimates for certain subgroups in which the use of only a cell phone is more common. This concern is particularly relevant for young adults. According to the most recent government estimate, more than 25% of those under age 30 use only a cell phone. An analysis of young people ages 18-25 in one of the Pew polls found that the exclusion of the cell-only respondents resulted in significantly lower estimates of this age group's approval of alcohol consumption and marijuana use. Perhaps not surprisingly, excluding the cell-only respondents also yields lower estimates of technological sophistication. For example, the overall estimate for the proportion of 18-25 year olds using social networking sites is 57% when the cell-only sample is blended with the landline sample, while the estimate based only on the landline sample is 50%.

[106]Details about the four studies—The Cell Phone Challenge to Survey Research; Online Papers Modestly Boost Newspaper Readership; A Portrait of Generation Next; and Cell-Only Voters Not Very Different—are available through the Pew Research Center's Internet site; see *http://pewresearch.org/pubs/*.

Including a cell-only sample with a traditional landline-based poll is feasible, as the four studies conducted last year indicate. But even if feasible, cell-only surveys are considerably more difficult and expensive to conduct than landline surveys. Federal law prohibits the use of automated dialing devices when calling cell phones, so each number in the cell phone sample must be dialed manually. It also is common practice to provide respondents with a small monetary incentive to offset the cost of the airtime used during the interview. And the screening necessary to reach cell-only respondents among all of those reached on a cell phone greatly increases the effort needed to complete a given number of interviews. Pew estimates that interviewing a cell-only respondent costs approximately four to five times as much as a landline respondent.

Pollsters recognize that some type of accommodation for the cell-only population will have to be made eventually, as was clear from the large amount of research on the topic presented at the AAPOR conference [in 2007].[107] In addition to the use of so-called "dual frame samples," such as those described above (calling both a cell phone sample and a landline sample), practitioners are discussing other alternatives, including the establishment of panels of cell-only respondents that can be surveyed periodically to track their opinions, and employing mail or Internet surveys to reach the cell-only population.

Acknowledgment. This article draws on research presented at the [2007] AAPOR conference. What's Missing from National RDD Surveys? The Impact of the Growing Cell-Only Population, [is] by Scott Keeter (Pew Research Center), Courtney Kennedy (University of Michigan and Pew Research Center), April Clark (Pew Research Center), Trevor Tompson (The Associated Press), and Mike Mokrzycki (The Associated Press).

References

Blumberg, Stephen J. and Luke, Julian V. (2007). Wireless Substitution: Early Release of Estimates Based on the National Health Interview Survey, July–December 2006, report by the U.S. Centers for Disease Control and Prevention, available on the Internet at: *www.cdc.gov/nchs/data/nhis/earlyrelease/wireless200705.pdf*

Keeter, Scott (2006). Cell-Only Voters Not Very Different: Fewer Registered, More First-time Voters, Pew Research Center, available on the Internet at *http://pewresearch.org/pubs/80/cell-only-voters-not-very-different*.

Pew Research Center (2006). The Cell Phone Challenge to Survey Research: National Polls Not Undermined by Growing Cell-Only Population, available on the Internet at: *http://people-press.org/reports/display.php3?ReportID=276.*

Pew Research Center (2006). Online Papers Modestly Boost Newspaper Readership: Maturing Internet News Audience Broader Than Deep, available on the Internet at: *http://pewresearch.org/pubs/238/online-papers-modestly-boost-newspaper-readership.*

[107] [Another source of information on telephone surveys is the monograph from the 2006 International Conference on Telephone Surveys. Readers interested in this topic may want to see Lepkowski et al. (2007). See also the just-released special issue of *Public Opinion Quarterly*, which focuses on cell phone numbers and telephone surveying in the U.S. (AAPOR, 2007)]

Pew Research Center (2007). A Portrait of Generation Next, available on the Internet at: *http://pewresearch. org/pubs/278/a-portrait-of-generation-next.*

Keeter's concerns, above, about recent technological changes that impact voter response are a cautionary tale that may well become more serious as time goes on. Furthermore, cell phones are not the only area of concern. Several papers in earlier chapters of this volume have also touched on the use of the Internet for election polling. As already noted, one of the primary issues is Internet suitability, given that most Internet data collection remains based on nonprobability samples. Two additional studies related to Internet surveys are described by Atkeson et al. (2007) and Crampton (2007). Atkeson et al. used a probability sample of general election voters to compare Internet and mail responses in a mixed mode survey and found that the responses of the Web users compared somewhat favorably to those of the mail respondents, though the authors cited some limitations. In another write-up, Crampton discusses the success of Web polls in selected European countries, where incentives are used to encourage participation. He suggests a similar approach be considered in the U.S. Despite these apparent successes, the consensus is that pollsters would do well to approach these new technologies with caution.

The reading below, by Chang, Holland, and Piehl (2007) addresses a nonresponse issue that has not been covered in other papers in this book—it examines the effectiveness of probes to obtain responses to questions that initially do not yield a usable answer. Because of the sensitivity of asking respondents about voter preference, this is an issue that bears consideration.

6.3.3   Evaluating Follow-up Probes to *Don't Know* Responses in Political Polls[108]
LinChiat Chang, Keating Holland, and Jeri Piehl

Why Do Respondents Say "Don't Know"? "Don't know" responses are most common among respondents with low knowledge or interest in the survey topic. That is, respondents with lower education, cognitive skills, prior exposure, experience, perceived importance of survey topic, etc. are most likely to say "don't know" (Ehrlich, 1964; Klopfer and Madden; 1980; Krosnick and Fabrigar, 2006). In many of these instances, "don't know" is a valid and accurate response, because these respondents genuinely lack an opinion on the issue at hand.

However, other respondents hold a valid opinion but are withholding it for a variety of reasons. Why do some respondents say "don't know" when they, in fact, have a valid opinion? Common reasons include:

1. Some respondents do not feel completely sure of their response; hence they are not comfortable or not confident enough to pick a valid answer (Hippler and Schwarz, 1986; McClendon, 1986; McClendon and Alwin, 1993).

---

[108]Evaluating Follow-up Probes to Don't Know, by LinChiat Chang, Keating Holland, and Jeri Piehl, reprinted with permission. Copyright ©2007: Opinion Research Corporation. All rights reserved.

2. Some respondents' beliefs, attitudes and feelings on the issue are contradictory. This state of ambivalence can result in "don't know" responses, because the conflicting attitudes and feelings preclude a clear stand on the issue (Feick, 1989; Klopfer and Madden, 1980; Ready, Whitehead, and Blomquist, 1995).

3. Some respondents are not comfortable giving out an answer that they would prefer to keep private (Berinsky, 1999, 2002; Glynn and Park, 1997; Schaeffer and Bradburn, 1989). For example, given the apparent decline in Bush approval ratings, pro-Bush respondents may be more reluctant to state their opinions; but some of them will do so upon probing. Indeed, data from our field experiment below showed more pro-Bush and pro-Republican responses among responses elicited by probes, implying at least some support for this reasoning.

4. The survey question is unclear or difficult to understand (Faulkenberry and Mason, 1978; Converse, 1976; Knäuper, Belli, Hill, and Herzog, 1997).

5. The response options provided are a poor match of respondents' opinions, so it is frustrating and difficult to pick an appropriate answer (Borgers and Hox, 2001; Coombs and Coombs, 1975; Converse, 1976; Ehrlich, 1964; Feick, 1989).

6. Interviewer expectations have a direct impact on the prevalence of "don't know" responses. Interviewers who expect to get valid answers are more likely to do so than interviewers who expect respondents to have no opinion on the subject (Singer, Frankel, and Glassman, 1983; Pickery and Loosveldt, 1998). In addition, many interviewers and field supervisors have attested to the phenomenon where respondents say "don't know" when they are merely stalling while they think, and these respondents will often provide a valid answer if the interviewer waits or probes (Bradburn and Sudman, 1988).

In other words, there are many reasons why respondents say "don't know." Some respondents genuinely have no opinion; some are reticent about sharing their opinions, while others are obstructed from stating their opinions due to flaws in question design or interviewer training. Hence, it seems worthwhile to probe for valid responses that would not otherwise be captured.

To Probe or Not to Probe?   The main objective of using follow-up probes to "don't know" responses is to elicit valid answers that would otherwise be lost. However, a potential problem is that probes might induce respondents to select a valid response when they actually have no opinion. Past research has demonstrated that when survey questions are phrased in ways to suggest that respondents ought to have opinions on the issue, some respondents would actually make up an arbitrary answer, because they do not wish to appear ignorant (Bishop, Tuchfarber, and Oldendick, 1986; Bishop, Oldendick, Tuchfarber, and Barnett, 1980; Hawkins and Coney, 1981; Schuman and Presser, 1981). When a probe was used in such studies, the effect was magnified. For example, before any probe was administered, nearly half of a national sample of adults said that they either agreed or disagreed with repealing the "1975 Public Affairs Act," a bogus issue. When respondents who initially gave no opinion were

Table 6.7   Operation variables

	No Probe	Hard Probe	Soft Probe
Number of Completions	424	434	428
Response Rate (AAPOR formula #3)	26%	21%	23%
Cooperation Rate (AAPOR formula #3)	57%	48%	51%
Refusal Rate (AAPOR formula #1)	3%	3%	3%
Average Length of Survey (in minutes)	4.3	4.7	4.5
Number of Completions per Interviewing Hr	2.5	2.1	2.4
Average Interviewer Quality Score (0-100)	83	85	85

[AAPOR refers to the American Association for Public Opinion Research.]

subjected to a follow-up probe, an additional 10% of respondents provided a valid response (Morin, 1995).

However, researchers rarely design surveys around bogus issues to trick their respondents. On the contrary, researchers strive to design surveys to obtain the most accurate gauge possible of legitimate variables of interest. In the above studies, survey items were intentionally phrased to suggest that respondents *should* know of those bogus issues and *should* have opinions on those issues, thus increasing pressure on respondents to provide an answer when they have none.

Nonetheless, if probes can elicit opinions on bogus issues, then it is possible that probes can also elicit bogus opinions on legitimate issues. If so, including responses elicited by probes *can* undermine data quality. To date, research comparing data with and without probes has not supported this reasoning. Instead, current evidence has consistently shown that *including responses elicited by probes produce data of higher validity* than simply accepting "don't know" responses as is (e.g., Das and Deshmukh, 1997; Gilljam and Granberg, 1993; Visser, Krosnick, Marquette, and Curtin, 2000). For example, when predicting final vote choice on electoral candidate races and issue referenda, analyses that included responses elicited by probes consistently yielded higher predictive validity than analyses that depended only on responses offered without probes (Visser et al., 2000).

Our Field Experiment.   In order to investigate the impact of probing and effects associated with specific probe wording, Opinion Research USA commissioned a field experiment in which the same nationwide political poll (based on standard CNN Poll items) was conducted in three different versions: a control condition with no probe and two conditions with different probe wording. One probe was worded strongly ("if you had to choose. . .") while the other probe was worded as a mild request ("we are interested in your general inclination. . ."), in order to explore the impact of hard vs. soft wording.

Items included standard polling questions on whether [the] respondent approved or disapproved of the way George W. Bush is handling his job as President, terrorism, the situation in Iraq, the economy, and immigration. To check for possible approval biases in response, we also added an item on whether [the] respondent approved or disapproved of the way Bill Clinton handled his job as President when he was in office. A congressional vote choice item asked "If the elections for Congress were being held today, which party's candidate would you vote for in your Congressional district?" A second item related to the congressional elections was "If George W. Bush supported a candidate for political office in your area, would you be more likely or less likely to vote for that candidate?"

All three polls were conducted on the same days in June 2006, by the Opinion Research USA call center located in Tucson, AZ, based on RDD [Random Digit Dialing] samples of American adults age 18 or older. We attained a completed sample of 424 respondents in the control condition, 434 respondents in the hard probe condition, and 428 respondents in the soft probe condition.

To assure equivalence, the three completed samples were compared in terms of demographic and operation variables. All three poll samples showed highly similar distributions on age, gender, race and ethnicity, highest education attained, annual household income, geographical region, voter registration status, and political ideology (liberal-conservative continuum). The only exception was that there were slightly more Democrats in the Hard Probe condition, compared to the other two conditions. Table 6.7 displays the key operation variables. As shown, the three polls had comparable response and cooperation rates, as well as interviewer quality scores. The average length of survey was under 5 minutes in all three conditions. In sum, the three conditions were roughly equivalent.

Reduction in Item Nonresponse.   Table 6.8 shows the proportion of "don't know" responses across the three conditions. As shown in the bottom row in Table 6.8, the average proportion of "don't know" responses was 8% in the "no probe" condition, 4% in the hard probe condition, and 3% in the soft probe condition. Hence, both probes were effective in reducing item nonresponse.

We ran various statistical tests to assess the relative effectiveness of the hard vs. soft probes, and found no statistically significant difference between the two probes in terms of reducing item nonresponse. In other words, both probes were equally effective in eliciting valid responses after respondents said "don't know."

Impact on Substantive Poll Estimates.   Ultimately, the objective of a nationwide political poll is to provide survey estimates to gauge the current sentiments of the nation. Hence, the next step was to assess whether the overall weighted poll estimates changed significantly if we excluded vs. included responses that were elicited after probing. Each of the three samples was weighted to match census parameters for age, gender, race, education, and geographical region. As shown in Table 6.9, follow-up probes had minimal impact on substantive poll estimates; none of the shifts in poll estimates before vs. after probing reached statistical significance.

Table 6.9 also revealed a marginal but consistent trend on the approval ratings, such that slight changes, if present, were always in increments in approval. This trend, which was present for both Bush and Clinton ratings, suggests that respon-

Table 6.8   Proportion of "don't know" responses

	No Probe	Hard Probe	Soft Probe
Bush Approval: handling job as President	10%	5%	4%
Bush Approval: handling terrorism	7%	4%	2%
Bush Approval: handling the situation in Iraq	6%	3%	1%
Bush Approval: handling the economy	9%	3%	3%
Bush Approval: handling immigration	14%	7%	7%
Clinton Approval: handled job as President	5%	2%	2%
Vote Choice: Candidate for Congress	9%	5%	2%
More or Less likely to vote for			
Bush-supported candidate	5%	4%	1%
Average Proportion of DK Responses:	8%	4%	3%

dents who converted from a "don't know" [DK] to a valid answer upon probing (DK converts) were more likely to say "approve" rather than "disapprove." Further, there was a marginal but consistent pro-Republican bias among DK converts in the Hard Probe condition. We monitored these biases in our subsequent poll, and replicated the approval bias, but not the pro-Republican bias. In short, DK converts are occasionally more likely to answer "approve" rather than "disapprove."

Differential Effectiveness of Probes across Demographic Groups.   The two probes did not work equally well across different demographic groups. Regression models were constructed to explore which demographic categories were most or least responsive to the respective probes. Analyses could not be conducted on each item, per se, due to the small number of DK converts; hence, all DK conversions across all items were combined for analyses. Logistic binary regressions were run such that the dependent variable was coded *1* for successful conversion by probe, *0* for unsuccessful conversion (i.e., remained a DK response); and the independent variables were a series of dummy variables corresponding to demographic groups.

Both probes [were] more effective on women than men; the hard probe was more effective in Northeast and West, while the soft probe was less effective among [the] 45-64 age groups and respondents with 4-year college degrees, but more effective among respondents with household income between $50,000–75,000. Note that all these differences were significant when controlling for the impact of all available demographic and political variables.

Conclusion.   Regardless of whether the probe was worded more or less assertively, both probes were effective in eliciting valid answers from respondents who initially said "don't know." Although inclusion of these probe-elicited responses did not have a significant impact on any estimate in this particular poll, the probes did elicit valid responses that would otherwise have been lost. Given the findings from this research, we have implemented the hard probe in all CNN Opinion Research Polls to-date.

Table 6.9 Poll estimates before and after probe

	Hard Probe		Soft Probe	
	Before Probe	After Probe	Before Probe	After Probe
Bush Approval: handling job as President	35%	37%	37%	38%
Bush Approval: handling terrorism	43%	43%	51%	51%
Bush Approval: handling the situation in Iraq	35%	35%	39%	39%
Bush Approval: handling the economy	36%	37%	36%	37%
Bush Approval: handling immigration	29%	32%	35%	36%
Clinton Approval: handled job as President	65%	66%	62%	62%
Vote for Republican candidate for Congress	35%	36%	34%	34%
More likely to vote for Bush-supported candidate	24%	25%	28%	28%

Although both probes were equally effective, the hard probe was selected over the soft probe because the hard probe exhibited somewhat more consistent impact across different demographics than the soft probe.

References

Berinsky, A. J. (1999) The Two Faces of Public Opinion, *American Journal of Political Science*, 43, 1209-1230.

Berinsky, A. J. (2002). Silent Voices: Social Welfare Policy Opinions and Political Equality in America, *American Journal of Political Science*, 46, 276-287.

Bishop, G. F.; Oldendick, R. W.; Tuchfarber, A. J.; and Bennett, S. E. (1980). Pseudo-opinions on Public Affairs, *Public Opinion Quarterly*, 44, 198-209.

Bishop, G. F.; Tuchfarber, A. J.; and Oldendick, R. W. (1986). Opinions on Fictitious Issues: The Pressure to Answer Survey Questions, *Public Opinion Quarterly*, 50, 240-250.

Borgers, N. and Hox, J. (2001). Item Nonresponse in Questionnaire Research with Children, *Journal of Official Statistics*, 17, 321-335.

Bradburn, N. M. and Sudman, S. (1988). *Polls and Surveys: Understanding What They Tell Us*, San Francisco, CA: Jossey-Bass.

Converse, J. M. (1976). Predicting No Opinion in the Polls, *Public Opinion Quarterly*, 40, 515-530.

Coombs, C. H. and Coombs, L. C. (1975). 'Don't know': Item Ambiguity or Respondent Uncertainty? *Public Opinion Quarterly*, 40, 497-514.

Das, A. and Deshmukh, J. (1997). What We Know about the "Don't Knows" (and What We Know about What to Do with Them), *WAPOR Newsletter*, 12-13.

Ehrlich, H. J. (1964). Instrument Error and the Study of Prejudice, *Social Forces*, 43, 197-206.

Faulkenberry, G. D. and Mason, R. (1978). Characteristics of Nonopinion and No Opinion Response Groups, *Public Opinion Quarterly*, 42, 533-543.

Feick, L. F. (1989). Latent Class Analysis of Survey Questions that Include Don't Know Responses, *Public Opinion Quarterly*, 53, 525-547.

Gilljam, M. and Granberg, D. (1993). Should We Take "Don't Know" for an Answer? *Public Opinion Quarterly*, 57, 348-357.

Glynn, C. J. and Park, E. (1997). Reference Groups, Opinion Intensity, and Public Opinion Expression, *International Journal of Public opinion Research*, 9, 213-232.

Hawkins, D. and Coney, K. (1981). Uninformed Response Error in Survey Research, *Journal of Marketing Research*, 18(3), 370-374.

Hippler, H. and Schwarz, N. (1986). Not Forbidding Isn't Allowing: The Cognitive Basis of the Forbid-Allow Asymmetry, *Public Opinion Quarterly*, 50, 87-96.

Klopfer, F. J. and Madden, T. M. (1980). The Middlemost Choice on Attitude Items: Ambivalence, Neutrality or Uncertainty? *Personality and Social Psychology Bulletin*, 6, 91-101.

Knäuper, B.; Belli, R.F.; Hill, D. H.; and Herzog, A.R. (1997). Question Difficulty and Respondents Cognitive Ability, *Journal of Official Statistics*, 13 (2).

Krosnick, J.A. and Fabrigar, L.R. (2006). *Questionnaire Design for Attitude Measurement in Social and Psychological Research*, New York, NY: Oxford University Press.

McClendon, M. J. (1986). Unanticipated Effects of No Opinion Filters on Attitudes and Attitude Strength, *Sociological Perspectives*, 29, 379-395.

McClendon, M. J. and Alwin, D. F. (1993). No-opinion Filters and Attitude Measurement Reliability, *Sociological Methods and Research*, 21, 438-464.

Morin, R. (1995 April 10-16). What Informed Public Opinion? A Survey Trick Points out the Hazards Facing those who Take the Nation's Pulse, *The Washington Post National Weekly Edition*, e36.

Pickery, J. and Loosveldt, G. (1998). The Impact of Respondent and Interviewer Characteristics on the Number of "No Opinion" Answers, *Quality and Quantity*, 32, 31-45.

Ready, R.C.; Whitehead, J. C.; and Blomquist, G. C. (1995). Contingent Valuation when Respondents Are Ambivalent, *Journal of Environmental Economics and Management*, 21, 181-196.

Schaeffer, N.C. and Bradburn, N.M. (1989). Respondent Behavior in Magnitude Estimation, *Journal of the American Statistical Association*, 84, 402-413.

Schaeffer, N.C. and Bradburn, N.M. (1989). Respondent Behavior in Magnitude Estimation, *Journal of the American Statistical Association*, 84, 402-413.

Schuman, H. and Presser, S. (1981). *Questions and Answers in Attitude Surveys: Experiments on Question Form, Wording, and Context,* New York, NY: Academic Press.

Singer E.; Frankel, M.R. and Glassman, M.B. (1983). The Effect of Interviewer Characteristics and Expectations on Response, *Public Opinion Quarterly,* 47, 84-95.

Visser, P. S. ; Krosnick, J. A. ; Marquette, J. ; and Curtin, M. (2000). Improving Election Forecasting: Allocation of Undecided Respondents, Identification of Likely Voters, and Response Order Effects, *Election Polls, the News Media, and Democracy*. Lavrakas, P. and Traugott, M. (eds.), New York, NY: Chatham House.

From the general survey literature, it is important to cite work such as that found in Singer (2007). Singer has conducted research to examine issues related to surveys of sensitive topics. In her 2007 talk, she focused on measurement error in a Web experiment and found that socially undesirable responses tend to be under-estimated. She considered a number of issues that affect response bias and found that general issues about privacy and confidentiality, attitudes towards surveys, and subjective concerns about trust influence replies. She noted that it is up to the researcher to find ways to minimize these influences.

Frankovic (2007a) reported on a somewhat related data quality issue—lying to the interviewer. She cites a number of circumstances that can affect the response to political polls. Some may be purposeful and some inadvertent, but the end result is the same—the response does not represent the respondent's true behavior or sentiments. Such problems can bias the results, as well, especially when race (or gender) bias could be a factor, as it was some years ago when Doug Wilder (an African American) ran for governor in Virginia.

6.4   GUIDELINES FOR QUALITY POLLING

The widespread use of exit polls for projecting election outcome results, analyzing characteristics of voters, and providing general insights about election practice—including the potential legitimacy of the voting process—both in the U.S. and worldwide, also raised broader ethical and procedural issues about the conduct of this kind of research. (As we have noted in this volume, some of the more recent exit polls have generated considerable controversy about U.S. presidential elections. Controversies have also resulted from some state and local elections and from elections in other countries.[109])   On May 14, 2005, the American Association for Public Opinion Research adopted Standards for Minimal Disclosure as part of the amendments to its Code of Ethics and Professional Practices. Those standards are listed below. They apply to statistical results that are released to the public.

6.4.1   AAPOR Standards for Minimal Disclosure
The American Association for Public Opinion Research

AAPOR believes that the following information should accompany published polls to help readers better judge the poll findings and accuracy:

- Exact question wording and response options, including any instruction and explanation texts that might reasonably be expected to affect the response.

- Definition and description of the sampling frame used to identify this population.

- Method by which respondents were selected.

- Size of sample and, if applicable, information on eligibility criteria, screening procedures, and completion rates.

- Method, location, and dates of interviews.

- When results are based on parts of the sample, instead of the whole.

- Who sponsored and conducted the poll.

- Information about the precision of the findings, including, if appropriate, estimates of sampling error, and a description of any weighting or estimating procedures used.

---

[109] One notable example is the referendum election to recall President Hugo Chavez in Venezuela, where election audits were introduced for the purpose of safeguarding against fraud. Hausmann and Rigobón (2004) present an argument that the election was manipulated and the audits were circumvented; Weisbrot, Rosnick and Tucker (2004) question their assumptions and deny their conclusions. Another paper that deals with a controversial election overseas is that of Durand et al. (2004), who discussed pre-election polling in the 2002 French presidential election.

A paper in this volume by Slater and Christensen (2002) adapts these guidelines to fit exit polling.[110] Blumenthal (2008) points out that transparency in reporting polling results would go a long way to helping users understand why major discrepancies occur between pre-election predictions and official vote counts. While the adoption of these guidelines strengthens polling procedures, however, they are only effective if they are followed. Currin-Percival (2007) analyzed results of polls reported by selected print and television media and found that compliance with these AAPOR standards tends to be mixed, at best. Both Blumenthal and Currin-Percival suggest that this failure to conform to the minimal standards may well be undermining cooperation with public opinion polls and acceptance of their results—making the long run downward trends in response even worse.

As a follow-on to the AAPOR effort, in 2006 the International Social Science Council of the United Nations approached the World Association for Public Opinion Research (WAPOR) to examine current principles and practices and develop suggested guidelines and benchmarks for the conduct and disclosure of information from exit polls. Those recommended standards, developed by the WAPOR Exit Poll Committee, are presented below.

6.4.2   WAPOR Guidelines for Exit Polls and Election Forecasts
WAPOR Exit Poll Committee (Approved by WAPOR Council, October 12, 2006)

Public opinion is a critical force in shaping and transforming society. Properly conducted and disseminated, survey research provides the public with information from the measurement of opinions and attitudes and allows its voices to be heard. This document provides guidelines for conducting exit polls and making election forecasts on election days.

Evaluating Exit Polls and Election Forecasts.   Exit polls are polls of voters, interviewed *after* they have voted, and *no later* than Election Day. They may include the interviewing *before* Election Day of postal, absentee, and other early voters. In some countries, Election Day polls cannot be conducted at the polling place, but in most cases, interviewing takes place at the polling location.

Exit polls can serve three different functions that are not mutually exclusive: predicting election results; describing patterns of voter support for parties, candidates, and issues; and supporting extensive academic research efforts. The main difference between these may be the speed with which the results are formulated and disseminated.

Exit polls used for projections should be reported as soon after the polls close as practical. Any delay in disseminating the results will inevitably raise questions about the legitimacy of the effort, especially with regard to estimating the outcome of the election. If analysis is the only purpose of the exit poll, prompt release is less important.

---

[110] See 2.3.1 in Chapter 2.

In some countries, election laws prohibit the publication of exit poll data until after the polls have closed. WAPOR and ESOMAR [the World Association of Research Professionals] oppose regulation of the conduct and reporting of polls in principle. However, no statement about the outcome of an election based on exit polls should be published before all the polls in the contest have closed. In national elections, this means polls relating to election results for elections in smaller voting units can be reported when all the polling places have closed in those locations, rather than waiting until all polling places used for voting that day have closed. Descriptive information other than voting behaviour may be published before the polls have closed.

Ethical Principles.   Survey researchers, in general, and those conducting exit polls in particular, need to follow certain broad principles in conducting their research:

1. Exit polls conducted for public consumption should be impartial and non-partisan. Exit polls are scientific research designed to collect data and report information on electoral outcomes. They are not tools for partisan advocacy.

2. Methods should be transparent, public, and well-documented. These goals can be achieved by publicly describing the methods prior to conducting the exit poll and by adhering to the standards of minimal disclosure delineated in this document. It is also recommended that when the exit poll is used for analysis, the data set (without individual identifiers), along with appropriate survey documentation, be deposited in public archives and/or on Web sites for general access.

3. Data collectors must adopt study designs for their exit polls that are suitable for producing accurate and reliable results and that follow specific procedural and technical standards stipulated in this document.

4. When reporting results from exit polls, data collectors and analysts must be careful to keep their interpretations and statements fully consistent with the data. Speculation and commentary should not be labeled as data-based reporting. Limitations and weaknesses in the design of an exit poll, its execution, and the results must be noted in all reports and analysis. Results should be released to the public and other interested parties through the general media and simultaneously made accessible to all.

5. The identity of respondents in exit polls must be protected. No identifying information (e.g., name, address, or other IDs) should be maintained with the voter-level records, and the data set should not allow deductive disclosure of respondents' identity. To limit the chances of deductive disclosure, small-area geographic details, such as the specific polling place in which votes were cast, should not be revealed.

Exit Poll Methods and their Disclosure.   Poll methods must be generally accepted as good survey practice and must be disclosed in advance of the conduct of the exit poll, as well as with any projection or analysis or subsequent public release of the data set.

Items for Minimal Disclosure. These items should be disclosed with any exit poll report or when any projection is made. Good practice would be to disclose as much of the methodology in advance as possible, particularly those items marked with an asterisk, which should be disclosed *before* Election Day.

* Sponsor of the exit poll
* Name of the polling company or principal researcher; prior experience (if any) in exit polling; and whether the data collector has any business or personal ties to political parties, candidates, political organizations or governmental bodies.
* Name of the organization responsible for analysis and projections, if different.

  Number of interviews
* Number of sampling points
* Sampling frame
* Geographic dispersion and coverage
* How sampling points are selected
* Where interviews are conducted—at polling places, in person at homes, by phone, etc.
* Any legal limits on data collection that might affect polling accuracy (e.g., minimum distance of interviewers from the polling place)

  Time of day of interviewing

  Whether interviewers are part of a permanent field staff or hired for the occasion
* How respondent anonymity is guaranteed (paper questionnaires, etc.)

  The interview schedule or questionnaire and instructions

  Which results are based on parts of the sample, rather than the whole sample

  A description of the precision of the findings, including estimates of sampling error

  Monitoring and validation procedures (if any)

  Weighting procedures

  Response rates (using one of the definitions in the AAPOR/WAPOR *Standard Definitions: Final Dispositions of Case Codes and Outcome Rates for Surveys*) and item nonresponse on vote questions

  Any known nonresponse bias

  General description of how estimates are made and the kinds of variables that are being used, and whether adjustments for nonresponse have been made

  Known design effects.

Political parties may sometimes make claims about private data. These claims also require documentation. Any public statement referring to exit poll results should abide by these disclosure principles and requirements.

Good Practices. Those conducting exit polls should always use generally accepted scientific methods. However, there are a number of good practices that apply specifically to exit polls.

Exit polls typically employ clustering in their sample designs. Because of the possibilities that various groups might attempt to influence voters and/or exit poll respondents, exit poll researchers are *not* expected to disclose the actual sample points or locations.

Exit polls should collect information across the whole of the polling day. Probability sampling (or full census) for interviews conducted at the polling place is the only acceptable selection method. Quotas are not appropriate for sampling at the polling place.

A national exit poll should represent the entire country, with 95% of the target population included in the sampling frame. If the sampling frame covers less than 95% of the target population, there should be an explanation for that decision.

Exit pollsters should keep in mind the relationship between small units for which the votes are tabulated and that can also serve as clusters for exit poll interviews. One way to evaluate an exit poll is to compare the actual election results and the estimates derived from the exit poll interviews for these same units. This comparison of small unit accuracy, typically at the precinct or polling place level, is one of the best ways to understand the exit poll's success. But there are situations where this will not be possible, either because no tabulations are reported at the smallest voting unit level or because the sampling units do not coincide with voting units.

Election Projection Methods and their Disclosure.   Election projections can be made in other ways than by interviewing voters as they exit the polling place. While most projections are based on interviews with voters after they have voted at a polling place, other forecasting models may include:

- Interviews in person, by telephone, or by other means of communication with voters after *or before* having cast their votes

- Counts of official votes in a sample of precincts, often known as *quick counts*

- A mix of methods.

A projection is an estimate that leads to a conclusion about the outcome of an election in a jurisdiction such as a nation, a state, or a district. This may occur in two different situations:

- If the winner is based on the popular vote for an office or a party, then a projection of the division of that vote is a projection of the outcome in the jurisdiction.

- If the winner is based on the vote in multiple jurisdictions, such as election of a Parliament where votes are cast in districts or of a President where votes are accumulated based on victories won in many jurisdictions, a conclusion about which party has a plurality of seats in the new Parliament or which presidential candidate has a winning number of votes is a national projection.

The projection need not reach a conclusion about each sub-jurisdiction. It need only reach a conclusion about the jurisdiction outcome.

The objective of any projection is a conclusion about an election for some jurisdiction. A sample of that jurisdiction must be adequate to reach an unbiased conclusion with sufficient/appropriate confidence in the estimate. A national projection typically requires the coverage of the entire country, with at least 95% of the target population in the sampling frame.

There will be times that a subset of the country will be used (for example, only competitive districts). But if a sampling frame is used that includes something less than the entire voting population of a jurisdiction, then the pollster should define what is and is not included in the sampling frame in a disclosure statement. The pollster also must publish a rationale to justify the pollster's ability to make an unbiased conclusion about the election outcome based upon collecting information from a subset of all jurisdictions.

WAPOR Exit Poll Committee Members

Chair: Kathleen Frankovic, former WAPOR President, U.S. Director of Surveys, CBS News

Nick Moon, United Kingdom Director, GfK NOP Social Research

Richard Hilmer, Germany Managing Director, Infratest dimap

Mahar Mangahas, Philippines President, Social Weather Stations

Alejandro Moreno, Mexico Department of Public Opinion Polling Head, Reforma

Anna Andreenkova, Russia Co-Director, CESSI

Warren Mitofsky, U.S. (until his death in September) President, Mitofsky International

Ex officio: Tom Smith, WAPOR Professional Standards Chair

Michael Traugott, WAPOR Vice President

> Both Mathiowetz (2007) and Frankovic (2007b) also stress the importance of disclosing information about statistical design and methods, in order to understand and accurately use public opinion poll results.

6.5 NEXT STEPS

> Improving data quality and, hence, the accuracy and usefulness of elections and exit polling results, requires statistical efforts like those cited in this volume. However, polling cannot be fully enhanced without changes to the voting system, itself. This section looks at a few recommendations for overcoming deficiencies and limitations in the voting process.[111]

[111] Election reform is not a new issue. Shaw and Ragland (2000) and Shapiro (2003) explored public opinion towards term limits and campaign financing. Panagopoulos (2004) presents a historical look at public opinion on election reform.

We begin with Rotz and Gracely (2007), who summarize the findings of the American Statistical Association's (ASA) Working Group on Fair and Accurate Elections, which reported for the ASA Special Interest Group on Volunteering in December 2007.

6.5.1   General Principles for Statistical Election Auditing
Wendy Rotz and Edward Gracely

After the 2000 presidential election rocked Florida and the U.S., followed by similar public concern in Ohio after 2004, many state and federal initiatives were undertaken to improve the accuracy and voter confidence in the integrity of America's elections. Statistical methodology applied to the election process can help these efforts by:

- Efficiently determining when a complete manual recount is/is not statistically necessary in a close race,

- Efficiently detecting accidental errors and/or deliberate fraud,

- Discouraging fraud with random recounts,

- Routinely identifying quality and efficiency improvements for the future elections, and

- Increasing public confidence in the integrity and accuracy of the U.S. election process.

Without statistical approaches, the number of precincts sampled all too often will represent either a wasteful review of more than is needed to establish the correctness of the election outcome or (worse) use of too few to achieve that goal.

With statistical approaches, election officials can establish with a pre-specified and high degree of confidence that even if there were errors in un-reviewed precincts, there is a quantifiably low probability that the errors are large enough to change the outcome of the election.

In this chapter, general guidelines are provided for developing statistically sound recount procedures. Naturally, the specific design for each election may need to take local factors into consideration. It is furthermore not intended that election officials would be able to perform all of the methodologies discussed in this chapter. Many are technical matters that require specific statistical listsexpertise. We highly recommend that a statistical consultant be used to prepare the details of individual audit plans.

There are a number of general principles applicable to statistical samples in election auditing, notably:

- Plans should be established prior to elections and *every election should be audited* as due course in the election process, regardless of whether any particular races are close.

- An appropriate sampling unit must be chosen.

- Samples need to be large enough to detect problems.

- Sample selections must be random.

- Certain types of precincts should always be recounted ("certainty precincts").

- Selections should cover key demographics.

- A verifiable audit trail is required for a meaningful recount.

- Recount procedures should listsminimize human bias.

- Plans should include other common-sense checks.

Each of these recommendations is addressed in the sections below.

Plan to Audit Every Election.   For quality control and improvement, to deter fraud, and to aid the confidence of voters in the election integrity, every election should be subjected to random auditing, not just close races.

Random testing can be a fraud deterrent in and of itself. Moreover, even in well-run elections, quality improvements to aid future elections can be identified from random testing. The principles of Six Sigma, operational research, or other quality control methods, can be applied to the election process, as well.[112] These require random testing of all elections, not just those that are close or those with perceived problems.

Choose an Appropriate Sampling Unit.   In many counties, recounts are conducted at the precinct level. Current legislation has consistently specified this level—largely because this is the paradigm and, logistically, counties are accustomed to collecting and compiling data at a precinct level.

A more resource-efficient alternative is to sample and verify counts by machine. This has several advantages. Assuming that at least some precincts have multiple machines, auditing say, 60 machines, involves far fewer actual votes than auditing the same number of precincts. In addition, there will likely be less variability among total machine counts than total precinct counts, potentially resulting in a sample size reduction. There will be more machines than precincts in the county or district but the requisite number of sampling units is not increased very much by that larger number and, indeed, may be offset by the reduction in variance. The net effect is that a machine-based sampling unit may be a better use of limited resources.

Unfortunately, a machine-based approach requires comprising a listing of all machines used in an election and uniquely identifying each one for the purpose of drawing a random sample. Some counties may not be able to easily compile such a listing. For

---

[112][*Six Sigma* at many organizations simply means a measure of quality that strives for near perfection. Six Sigma is a disciplined, data-driven approach and methodology for eliminating defects (driving towards six standard deviations between the mean and the nearest specification limit) in any process—from manufacturing to transactional and from product to service. The statistical representation of Six Sigma describes quantitatively how a process is performing.]

the sake of simplicity in this chapter, we will assume the recounts are on a precinct-by-precinct basis. However, most of the discussion is applicable to machine-based testing.

Determine the Sample Size (number of machines or precincts to sample). As in all audits, when considering the appropriate sample size, the intended use of the data and the required level of accuracy are factors contributing to the determination of the number of units to draw. The sample data could be used to trigger a complete manual recount, to detect processing errors, or to gather detailed information for quality improvement efforts. Often, the sample sizes for detailed quality improvement efforts are larger than those required to establish the accuracy of an election. The resources for such efforts then need to be weighed against the increase in sample they require. At a minimum, sample sizes should be sufficiently large to reasonably be able to detect problems according to the laws of probability.[113]

Determination of an appropriate sample size for election auditing also depends upon the action to be taken when discrepancies are found. Historically in some counties, the identification of a single discrepancy is cause for a complete manual re-enumeration. That may be overkill when random testing is conducted regardless of whether elections are close. An alternative is to establish a threshold that must be surpassed before a complete recount is initiated. These two approaches are discussed in more detail below.

Approach when a discrepancy larger than a threshold initiates a complete enumeration. The mere discovery of small discrepancies in isolated precincts do not necessarily require a complete recount if the discrepancies stem from non-systemic causes and would not change the outcome of the election, even if repeated in most of the non-sampled precincts. There may be little cause to doubt the correctness of the election outcome in this situation.

Instead, a threshold may be applied. The threshold should be based primarily upon the difference between the vote leader and the closest challenger. A complete enumeration should be conducted only when the sample indicates the election outcome may be altered.

Naturally, sample sizes would need to be large enough to determine whether small discrepancies could alter the election outcome and closer races would require larger sample sizes. Very close races may require samples so large that nearly every precinct would be manually recounted. In these cases, a complete enumeration makes sense.

Therefore, a threshold approach requires a statistician to determine sample sizes based on the difference between candidates, indicating when elections are so close they require a complete manual recount from the start and, for less close elections, determining the sample size and decision process for deciding if identified discrepan-

---

[113] [For a discussion of percentage-based audits vs. statistical sample-based audits, see Stanislevic, McCarthy, and Lindeman, (2007) and McCarthy et al. (2008). The latter is presented in Chapter 7 of this volume (7.4.3).]

cies are large enough so as to put the election outcome at risk and, therefore, should trigger a complete manual recount despite the outcome.

If a county or state uses a threshold of difference to determine the need for a complete recount, a statistician should prepare the statistical formulas and sample plan in advance of the elections. The statistician would also need to be prepared for the day after elections to quickly determine the sample size based on the spread of the elections results. Alternatively, the statistician could provide a table with sample sizes required for varying contingencies of the percent difference between candidates. To avoid the perception of bias after the fact, such a pre-specified model may be preferable.[114]

Note also, that certain kinds of discrepancies, such as ones that suggest the computer software may have been compromised, should *always* trigger a complete recount, even if the observed discrepancies are not great enough to suggest a change in the election outcome.

Approach when a single discrepancy initiates a complete enumeration. When the audit plan (or state law) calls for a complete recount of the relevant district, county, or state if a single discrepancy occurs, the audit plan should also call for random selections of enough precincts to provide a high probability (at least a 95% chance) that the recount will detect a problem with low prevalence (say, a problem occurring in as few as 5% of the precincts). A statistician can readily calculate that figure.[115]

A general idea of the number of precincts this approach could require is provided in the following examples. A county with 200 precincts requires a recount of 55 randomly selected precincts to have a 95% chance of detecting a problem occurring in as few as five percent of the county's precincts. A county with 1,000 precincts will need a sample of 60 for the same requirements.[116] Designing the plan so as to further reduce the number of problem districts comes with a price—more districts have to be counted. For example, to detect a problem occurring in as few as 5 out of 1,000 precincts, a sample of 450 precincts is required.[117] That may require quite an extensive effort. A statistician (with input possibly from county lawyers) may help election officials obtain the right balance between resources and public comfort level in an audit.

---

[114]A statistician may also work closely with a state or county to develop efficient but more complex plans, such as "stratified" or "PPS"[percent proportional to size] type sampling approaches, which can obtain a higher degree of precision from smaller precinct sample sizes. These kinds of plans should only be used when a statistician may be heavily involved in every stage from the planning, to selection of precincts, and interpretation of the recount findings. To take advantage of efficiencies of more complex forms of sampling, it is important that regulators do not mandate a number or percentage of precincts to audit, but instead mandate a high level of confidence (say 95%) and an election-dependent requirement for the required accuracy (for example, being able to detect a discrepancy as small as half the difference between the candidates in the closest election).

[115]Sample sizes may also be obtained from auditing tables, for example, see the textbook by Wilburn (1984)—*Practical Statistical Sampling for Auditors.*

[116]That is, 60 precincts are required to have the same 95% chance of detecting a problem occurring in as few as 5% of the precincts.

[117]Wilburn (1984), Appendix C.

Note also that:

1. The capacity of the recount process to catch errors is much more closely related to the *number* of precincts recounted than to the *percentage* of precincts recounted (failure to realize this is a common layman's misunderstanding of sampling). In the example above, a sample of 55 (or 28%) is sufficient for a county with 200 precincts, yet for the same probability and low-prevalence percentage threshold, only five more precincts for a sample of 60 (or 6%) is needed in a county with 1,000 precincts.

2. In evaluating sample sizes, it is important to understand that the *number of precincts* tested is the key, not the *number of votes* recounted. For example, the election official is more likely to find a problem if the audit recounts 50 precincts totaling 5,000 ballots than 10 precincts totaling 10,000—even though more ballots are recounted in the latter. The key is that more precincts are counted and verification of 50 total precinct counts is more revealing than verification of only 10 precinct counts.

3. A smaller, low-prevalence percentage threshold may be required (leading to larger sample sizes) when races are close. A statistician may aid the election authorities in preparing for this contingency.

4. Do not use any of the precincts selected with certainty towards fulfilling this random part of the audit.

More sample size factors to consider.

- **Extremely tight elections.** Many states have laws dictating that a complete manual recount must be conducted if the two leading candidates differ by less than a small percentage. A statistician can help the state or county determine a reasonable threshold to set.

- **Coordination of county audit plans for state and national races.** For state-wide or national races, the sampling approach may require coordination with other counties in the state, especially where computer voting machines are used. Without a coordinated plan, election officials could be subject to criticism that they allowed small known discrepancies to go unchecked on a county by county basis—when, in summation, the discrepancies could have been large enough to change state-wide results. This can occur when small errors are made more frequently in one candidate's favor during a close race, especially if the small errors occur in counties in which one candidate wins a clear majority.

   In addition, there are often state and national legislative races in which districts cross county lines, sometimes extending into several counties. Such common scenarios require careful attention. It may be necessary to sample precincts based on specific races, rather than primarily on a county-by-county basis. To take such situations into account, choice of sampling methods and procedures should ideally be determined at the state level, rather than the local or county level.

- **Complex elections.** In most elections, there are races at several levels, such as statewide, countywide, and local. Thresholds for complete counts may be determined separately on a race-by-race basis.

  Audit plans for the purpose of quality improvement or detection/deterrence of fraud may include every race, but that is labor intensive. Multi-stage sampling may be a cost saving alternative. For example, first, randomly select which races to audit and, then, randomly select precincts to recount in those races. Such an approach, however, would have to include a stipulation to select with certainty the races that are close. These close races should always be subjected to sampling or, perhaps, a complete manual recount if the race is extremely close. The same sampling procedures described in the section below may be used to select the non-certainty races.

Determine How to Randomly Select the Sample. The mathematical foundation supporting any statistical sampling plan relies upon random selection. A qualified independent statistician should select the sample. The random sample selection method should be transparent, but the actual choices must be unpredictable in advance.

Even with an outside independent listsexpert, random draws could receive the same public distrust as software programs, unless the selections are repeatable and verifiable following a highly transparent process, with no chance for human manipulation.

Therefore, it is recommended to draw the sample via a publicly available random sampling package[118] that can provide a reproducible and readily verifiable random sequence that cannot be known prior to the elections and cannot be manipulated after the elections.[119]

As an alternative to software, some kind of random draw with multiple-sided dice has the advantage of being easily understandable.[120] But most districts will probably use software to do the sampling, so we focus on that method here. In truth, discussions about the best way to do such sampling are ongoing among statisticians, with some passionately supporting each method. The issues are not statistical, but political and pragmatic—any of these methods can result in good, useful samples, but these are undermined if the public does not believe them or if there is a potential for fraud.

It is helpful to understand a bit about how typical computerized random number generators work to draw a sample. They start with a certain value, the "seed," and perform a series of mathematical calculations based on the seed to generate a sequence of "random" numbers. The numbers are not truly random—they are a mathematical sequence that is fully determined by the algorithm and the seed.

---

[118]For example, the RAT-STAT package from the U.S. Department of Health and Human Services has been carefully tested by the National Institute of Standards and Technology and is available free online at *http://oig.hhs.gov/organization/OAS/ratstat.html*.

[119]An example of manipulation is repeatedly redrawing the sample until the one drawing the sample is happy with the precincts selected.

[120]See Cordero, Wagner, and Dill (2006).

The key is that while (for transparency) the software may be known in advance, the seed must not be. Otherwise the precincts to be selected can be determined prior to elections, leaving the election officials subject to suspicion of a "rigged" recount. Yet the seed must also not appear vulnerable to manipulation, negating the possibility of allowing the computer program to choose the seed. It would be too easily criticized that the sampler merely kept redrawing until a desirable sample was selected.

An interesting approach to obtain a listsunique and unpredictable, yet verifiable, seed is to assign the seed to be the national total vote count (or an appropriate state count if there is no national election). This cannot be known before the election. It would be important to choose a total vote count that no election official (or computer programmer) could possibly manipulate. The next winning lottery number is another possible seed.

In addition to an unknown seed that is free from manipulation, the audit plan needs to include the order in which the random sequence is to be applied. This dispels concerns regarding after-the-fact reordering of a precinct listing to position favorable precincts where they will be selected by the random sequence. The simplest listssolution is to specify listing the precincts in alphabetical order by town within the county, then by number within towns or some other logical fashion. If the precincts are numbered throughout the county, that number sequence can be used as well.

If machines are sampled from a list, then care must be given to assign uniquely identifiable tracking numbers to each machine and to listsassure that one machine's results are not replaced with another's at the same precinct.

To prevent manipulation of random selection through human choices in certainty selections, it is best that certainty cases be included in the list from which to sample (see the next section). If one of the randomly identified precincts is already a certainty case, simply move on to the next precinct according to the random order of selection. This process will be reproducible, verifiable, and not subject to change from a human decision process in the certainty selections

Review "Certainty Precincts."   Some precincts should always be manually verified. A few are mentioned below.

Precincts with known problems.   For example, precincts where machines broke down on Election Day or voters reported the machines were not appearing to accept their votes. Precincts with problems in previous elections might be pre-specified for a recount, as well.

Extremely large precincts.   If a precinct is so large that a small percentage change in its total could change the results of a close election, the precinct counts should be manually verified. The rules for identifying these precincts should be established in advance.

"Challenge Precincts."   An interesting proposal is to also allow each major party to choose a fixed number of precincts to audit with certainty. Individual candidates and their campaigns are often the most acutely aware of where voting problems may occur, particularly where fraud might be most likely. One way to take advantage of such knowledge would be to invite each candidate with a minimum number of votes to designate a certain set of precincts to be audited immediate following the election.[121]

These precincts should not be included towards the random portion of the audit, but instead evaluated separately.

Check that Key Demographics Have Been Covered.   In addition to the certainty precincts, and the randomly selected precincts, an audit plan should also listsensure specific *types* of precincts have been checked. For example, has the audit covered:

- Each type of voting machine?

- A precinct that switched from Democratic to Republican and vice versa?

- An ethnically mixed precinct?

- An urban precinct?

- A rural precinct?

- A precinct from a polling location responsible for multiple precincts?

If the random selections or certainty selection do not include at least one instance of these types of precincts, additional precincts may be randomly selected—as a separate study.[122] Election officials, lawyers, and the statistical consultant should discuss this prior to the election and specify in the audit plan which types of precincts should have additional selections to ensure coverage.

A Verifiable Audit Trail Is Essential.   There is no value in asking a computer to recount its tally—it should always return the same number. Recounts need to be manual. Running the same punch cards through the same machine will not detect a problem with the machine accurately reflecting the count. With paper ballots, the ballot itself is the audit trail.

With computerized voting machines, a paper ballot is still necessary to listsensure the integrity of the election. The paper recording method needs to be independent of the computerized data and should be correct even if the software is defective. That is, regardless of the votes stored on the system, the printed ballot should be voter verified. However, it should also be independent of processing error by the voter. For example, it should not rely on the voter remembering to turn in a ballot or paper print out.

As the United States moves more and more toward computerized voting and tallying equipment, routine recounts of vote tallies become more and more critical. By design, the public, and often even the election officials, does not know exactly how

---

[121] See discussion of this topic at *https://vvf.jot.com/WikiHome/ChallengeAudits*.

[122] It would be possible for a statistician to develop a plan that would ensure coverage of most of these kinds of precincts. However, that plan would most likely be quite complex and cumbersome.

such systems work. Thus, even more so than in the days of manual paper ballots, the public needs assurance that votes are counted correctly. This requires an auditable paper trail.

Minimize Human Bias in Recounts.   It is human nature to conform to the norm. Therefore, the counters should not know what the machine tally was or, indeed, results of other counters for the same stack of votes. Furthermore, the counters should be randomly assigned to precincts/machines and should not be informed of their assignments until just prior to the recount. Reasonable caution must be employed to listsensure counters do not bring precinct listings with them.

A stricter approach has a three-person team for each precinct, ideally one person from each party and a randomly assigned mediator. The two partisan counters work alone counting the entire precinct without discussion. Their results are presented to the third party. If the results agree, and also agree with the machine, all is well. If the counters disagree, they repeat the process. Neither is told what the other reported until they agree.

Employ Common Sense Sanity Checks.   Recounts of votes only test one stage in the election process. There are many others. A total quality management approach would outline the entire election process from voter registration, machines used, ballot layout, voting location, delivery of machines/ballots/supplies, sufficiency of machines, ballots, supplies, through vote casting, machine tallies, reporting to county, to reporting to state, ... All steps are important in listsassuring a fair and accurate record of the voters' intent. At each step there should be some form of verification of an accurate process and assessment of areas for improvement. A useful exercise prior to elections is to trace a ballot from printing/generation through the time when a vote is cast to how it is tallied and how it is counted and reported up to the state level. Some basic sanity checks that could be incorporated in the recount audit plan include:

- Were more votes cast than voters registered in a precinct?

- Were more votes recorded than number of voters turning out at a precinct?

- Do the precinct counts of voter turnout match the number of ballots cast in the precinct?

- Are all ballots (used, unused, and destroyed) delivered to a precinct accounted for?

- Do any of the machines in a precinct show a very different percentage from the others?

These very simple questions could identify problems a random manual recount may not.

Conclusions. There are many other applications of statistics that can improve the quality and efficiency of the election process, but it is beyond the scope of this chapter to do them justice. A good reference is the Brennan Center Report.[123]

Sound statistical principles can be used to improve election quality on many levels. For the sake of transparency, most of the guidance in this chapter is best applied before an election, however, much is still applicable and sound advice, even after an election has taken place.

Even when elections are not close, statistically-based quality improvement samples and processes may give election officials the information they need to steadily improve each election's quality and address public concerns regarding integrity.

Transparent, random recounts applied with a statistical approach can be a powerful tool for demonstrating the integrity of an election or detecting problems requiring a complete recount. Election officials may rely on random recounts to provide public trust in the election process.

Acknowledgments. This paper is a product of the U.S. Elections working group within the American Statistical Association's (ASA) Special Interest Group (SIG) on Volunteerism. Special thanks to the input and comments from the entire working group, especially those of Arlene Ash, John McCarthy, Judy Tanur, Mary Batcher, Howard Stanislevic, and Mark Lindeman. This article solely reflects the views of its authors and not of their professional affiliations.

References

Cordero, Arel; Wagner, David; and Dill, David (2006). The Roll of Dice in Election Audits —Extended Abstract, IAVOSS Workshop on Trustworthy Elections (WOTE 2006), June 29, 2006, available on the Internet at: *http://www.cs.berkeley.edu/~daw/papers/dice wote06.pdf*.

Wilburn, Arthur J. (1984). *Practical Statistical Sampling for Auditors*, 52, New York, NY: Marcel Dekker, Inc.

Additional Recommended Readings

Cochran, William G. (1977). *Sampling Techniques,* 3rd ed., New York, NY: John Wiley & Sons.

Guy, Dan M.; Carmichael, D.R.; and Whittington, Ray (2002). *Audit Sampling*, 5th ed., New York, NY: John Wiley & Sons, Inc.

Roberts, Donald M. (1978). *Statistical Auditing*, New York, NY: American Institute of Certified Public Accountants, Inc.

After the November 2007 elections, the Connecticut Citizen Election Audit Committee observed 31 audits in Connecticut and determined the need for clear statistical thresholds to be established to assess if discrepancies require more rigorous review.[124] On January 19, 2008, the Governor of New Jersey signed into law a

---

[123] See *http://www.brennancenter.org/content/section/category/voting_technology*.
[124] See *http://www.CTElectionAudit.org*

measure that would require mandatory auditing of statewide, county and municipal elections. While the state has been given until June 3, 2008 to retrofit machines to provide a paper trail,[125] this is still a promising outcome for ensuring more accurate and trustworthy election results.[126]

The next reading takes a further step in making recommendations to improve elections nationwide. In it, Mulrow and Scheuren (2007) describe their plan for a National Election Voter Scorecard that would provide valuable insights into the integrity and validity of the election process.

6.5.2   Producing a 2008 National Election Voter Scorecard
Edward Mulrow and Fritz Scheuren

Introduction. Newspaper reports of major elections since 2000 suggest that the Help America Vote Act (HAVA) of 2002[127] is not working, and that the American public distrusts the election process. Many news reports have focused on the myriad of problems some voters have faced in past elections.

For example, the November 8, 2006 edition of *The Washington Post* featured an article entitled "Courts Weigh in After Voting Difficulties Emerge at the Polls."[128] The article describes Election Day problems faced by voters that potentially disenfranchised citizens in a number of states. These problems occurred in a non-presidential election year, when voter turn-out was not as high as it might be in 2008. If the system was stressed under low turnout, it will be stressed again—and even more—in 2008, unless appropriate system changes are put in place.

Exit polling has focused historically on the "horse race"—on who won or lost—and particularly on the characteristics exit polls were not used to assess polling problems in the U.S.—certainly not the kinds of problems occurring in Florida in 2000 that led to HAVA.

Naturally, given the controversy created by the Florida vote in 2000, there was an interest in the extent to which voters might see if HAVA was successful. Using an exit poll to partly assess this was judged to be of value. Therefore, in Albuquerque (really Bernalillo County), New Mexico, for the 2004 presidential election an experiment

---

[125] See Appel (2008). Also see Ash (2007) and Batcher and Scheuren (2007)—testimony favoring an early version of this legislation. In March 2008, the Board of Directors of the American Statistical Association issued a formal statement encouraging the government to establish tools and practices to ensure the statistical integrity of elections nationwide (American Statistical Association, 2008).

[126] Those interested in following these developments may find information on this topic on the Internet at: *http://vevo.verifiedvoting.org/mailman/listinfo/auditing*.

[127] The Help America Vote Act (HAVA) of 2002 authorized payments to states to off-set costs of improving the administration of elections and replacement of punch card and lever machines for voting. On the Internet, see *http://thomas.loc.gov/cgi-bin/query/F?c107:1:./temp/~c107ykUUgV:e1013:* for the text of the bill. (Last accessed December 12, 2007.)

[128] See Goldstein, Amy and Cooperman, Alan (2006).

was run to see if exit polls could learn about the quality of the voting process, as seen by the voter. What we found was that it could.[129]

Because the Albuquerque polling was staffed largely by volunteers, mostly from Common Cause, the effort was not expensive. Even so, it was hard to see how the survey we did in Albuquerque could be scaled up to work on a national basis, like the exit polling then run by Edison-Mitofsky. Instead, we felt that a mixture of methods might work. A few county-level exit polls, combined with a very large national telephone survey, seemed more promising.

Because Ohio Was So Important from an Electoral Perspective, to try out this combined option in 2006, we conducted *both* an exit poll and a Random Digit Dialing (RDD) telephone survey in Franklin County, Ohio (where the city of Columbus is located). While there were differences in the way the questions had to be asked, the results were quite close. In fact, the results were similar enough so that we went on to develop a full scale proposal for the 2008 election.

This paper describes the results we found in Franklin County and presents our early plans to scale up nationally what we learned—to produce what we call a National Election Voter Scorecard. The National Elections Voter Scorecard would be based on a nationwide survey of voters, conducted during a short time period just before and on Election Day 2008. The scorecard would present a broad view of the election process as seen by the voters and provide insights into systematic problems in the election process that might need to be addressed at a national level.

The rest of this paper describes the pilot test employed in Franklin County and presents results from the survey's assessment of the election process. Then, we more fully describe what we recommend for the 2008 presidential election.

Franklin County Pilot. For the November 2006 election, staff of the National Opinion Research Center at the University of Chicago (NORC) administered a short telephone questionnaire which took respondents less than 10 minutes to complete. The questionnaire first determined whether the respondent or anyone in the household was eligible to vote. If the respondent was a voter, the questionnaire asked whether the person voted in-person or absentee. It then went on to elicit information on any problems the respondent had with voting. Broadly, problems were divided into issues with getting to the point of using the ballot, difficulties using the ballot, as well as general comments on the process.[130]

Problems with getting to the point of using the ballot typically dealt with potential barriers of access to the electoral process. These included whether respondents had difficulty finding the polling location or meeting the identification requirements, were incommoded by the length of the wait to vote,[131] or could not be located on the list of

---

[129] For details, see Liu et al. (2004).

[130] Results of the survey were compared to those by the exit poll, conducted by a group from Brigham Young University (BYU). They were similar in their outcomes, except the BYU results did not obtain data from early voters. The BYU sample was drawn by Warren Mitofsky at no cost. See Grimshaw et al. (2004) for information on the Utah Colleges Exit Poll, in general; see Monson (2007) for more on this effort.

[131] See Allen and Bernshteyn (2006) for information on waiting lines in Franklin County, OH.

registered voters. The section of the questionnaire dealing with difficulties in using the ballot asked respondents whether they had trouble with understanding how to cast their votes and if there was a lack of assistance when voting. It also assessed their confidence about whether their vote would be counted correctly. Respondents were then invited to make open-ended comments about the election process generally.

The questionnaire was administered in two counties in Ohio, one of which was Franklin County.[132] The sample was selected using 1+ list-assisted RDD.[133] Because telephone exchanges do not perfectly match voting precincts and overly-inclusive procedures were judged to be better than under-inclusion, the sample was drawn using telephone exchanges that completely covered Franklin County. Records were kept for each exchange, with the overall finding that 85% of the respondents were potential voters in Franklin County. Manually dialed calls were made the day of the election, as well as the following day, and were skewed towards the evening hours in order to catch many potential respondents at a convenient time. Interviewers recorded respondents' answers via paper and pencil. The data and final dispositions were coded into Excel spreadsheets.

Results. Of the 363 responders, 274 had voted and 89 had not voted. Of voters, 207 voted in person, while the remaining 67 voted via absentee ballot. Of the 89 people who did not vote, two people provided data only for the first 4-5 (household) questions, 5 planned to vote later in the day, and 82 did not vote, but did provide the additional information requested.

Presented in the tables below are demographic information for the 67 absentee voters, the 207 who voted in person, and the 82 non-voters who provided data. As can be seen in Table 6.10, the bulk of the sample fell between the ages of 25 to 54 and 55 or over, with only 5% belonging to the under 25 age group. As has been long documented in election research , younger cohorts were disproportionately less likely to vote, while the 55 and older group was disproportionately more likely to vote. The 55 and older group made up a disproportionately large proportion of those voting absentee. The distribution of respondents whose age is unknown most resembles the pattern seen among younger voters.

Tables 6.11 and 6.12 present the distribution of voter category by sex and race of respondent. Six percent of the sample did not respond to the sex of respondent question. Even assuming those who did not answer the sex of responder question were all male, responders are far more likely to be female than male. Female respondents constituted a disproportionately high percentage of those who voted absentee, which may be partially explained by the fact a larger portion of the over 55 population is female.

---

[132] The other county was Cuyahoga, but for Cuyahoga no exit polling was done by BYU.

[133] Random Digit Dialing (RDD), as usually practiced, samples telephone numbers within 100 blocks made up of the last two digits taken by stratifying on area codes, then sampling a 100 block at random, and selecting a fixed number of numbers within the block. In this case, we confined attention just to 100 blocks with at least one listed residential telephone number. This approach is the way that the Mitofsky-Waksberg RDD telephone sampling method is currently carried out. For more information on this technique, see Brick and Tucker (2007).

Table 6.10 For each category of voters, distribution of responders by age

Age	Voted Absentee	Vote in Person	Did not vote	Overall
Under 25	1%	3%	11%	5%
25 to 54	19%	53%	49%	46%
55 or over	76%	42%	18%	43%
Unknown	3%	3%	11%	7%
Total	100%	100%	100%	100%

Table 6.11 For each category of voters, distribution of responders by sex

Sex	Voted Absentee	Vote in Person	Did not vote	Total
Female	73%	57%	49%	58%
Male	24%	43%	30%	36%
Nonresponse	3%	*	21%	6%
Total	100%	100%	100%	100%

* Indicates less than 1%, but greater than 0.

Table 6.12 For each category of voters, distribution of responders by ethnicity

Ethnicity	Voted Absentee	Vote in Person	Did not vote	Total
Caucasian/White	78%	79%	48%	71%
African American	10%	12%	20%	13%
All other*	7%	6%	10%	8%
Nonresponse	4%	3%	22%	8%
Total	100%	100%	100%	100%

* Asian/Pacific Islander.; Hispanic; Native American; Other

Those who did not respond to the sex of responder question were disproportionately less likely to vote.[134]

Both item and unit nonresponse pose potential threats to the external validity of survey research. This is because people who do not respond to specific items of a survey—or who refuse to cooperate with the survey entirely—are often unlike those who agree to participate in the survey and answer the questions. When studies simply use data from first round respondents and ignore item and unit nonresponse, this makes the generalization of sample characteristics to the population of interest problematic, at best. Further, while analytical techniques do exist to try to correct for some of the problems associated with missing data, such techniques are post hoc and constitute no replacement for obtaining the data in the first place.

In an effort to address this problem, we tried using capture/recapture techniques with regard to the Franklin County data. Capture/Recapture techniques are one of the best methods available to handle missing data problems (Scheuren, 2007). These deal with nonresponse by directly sampling nonrespondents (Hansen and Hurwitz, 1946), while at the same time re-sampling a group of people who responded in the original study. By sampling previous respondents and nonrespondents at a later date, researchers can compare the characteristics of nonrespondents to original respondents, as well as second time respondents to original respondents, thus disaggregating the effects of second-round sampling from the characteristics associated with nonresponse. The technique is, thus, an attempt to triangulate the effects of nonresponse on the sample characteristics and, hence, obtain—with a reasonable degree of confidence—the true characteristics of the population. This study implemented the capture/recapture technique for one day in December 2006, obtaining responses from both households originally in the survey and households which did not initially respond. The questionnaire consisted of a very short set of questions about a Franklin County congressional House race which had not yet been decided. Unfortunately, the results of the capture/recapture test were not conclusive, as potential respondents were no longer interested in the topic by December and, hence, the second survey was judged unusable. Nonetheless, the approach is a valid one that bears consideration for the future.

During Election Day in November 2006 and on the following day, 8,985 phone numbers were called by NORC staff, most twice. From these initial contacts, 364 completed questionnaire responses were obtained. Of the nonresponders, 6,087 appeared to be outside of the scope of the survey (e.g., businesses, dormitories or non-working numbers). There were 2,534 nonresponders who were probably in-scope (i.e., residential working phone numbers). Of these "in-scope" nonresponders, 1,541 were answering machines (some of which may have been businesses and, hence, really were out-of-scope), 205 were busy (some of which may have been out-of-scope), and 788 were refusals, hang-ups, or "call back later." Of the 364 questionnaire responses

---

[134]We expect to have an over-representation of females in the 2008 sample, just as we had here. The data will be reweighted to eliminate this potential bias, however.

Table 6.13 Who had problems? Percentages by sex of those voting in person

Sex	Problems/Concerns		
	Getting to Use Ballot	Using the Ballot	General
Male	10.23%	3.41%	5.68%
Female	10.17%	0.85%	10.17%
Unknown	0%	0%	0%

Table 6.14 Who had problems? Percentages by race of those voting in person

Ethnicity	Problems/Concerns		
	Getting to Use Ballot	Using the Ballot	General
Caucasian/White	7.05%	1.92%	6.41%
African American	29.17%	4.35%	16.67%
All other*	13.64%	0%	14.29%

*Asian/Pacific Islander.; Hispanic; Native American; Other

obtained, one was classified as a refusal, because the only data obtained were for age, gender and ethnicity.[135]

Some highlights on problems found. The most important highlight is that about one-sixth or 35 of the 207 voters who voted in person had one or more problems. There were 12 individuals who had difficulty finding the polling location. There were 4 respondents who said that their name was not on the list of registered voters, 2 reported their identification was not accepted, and 4 responded that their right to vote was challenged in some other way.

The tables above [6.13 and 6.14 and Table 6.15 on the following page] provide a breakdown by demographic characteristics of who (of 207 people voting in person) had problems voting. For example, of the 88 males who voted in person; Table 6.13 (by gender) indicates that 10.23% of them had problems getting to use the ballot. Table 6.14 (by race) shows 29.17% of the African Americans voting in person had problems getting to use the ballot (i.e., 7 out of 24 African Americans respondents). Of course, all of these numbers, except the overall estimate of voters with problems (one in six) are too small to be used alone, but if the survey had been national might

---

[135] The cooperation rate was, thus, 363/(363 + 788) or about one-third. Exit poll cooperation rates were somewhat higher, but still less than 50%.

Table 6.15 Who had problems? Percentages by age of those voting in person

	Problems/Concerns		
Age	Getting to Use Ballot	Using the Ballot	General
Under 25	14.29%	0%	0%
25-54	12.73%	1.82%	10.91%
55+	7.06%	2.35%	5.88%

have constituted something statistically significant. They are only shown here for illustrative purposes.

Scaled-Up Scorecard Approach. The National Election Voter Scorecard Survey proposed would be conducted as a telephone survey, backed-up by a limited number of well-designed exit polls. Plans are for NORC to conduct the telephone survey and BYU to carry out a limited number of exit polls across the country to give context and detail to the larger RDD effort.

The telephone surveying would be conducted in two periods:

1. Beginning Sunday evening and then throughout on the day before the 2008 election,[136] and

2. On Election Day, itself, the National Election Voter Scorecard Survey would sample from the remaining 60% to 70% of the voters.

If BYU is also able to conduct exit polling nationwide, these results would be integrated in and used to confirm the RDD results. Some key parts of the approach:

- Using the vote counts by the media—available on public Websites maintained throughout the election—we would reweight the data by gender, age, and possibly race.

- Plans are to report preliminary national voter National Scorecard results on Wednesday, right after the official election results are known.

- Methods will be based on those used in the 2006 election study in Ohio, where the comparability of exit poll results with telephone surveys conducted the day of the election was established.

- Questionnaire design is to be based on the third generation of the survey developed for the Albuquerque, New Mexico, 2004 exit poll study and the second generation of the Ohio telephone survey, still in preparation.

---

[136]This would be to catch early voters (expected to be between 30% and 40%) and to monitor turnout, including such issues as voter concerns about anticipated long lines.

- A large random sample of household telephone number will be selected, in order to obtain at least 11,000 completed surveys.

- Results from well-designed exit polls will be used with the telephone survey results to address potential undercoverage of voters by the telephone sampling frame, as the sampling of cell phone only households will not probably be undertaken.

- The survey telephone process will provide reliable national and state representative information regarding voter satisfaction with the election process.

With this National Scorecard Survey, we should be able to ascertain whether polling place problems that occur are systematic or if they are due to extenuating circumstances that are limited to certain local areas.

Concluding Remarks. We believe that a National Election Voter Scorecard approach will provide a better backdrop for evaluating and fixing election problems compared to the "knee-jerk" reaction approach employed by political observers—including national, state, and local officials—in response to anecdotal media stories. Of course, we realize that the National Scorecard approach will have its own problems—in particular, a high nonresponse rate—and that statistical methods may be necessary to adjust for them.

Still, the National Scorecard could aid in achieving the needed broader look at the election process. It would also, in itself, help develop metrics that make the election process more transparent and present a valid picture of the overall voting system in the U.S. How common are problems such as long lines at polling places and disenfranchisement of voters?[137] We could replace anecdote with statistical data, if we are able to mount a National Election Voter Scorecard Survey.

In our view, the bottom line is that making system-wide changes to the nation's elections based on isolated incidents does not provide the quickest path to quality improvement. Instead, a broader review of the voting system is needed in order to determine how to address problems in the election process. Letting voters keep score would be one part of the effort to continue improvements.

Acknowledgments. This paper profited from an early draft of the discussion of the Ohio experiment and its results, provided by Dr. Clifton Emery, when he was an intern at the National Opinion Research Center completing his doctoral studies. Important contributions were also made by our colleagues at Brigham Young University, notably Professor Quin Monson.

References

Allen, Theodore and Bernshteyn, Mikhail (2006). Mitigating Voter Waiting Times, *Chance*, 19, 14, 25-34.

---

[137] Allen and Bernshteyn (2006) discuss long waiting times with regard to the 2004 election in Franklin County, OH.

Brick, J. Michael and Tucker, Clyde (2007). Mitofsky-Waksberg: Learning from the Past, *Public Opinion Quarterly*, 71(5), 703-716.

Grimshaw, Scott D.; Christensen, Howard B.; Magleby, David B.; and Patterson, Kelly D. (2004). Twenty Years of the Utah Colleges Exit Poll: Learning by Doing, *Chance*, 17, 2, 32-38.

Goldstein, Amy and Cooperman, Alan (2006). Courts Weigh In After Voting Difficulties Emerge at the Polls, *The Washington Post*, A35, November 8, 2006.

Hansen, M. and Hurwitz, W. (1946). The Problem of Nonresponse in Sample Surveys, *Journal of the American Statistical Association*, 41, 516-529.

Liu, Yan et al. (2004). *Albuquerque Exit Polling Codebook*, available on the Internet at: *www.votingsystems. us*.

Monson, Quin (2007). 2006 Voter Problems and Perceptions: A Combined Exit and RDD Approach in Franklin County, Ohio, paper presented at the American Association for Public Opinion Research in Anaheim, CA.

Scheuren, Fritz (2007). Eight Rules of Thumb for Understanding Survey Error, a paper presented at the Research Triangle Institute, March 23, 2007, Research Triangle Park, NC.

6.6   SUMMARY OBSERVATIONS

As this book goes to print, the 2008 U.S. presidential primary season is well under-way. Pre-election polls, entrance polls for caucuses and exit polls for primaries are being used to predict and analyze both state caucuses and voting primaries across the country. In some of those contests, the data have been "right on," predicting the winner with relative clarity and precision. In others, the data have yielded forecasts that differ considerably from official vote counts. One result is an almost immediate flurry of discussions on the Internet—particularly on *AAPORnet* and *Pollster.net*—and in the media about "what went wrong." [138]

The intense interest and concern clearly has its pluses and minuses. On the down side, discrepancies between predictions and official counts in New Hampshire and South Carolina raise concerns about the validity of polling as a predictor. When combined with anecdotal concerns about voting problems, they open the entire U.S. election system to question about the validity of any and all results. These are serious repercussions. Nonetheless, closer scrutiny will, no doubt, also make any efforts to commit fraud more difficult to get away with. Further investigation will also lead to improvements in polling and in the actual voting process. Important here will be methods standardizing the procedures and placing greater emphasis on quality control to ensure more accurate data. [139] The greater scrutiny in this election round promises to provide all of us with a more solid understanding about what factors are biasing polling results, so that sampling and weighting methods can be enhanced.

This chapter essentially brings this historical look at the first decade of the 21st century's election and exit polling to a close. The next chapter contains some

---

[138] See, for example, Dopp (2008); Erikson and Wlezien (2008); Lenski (2008)

[139] Blumenthal (2008) urges pollsters to follow AAPOR guidelines; AAPOR (2008) subsequently announced the formation of an ad hoc committee to evaluate the New Hampshire methodology and publicize the results.

technical presentations and appendices to submissions that appear earlier in this volume, but our attempt to draw together recent developments in election polling research culminate with the materials provided here. Recommendations for future efforts to improve the quality of exit polling data and official guidelines to encourage transparency and full disclosure for exit pollsters serve as the foundation. The examples of research we provide—regarding better sampling of diverse populations, more thorough understanding of the impact of cell phones, and alternatives to overcome item nonresponse from inexplicit answers—are but a start in some of the exciting and innovative efforts already underway. Add to these examples improvements to understanding and interpreting the election process by ensuring collection of statistical data through paper trails, audits, and a National Election Voter Scorecard, and—despite the issues that have emerged thus far in the 2008 election primaries— we can be assured that current practitioners will be able to keep Warren Mitofsky's contributions to election surveying alive and well for many years to come.

Editors' Additional References

American Association for Public Opinion Research (2007). Special Issue: Cell Phone Numbers and Telephone Surveying in the U.S., *Pubic Opinion Quarterly*, 71 (5).

American Association for Public Opinion Research (2008). AAPOR Announces Formation of Ad Hoc Committee to Evaluate New Hampshire Pre-Election Primary Poll Methodology, news release, January 14, 2008, available on the Internet at *http://www.aapor.org*.

American Statistical Association (2008). Position on Electoral Integrity, statement adopted by the Board of Directors, March 7, 2008.

Appel, Heather (2008). No Paper Trail for Feb. 5 Primary, *Herald News*, January 11, 2008.

Ash, Arlene (2007). State of New Jersey 212th Legislature Testimony, Senate, 507, November 24, 2007.

Atkeson, Lonna Rae et al. (2007). Using Mixed Mode Surveys (Internet and Mail) to Examine General Election Voters, paper presented at the annual meeting of the American Association for Public Opinion Research in Anaheim, CA.

Barreto, Matt A.; Streb, Matthew J.; Marks, Mara; and Guerra, Fernando (2006). Do Absentee Voters Differ from Polling Place Voters? New Evidence from California, *Public Opinion Quarterly*, 70 (2), 224-234.

Batcher, Mary and Scheuren, Fritz (2007). Ballot Integrity Act of 2007, S. 1487, testimony for election auditing in the State of New Jersey, November 17, 2007.

Blumenthal, Mark (2008). The Secret Lives of Pollsters, February 7, 2008, available on the Internet at *http://www.nytimes.com/2008/02/07/opinion/07blumenthal.html*.

Chang, LinChiat; Holland, Keating; and Piehl, Jeri (2007). Evaluating Follow-up Probes to *Don't Know* Responses in Political Polls, paper presented at the annual meeting of the American Association for Public Opinion Research in Anaheim, CA.

Crampton, Thomas (2007). Polling Goes Online, with a Recruited Panel, *International Herald Tribune*, May 27, 2007, available on the Internet at: *http://www.iht.com/*.

Currin-Percival, Mary (2007). Reports about Polls in the Media: Variation in Adherence to AAPOR Disclosure Standards, paper presented at the annual meeting of the American Association for Public Opinion Research in Anaheim, CA.

Dopp, Kathy (2008). New Hampshire's Democratic Primary Election Results Are Suspicious, January 14, 2008, available on the Internet at: *http://electionarchive.org/ucvData/NH/ReleaseReNHPrimary2008.pdf*.

Durand, Claire; Blais, Andre'; and Larochelle, Mylène (2004). Review: The Polls in the 2002 French Presidential Election: An Autopsy, *Public Opinion Quarterly*, 68 (4), 602-622.

Edison Media Research and Mitofsky International (E/M) (2005). *Evaluation of Edison/Mitofsky Election System 2004*, January 19, 2005, *http://www.vote.caltech.edu/media/documents/EvaluationJan192005. pdf*

Erikson, Robert S. and Wlezien, Christopher (2008). Likely Voter Screens and the Clinton Surprise in New Hampshire, January 13, 2008, available on the Internet at: *http://www.pollster.com/blogs/likely_voter_ screens_and_the_c.php.*

Frankovic, Kathy (2007a). To Tell the Truth to Pollsters, Article available on the Internet at: *www.cbsnews. com/stories/2007/08/15/opinion/pollpositions/main3169223.shtml*, August 15, 2007 (last accessed December 3, 2007).

Frankovic, Kathy (2007b). Sticking to Standards Key with Polls, appeared on the Internet at: *www.cbsnews. com/stories/2007/10/24/opinion/pollpositions/main3401787.shtml*, October 10, 2007 (last accessed December 3, 2007).

Hausmann, Ricardo and Rigobón, Roberto (2004). In Search of the Black Swan: Analysis of the Statistical Evidence of Electoral Fraud in Venezuela, available on the Internet at: *http://ksghome.harvard.edu/ ~rhausma/new/blackswan03.pdf* .

Keeter, Scott (2007). How Serious Is Polling's Cell-Only Problem? The Landline-less Are Different and Their Numbers Are Growing Fast, paper presented at the annual meeting of the American Association for Public Opinion Research, Anaheim, CA.

Lenski, Joe (2008). Response to Hardball Citation, discussion on AAPORNET, January 10, 2008.

Lepkowski, James M.; Tucker, Clyde; Brick, Michael J.; deLeeuw, Edith D.; Japec, Lilli; Lavrakas, Paul J.; Link, Michael W.; and Sangster, Roberta L. (2007). *Advances in Telephone Survey Methodology*, New York, NY: John Wiley & Sons, Inc.

Mathiowetz, Nancy (2007). Pre-election Poll Methodology: What the Public Has the Right to Know, posted to the Internet at *www.huffingtonpost.com/nancy-mathiowetz/preelection-poll-methodo_b_74696.html*, November 29, 2007 (last accessed December 3, 2007)

McCarthy, John; Stanislevic, Howard; Lindeman, Mark; Ash, Arlene S.; Addona, Victtorio; and Batcher, Mary (2008). Percentage-Based versus Statistical-Power-Based Vote Tabulation Audits, *The American Statistician*, 42 (1), 11-16.

Mulrow, Edward and Scheuren, Fritz (2007). Producing a 2008 National Election Voter Scorecard, unpublished paper.

Panagopoulos, Costa (2004). Poll Trends: Electoral Reform, *Public Opinion Quarterly*, 68 (4), 623-640.

Pedraza, Francisco and Barreto, Matt (2007). Exit Polls and Ethnic Diversity: How to Improve Estimates and Reduce Bias Among Minority Voters, paper presented at the annual meeting of the American Association for Public Opinion Research, Anaheim, CA.

Rotz, Wendy and Gracely, Edward (2007). General Principles for Statistical Election Auditing, unpublished paper.

Shapiro, Robert Y. (2003). Public Attitudes Toward Campaign Finance Practice and Reform, *Inside the Campaign Finance Battle: Court Testimony on the New Reforms*, Corrado, Anthony; Mann, Thomas; and Potter, Trevor (eds.), Washington, DC: The Brookings Institution.

Shaw, Greg M. and Ragland, Amy S. (2000). The Polls—Trends: Political Reform, *Public Opinion Quarterly*, 64 (2), 206-226.

Singer, Eleanor (2007). Reducing Measurement Error in Surveys on Sensitive Topics, paper presented at the U.S. State Department Conference on Survey Methodology, July 16, 2007.

Slater, Melaney and Christensen, Howard (2002). Applying AAPOR Final Disposition Codes and Outcome Rates to the 2000 Utah College's Exit Poll, *American Statistical Association Proceedings*, AAPOR-ASA Section on Survey Research Methods, 540-545 (also in this volume as paper 2.3.1).

Stanislevic, Howard; McCarthy, John; and Lindeman, Mark (2007). A Simple 3-Step Audit Same Size Calculation for Election Contests, working paper, October 18, 2007.

Weisbrot, Mark; Rosnick, David; and Tucker, Todd (2004). Black Swans, Conspiracy Theories, and the Quixotic Search for Fraud: A Look at Hausmann and Rigobón's Analysis of Venezuela's Referendum Vote, Washington, DC: Center for Economic and Policy Research.

CHAPTER 7

TECHNICAL APPENDIX

7.1 INTRODUCTION

In compiling this volume, it was our hope that this book would appeal not only to statisticians and survey researchers, but also to a much broader spectrum of readers, including journalists, political scientists, and election reform advocates, as well as some interested people from the general public. In order to attract a variety of readers, we focused on contributions from the research community that were not too technical for the average reader. Nonetheless, there are some materials that contain more detailed methodological discussions that we felt should be included. Those papers, plus the appendices of a few of the submissions, are included in this chapter.

As with the rest of this book, the selections provided here are merely examples of some of the work currently underway. They are, by no means, meant to be fully representative of those efforts or even to cover the depth and breadth of work that is

Editors' Note: Text in gray shading is provided by the editors of this volume. Boxed-in text represents excerpts from interviews with Warren Mitofsky. The balance of the chapter is made up of papers prepared by contributing authors. [Text in square brackets was added by the editors.]

Elections and Exit Polling. Edited by Fritz Scheuren and Wendy Alvey
Copyright © 2008 John Wiley & Sons, Inc.

ongoing. We begin with three excerpts from papers that appear in earlier chapters in this volume. Then, we present an appendix which provides key questions for the Swedish survey described in Chapter 5. This is followed by three other papers that complement earlier readings in the book.

7.2 STATISTICAL METHODS FOR PAPERS IN THIS VOLUME

We begin with an excerpt that applies a data mining decision tree approach to the identification of "likely voters." Likely voter modeling is one of the most important areas of continuing research in attempting to improve the accuracy and usefulness of pre-election polls. The balance of this paper appears in Chapter 4 (4.2.2).

7.2.1 Supplemental Excerpt from: Predicting Likely Voters: Using a New Age Methodology for an Age-Old Problem
Gregg R. Murray, Chris Riley, and Anthony Scime

A "New Age" Methodology: Iterative Expert Data Mining. With the objective of finding a small and efficient number of variables that, together, accurately identify likely voters, we combine data mining with domain expertise into an Iterative-Expert-Data Mining methodology (Scime and Murray, 2007). Generally speaking, data mining is a process of inductively analyzing data to assess known relationships, as well as to find interesting patterns and unknown relationships. The term "data mining" encompasses a number of techniques and involves both human and computational resources (Scime and Murray, 2007; Hofmann and Tierney, 2003). Classification data mining is used to analyze new data and to predict future outcomes for cases with a set of common characteristics. That is, it is both predictive and explanatory (Osei-Bryson, 2004).

Classification algorithms construct decision tree models and rules by looking at the past performance of input variables (i.e., independent variables) with respect to an outcome variable (i.e., a dependent variable). The decision tree is constructed inductively from cases with known values for the outcome variable. Input variables are sequentially selected from the data set to construct the decision tree using a divide-and-conquer algorithm that is driven by an evaluation criterion. The input variable selected at a given point in the tree is the one with the strongest association with the outcome variable among the remaining cases. Using this selection process, the data are repeatedly sub-divided until a pre-determined stopping rule is met, which indicates that further subdivision is meaningless (Han and Kamber, 2001).

In this research, we construct our decision tree models using a post hoc predictive classification process known as Chi-squared Automatic Interaction Detection (CHAID) (Kass, 1980; Magidson,1988, 1994). CHAID is appropriate for models of categorical dependent variables and is designed to identify a set of input variables and associated interactions that optimally predict the values of the outcome variable (Magidson, 1994). Importantly, it results in mutually exclusive groups of individuals and classifies every case into an outcome category. The CHAID process analyzes an exhaustive series of cross-tabulations between each input variable and the outcome variable, repeatedly subdividing the cases based on the input-outcome variable combination with the largest statistically significant χ^2 value among the set of subdivided cases. It reduces the dimensions of the data by merging categories of an input variable when the merged categories produce statistically indistinguishable differences in the association between that variable and the outcome variable.

After the decision tree is constructed, each branch of the decision tree is converted into a rule, which identifies a group of respondents that exhibit a set of common characteristics.[140] A decision tree provides for all possible combinations of the input variables and their allowable values reaching a single, allowable outcome. The decision tree and rules can be analyzed to predict future behavior of new data.

Scime and Murray (2007), Hofmann and Tierney (2003), and Ankerst, Ester, and Kriegel (2000) propose iterative processes that combine data mining techniques with domain expertise (i.e., relevant knowledge accumulated over time through research and experience) to increase the usefulness and accuracy of predictions and/or to reduce the number of variables required to classify the outcome variable. The Iterative-Expert-Data Mining (IEDM) process (Scime and Murray, 2007) enables us to identify a small number of input variables with economical predictive accuracy in a data set. The technique, which iteratively includes domain expertise in both case selection and variable selection, is scientifically rigorous and theoretically informed.

Case selection involves expert review in order to select cases that fit the goals of the data mining project (e.g., theoretical or practical significance). Variable selection involves three reduction processes. First, the number of variables is reduced based on the expert's theoretically and empirically informed knowledge of the domain. Second, a data mining-based evaluation criterion is used to suggest a further reduction in the number of variables. Finally, the result of this reduction is reviewed by the expert for further reduction and/or enhancement based on theoretically and empirically informed domain knowledge.

Finally, it is appropriate to note that outcomes can be predicted and relationships in data can be evaluated using other methods, as well. For example, regression analysis can also be used to classify data. Andoh-Baidoo and Osei-Bryson (2007) have shown that decision trees are more insightful than regression in predicting the interaction of variables on the outcome variable. Specifically, decision trees hold at least four significant advantages over regression analysis. First, regression requires all important relationships to be identified before testing, which limits the discovery of unknown relationships and increases the effects of analytical bias through variable selection. Decision tree classification, on the other hand, identifies both known and unknown relationships and reduces variable-selection bias by including more variables in the analysis. Second, regression requires missing values to be estimated or data to be eliminated, whereas decision tree algorithms maintain the integrity of the data by accounting for missing data. Third, decision trees provide direct knowledge of how

---

[140] [See Figure 4.2 in the balance of the paper in Chapter 4 (4.2.2).]

changes to the dependent variables can change the result. Finally, a decision tree produces output that is easily converted into specific, actionable rules.

Applying the IEDM Method. The data mining process is most efficiently and effectively applied to large databases that include extensive information. In this study, we rely on the publicly available American National Election Studies (ANES), an ongoing, long-term series of public opinion surveys intended to produce research quality data on American elections. The ANES collects data on a wide array of subjects, such as voter registration and choice, social and political values, social background and structure, candidate and group evaluations, and opinions about public policy. The ANES has conducted interviews of a nationally representative sample of adults every election year since 1948, except for the mid-term election of 1950. The 1948-2004 ANES Cumulative Data File is composed of the pooled cases and variables from each of the studies conducted since 1948 ($N = 47, 438$). The file includes most of the questions that have been asked in three or more ANES surveys. It is composed, therefore, of more than 900 variables.

With the objective of finding variables that, together, are economically indicative of respondents' likelihood of voting, we applied the IEDM methodology to the 1948-2004 ANES Cumulative Data File to construct a decision tree model and resulting rules. We analyzed multiple elections to capture long-term predictors that have persisted and are more likely to persist, over time.

In the first step of the IEDM process, we used domain expertise primarily derived from scholarly literature (e.g., Dimock et al., 2001; Freedman and Goldstein, 1996; Petrocik, 1991; Rosenstone and Hansen, 1993; Wolfinger and Rosenstone, 1980) to reduce the data set to 146 input variables, in addition to the outcome variable. We excluded variables that appeared to be variations of the outcome variable that would be predictive mostly due to their high correlation with the outcome variable. This class of variables included measures such as the configuration of a respondent's split-ticket vote. Of the remaining variables, we retained those that held the most empirical and theoretical significance.

In terms of case selection, the domain expert limited the data to respondents in the ANES validated-vote studies in the presidential election years of 1976, 1980, and 1988.[141] This selection reflects an effort to reduce the effects of vote over-reporting (Bernstein, Chadha, and Montjoy, 2001) as well as the recognition that during presidential years more extensive surveys are administered and citizens are more attentive. The resulting data set includes 4,562 presidential-year records. Importantly, by data mining convention, the data set is divided into a "training set" and a "test set." The training set is composed of the data used to construct the decision tree model. The test set is composed of unused, independent data that are reserved to test the completed decision tree for model-to-data over-fit and generalizability. In this case, we randomly

---

[141] The ANES also included validated-vote studies in 1964 and 1984, but two of our primary predictors of turnout—voting in the presidential primary election and previous presidential vote—were not asked in 1964 and 1984, respectively. As such, we do not include those years in this research.

selected a subset of 3,060 respondents for the training set and 1,502 respondents for the test set.

Following the IEDM process, we next used the training set to estimate the strength of association of each input variable with the outcome variable, using the CHAID algorithm as implemented by Answer Tree 3.1. We then ordered the input variables by χ^2 value from highest to lowest. Next, we constructed a series of decision tree models using the Answer Tree implementation of CHAID. That is, we executed the classification algorithm repeatedly, with each execution producing a decision tree from progressively smaller sets of variables. We then compared all the resulting decision trees and used domain expertise to determine the optimal tree in terms of number of input variables and accuracy. In the final IEDM stage, we reviewed the tree and used domain expertise for further variable additions and deletions. This review process enables the domain expert to add or delete critical variables that are likely to increase the model's accuracy and/or usefulness.[142]

References

Andoh-Baidoo, Francis K. and Osei-Bryson, Kweku-Muata (2007). Exploring the Characteristics of Internet Security Breaches that Impact the Market Value of Breached Firms, *Expert Systems with Applications*, 32(3), 703-725.

Ankerst, Mihael; Ester, Martin; and Kriegel, Hans-Peter (2000). Towards an Effective Cooperation of the User and the Computer for Classification, *Proceedings of the Sixth ACM SIGKDD International Conference on Knowledge Discovery and Data Mining*, Boston, MA, 179-188.

Bernstein, Robert; Chadha, Anita; and Montjoy, Robert (2001). Overreporting Voting: Why It Happens and Why It Matters, *Public Opinion Quarterly*, 65(1), 22-44.

Dimock, Michael; Keeter, Scott; Schulman, Mark; and Miller, Carolyn (2001). Screening for Likely Voters in Pre-election Surveys, paper presented at the meeting of the American Association for Public Opinion Research Conference, Montreal, Quebec, Canada.

Freedman, Paul and Goldstein, Ken (1996). Building a Probable Electorate from Pre-election Polls: A Two-stage Approach, *Public Opinion Quarterly*, 60, 574-587.

Han, Jiawei and Kamber, Micheline (2001). *Data Mining: Concepts and Techniques*, Boston, MA: Morgan Kaufmann.

Hofmann, M., and Tierney, B. (2003). The Involvement of Human Resources in Large Scale Data Mining Projects, *Proceedings of the 1st International Symposium on Information and Communication Technologies*, Dublin, Ireland, 103-109.

Kass, G. (1980). An Exploratory Technique for Investigating Large Quantities of Categorical Data, *Applied Statistics*, 29, 119-127.

Magidson, J. (1988). Improved Statistical Techniques for Response Modeling, *Journal of Direct Marketing*, 2(4), 6-18.

———— (1994). The CHAID Approach to Segmentation Modeling: Chi-squared Automatic Interaction Detection, *Advanced Methods of Marketing Research*, Bagozzi, R.P. (ed.), Cambridge, MA: Basil Blackwell.

Osei-Bryson, Kweku-Muata (2004). Evaluation of Decision Trees: A Multicriteria Approach, *Computers and Operations Research*, 31(11), 1933-1945.

Petrocik, John R. (1991). An Algorithm for Estimating Turnout as a Guide to Predicting Elections, *Public Opinion Quarterly*, 55, 643-647.

[142][For the results of this study, see 4.2.2 in Chapter 4 of this volume.]

Rosenstone, Steven J. and Hansen, John Mark (1993). *Mobilization, Participation, and Democracy in America,* New York, NY: MacMillan.

Scime, Anthony and Murray, Gregg R. (2007). Vote Prediction by Iterative Domain Knowledge and Attribute Elimination, *International Journal of Business Intelligence and Data Mining,* 2(2), 160-176.

Wolfinger, Raymond E. and Rosenstone, Steven J. (1980). *Who Votes?* New Haven, CT: Yale University Press.

Appendix: ANES Survey Items. ANES data set variable numbers follow in parentheses.

- **Outcome Variable**

    - *Vote Validation.* Was respondent's vote validated? (VCF9155)

- **Input Variables.**

    - *Vote Intent.* "On the coming presidential election, do you plan to vote?" (VCF0713)

    - *Previous Vote.* "Do you remember for sure whether or not you voted in that [previous] election?" (VCF9027)

    - *Primary Vote.* "How about you—did you vote in that primary election/caucus?" (VCF9026)

> The next contribution is the balance of a paper that appears in Chapter 2 (2.2.1). It describes the statistical methodology used to assess coder reliability in the post-2000 presidential election's evaluation of the Florida voting systems.

7.2.2 Supplemental Excerpt from: Reliability of the Uncertified Ballots in the 2000 Presidential Election in Florida[143]
Kirk Wolter, Diana Jergovic, Whitney Moore, Joe Murphy, and Colm O'Muircheartaigh

Description of Coder Reliability. The objective of the coding of the uncertified ballots was to obtain an accurate (reliable and valid) record of the marks on the ballots. Knowing that no coding operation can ever be flawless, our objective was to assess the quality of the coding operation. This quality has two principal dimensions: reliability and validity.

Reliability is a measure of the consistency in the data; it can be described using a variety of measures, some of which we discuss in what follows. Bailar and Tepping (1972) and Kalton and Stowell (1979) discussed coder reliability studies. The first dimension, *validity* (average correctness), is much more difficult to assess, and there

---

[143] Reliability of the Uncertified Ballots in the 2000 Presidential Election in Florida, by Kirk Wolter, Diana Jergovic, Whitney Moore, Joe Murphy, and Colm O'Muircheartaigh, reprinted with permission from *The American Statistician.*

is little opportunity to measure validity from internal evidence within our data. We attempted to avoid invalid data by not hiring biased coders and by coder training and supervision.

[The next two sections] discuss reliability issues for undervote and overvote ballots, respectively. Throughout these sections, we examine reliability at the candidate/ballot or chad level.

Reliability of undervote ballots. For a given universe of ballots, we gauge the reliability of alternative ballot systems using four statistics defined as follows:

$$\overline{P}_2 = \sum_{j}^{J} \left(B_{jj+} + B_{j+j} + B_{+jj} \right) / 3B, \tag{7.1}$$

$$\kappa_2 = \frac{\overline{P}_2 - \sum_{j}^{J} P_j^2}{1 - \sum_{j} P_j^2}, \tag{7.2}$$

$$\overline{P}_3 = \sum_{j}^{J} B_{jjj} / B, \tag{7.3}$$

and

$$\kappa_3 = \frac{\overline{P}_3 - \sum_{j}^{J} P_j^3}{1 - \sum_{j} P_j^3}, \tag{7.4}$$

where B is the total number of ballots considered in the universe; j indexes the code (or variable) values; J is the total number of discreet code values; B_{ijk} is the number of ballots assigned the ith code value by the first coder, the jth value by the second coder, and the kth value by the third coder; and $P_j = (B_{j++} + B_{+j+} + B_{++j})/3B$ is the (assumed common) probability of assigning the jth value. See Fleiss (1965, 1971).

$\overline{P}_2$ is the proportion of agreement among all pairwise comparisons between coders. If three coders examine a given ballot, then there are three pairwise comparisons for each of the 10 presidential candidates, or 30 pairwise comparisons for the ballot overall. Similarly, $\overline{P}_3$ is the proportion of agreement among all three-way comparisons. If three coders examine the given ballot, then there are 10 such comparisons, one per candidate. If there are B such ballots, then there are $30B$ pairwise comparisons and $10B$ three-way comparisons over all ballots. Both $\overline{P}_2$ and $\overline{P}_3$ range from 0 to 1, with larger values signifying greater consistency in coding.

κ_2 and κ_3 convey similar, though not identical, information. κ_2 measures the excess of actual pairwise agreement over the agreement expected given a randomization model. The excess is normalized by the maximum possible excess over random agreement. κ_3 is similarly defined, but in terms of three-way agreement. Both statistics range from 0 to 1. Small values signify that actual agreement is little above random agreement, while larger values signify that actual agreement is greater than random agreement.

Throughout the balance of this section, we present various reliability statistics and, where possible, draw conclusions about the reliability of different types of undervote ballots. We emphasize that all statements about reliability refer to the process of hand counting (or coding ballots), given that a machine has already classified the ballots as undervotes. Overall, there are two dimensions of reliability that must be of concern to election administration:

1. The extent to which ballots are classified as undervotes, and

2. The consistency of coding ballots, given they are classified as undervotes.

The former may be assessed by the undervote error rate.[144] This section deals strictly with the latter dimension.

Table 7.1 gives the four reliability statistics for each of the ten named presidential candidates and for each of three outcome variables. The first variable—*all codes*—refers to the original coding by the coders. For this variable, agreement is measured at the level of the raw code value. The second and third variables are derived from this first variable. *Dimple or greater* is an indicator variable defined equal to 1, if the original code signifies a dimple or greater, and to 0 otherwise, while *two corners or greater* is defined equal to 1, if the original code signifies two corners detached or greater, and to 0 otherwise. For these variables, agreement is measured in terms of the recoded values, not in terms of raw code values. Dimple or greater corresponds to a voting standard advocated by the Gore team during the presidential election, while two corners or greater corresponds to a voting standard advocated by the Bush team (see, e.g., *Orlando Sentinel* or *Wall Street Journal*, both of November 12, 2001).

As expected, we find that agreement is at a relatively lower level for the all codes variable than for the dimple or greater variable than for the two-corners or greater variable. The all codes variable provides the most detailed description of a candidate/ballot, and agreement for this variable must be a relatively rarer event than agreement for either of the other two variables. Dimple or greater is possibly less obvious and relies more on coder judgment than two corners or greater. The concepts of dimpling and two-corner detachment apply most directly to punch card systems, including Votomatic and Datavote ballots. They do not have exact counterparts for optical scan ballots. To continue our analyses, we defined these outcome variables as any affirmative mark and a completely filled arrow or oval, respectively. We believe these definitions make the outcome variables reasonably, though not exactly, comparable across the three types of ballots.

---

[144][See Wolter et al. (2003), Introduction (2.2.1 in this volume).]

Table 7.1 Reliability statistics by candidate by outcome variable: total marked ballots

Candidate	All codes				Dimple or greater				Two corners or greater			
	$\bar{P}_2$	κ_2	$\bar{P}_3$	κ_3	$\bar{P}_2$	κ_2	$\bar{F}_3$	κ_3	$\bar{P}_2$	κ_2	$\bar{P}_3$	κ_3
Bush	0.81	0.68	0.73	0.65	0.89	0.77	0.84	0.77	0.98	0.85	0.97	0.85
Gore	0.80	0.65	0.71	0.63	0.88	0.73	0.81	0.73	0.99	0.81	0.98	0.81
Browne	0.98	0.55	0.98	0.58	0.99	0.47	0.99	0.47	1.00	0.48	1.00	0.48
Nader	0.99	0.65	0.98	0.67	0.99	0.62	0.99	0.62	1.00	0.63	1.00	0.63
Harris	0.99	0.60	0.99	0.62	0.99	0.51	0.99	0.51	1.00	0.51	1.00	0.51
Hagel	0.99	0.62	0.98	0.65	0.99	0.55	0.99	0.55	1.00	0.56	1.00	0.56
Buchanan	0.99	0.63	0.98	0.65	0.99	0.64	0.99	0.64	1.00	0.58	1.00	0.58
McReynolds	0.99	0.59	0.99	0.62	1.00	0.46	0.99	0.46	1.00	0.43	1.00	0.43
Phillips	0.99	0.62	0.99	0.64	1.00	0.43	0.99	0.43	1.00	0.49	1.00	0.49
Moorehead	0.99	0.61	0.99	0.63	1.00	0.51	0.99	0.51	1.00	0.47	1.00	0.47

Reliability is relatively similar for Bush and for Gore. For example, $\overline{P}_2$ is 0.89 and 0.88 and κ_2 is 0.77 and 0.73 for Bush and Gore, respectively, given the dimple or greater variable. Reliability is apparently at a much higher level for each of the remaining eight candidates. For example, under the dimple or greater standard, Browne's $\overline{P}_2$ is 0.99. This apparent higher reliability is due to the fact that few voters intended to vote for these candidates and most of the variable values for these candidates are 0.

In what follows, we only present reliability statistics for the Republican, Democratic, and Libertarian candidates: Bush, Gore, and Browne. Results for other candidates mirror those for Browne. To simplify the presentation, we present reliability statistics only for the dimple or greater and two corners or greater variables. Also, because of the similarities between them, we drop the κ statistics and only present the $\overline{P}$ statistics.

Reliability by type of ballot appears in Table 7.2. For the dimple or greater standard, the reliability of the two punch card ballots is lower than that of the optical scan ballots. For the two-corners or greater variable, the reliability for Datavote (punch card without chad) is lower than for the other two types of ballots. Indeed, the two variables are quite similar for Datavote ballots, tending to differ only for absentee ballots.

While Bush and Gore reliabilities are similar for Votomatic and optical scan ballots, Bush reliability is lower than Gore reliability for Datavote ballots. This finding is probably due to the fact, that only 299 Datavote ballots were marked at all; that 135 were marked for Gore; and that 168 were marked for Bush. Thus, more ballots were blank for Gore—on which there was complete agreement between the coders— leading to relatively higher Gore reliability.

Table 7.3 presents reliability statistics by the absentee status (absentee versus regular) of the ballots. Overall, absentee ballots are slightly less reliable than regular ballots. Given the dimple or greater standard, the Bush and Gore pairwise agreement statistics are 0.86 and 0.87 for absentee ballots and 0.90 and 0.88 for regular ballots, respectively. All four reliability measures reveal the slightly greater reliability of regular ballots. For the two-corners or greater standard, the reliability of absentee and regular ballots tend to converge.

Votomatic ballots tend to follow the pattern by absentee status observed overall. Datavote ballots, however, reveal larger differences, but still regular ballots are the more reliable. For example, given the dimple or greater standard, the Bush pairwise agreement statistics are 0.77 for absentee ballots and 0.98 for regular ballots, respectively. The differences even continue for the two corners or greater standard. Datavote absentee ballots also reveal a relatively sizable difference between Bush and Gore reliability. As noted earlier, this finding is probably due to the fact that more of these ballots were blank for Gore. Optical scan ballots reveal negligible differences between absentee and regular ballots.

Table 7.4 examines various design formats used in optical scan counties. Some counties used an arrow design, while others used an oval design. From these data, it appears the oval design is slightly the more reliable. For example, given the dimple or greater standard, the Gore and Bush pairwise agreement statistics are 0.95 and 0.92 for the arrow design and 0.97 and 0.95 for the oval design, respectively. As we have

Table 7.2   Reliability statistics by candidate by outcome variable by type of ballot: total marked ballots

Candidate	Dimple or greater		Two corners or greater	
	$\overline{P}_2$	$\overline{P}_3$	$\overline{P}_2$	$\overline{P}_3$
A. Votomatic ballots				
Bush	0.88	0.83	0.98	0.98
Gore	0.87	0.80	0.99	0.98
Browne	0.99	0.99	1.00	1.00
B. Datavote ballots				
Bush	0.79	0.70	0.87	0.81
Gore	0.86	0.79	0.92	0.88
Browne	0.99	0.98	1.00	1.00
C. Optical scan ballots				
Bush	0.96	0.94	0.98	0.97
Gore	0.94	0.91	0.98	0.97
Browne	0.99	0.98	1.00	0.99

seen before, observed differences for the dimple or greater standard tend to converge for the two-corner or greater standard.

About three-fourths of the oval-design ballots listed the presidential candidates in a single column, and about one-quarter split the candidates across two columns. One might expect the single column design to be the more reliable, yet the statistics reveal that both of these variations on the oval design are about equally reliable.

In Figure 7.1, we study reliability by county, plotting Gore pairwise agreement versus Bush pairwise agreement, given the dimple or greater standard. We have obtained similar results for the other reliability statistics and for the two corners or greater standard. Four counties—Calhoun, Dixie, Hendry, and Wakulla—did not have any marked ballots, and thus they are omitted from the figure. For Bush, agreement varies from just over 0.70 in Baker county to essentially 1.00 in Union and other counties. For Gore, agreement varies from just over, 0.70 in Jefferson county to essentially 1.00 in Union and other counties.

If Gore and Bush agreements perfectly tracked one another (high for Bush also high for Gore and low for Bush also low for Gore) then the various points, each of which represents a county, would fall on a straight line. One can see reasonably close tracking for the Votomatic counties: the counties (or points) do tend to fall along a straight line. The Datavote counties are generally quite small, and the pairwise agreement statistics tend to be a bit erratic, because they are based upon a relatively small number of ballots. The optical scan counties form a large cluster with very high

Table 7.3   Reliability statistics by candidate by outcome variable by absentee status: total marked ballots

Candidate	Dimple or greater		Two corners or greater	
	$\overline{P}_2$	$\overline{P}_3$	$\overline{P}_2$	$\overline{P}_3$
A. Total marked ballots				
A.1. Absentee ballots				
Bush	0.86	0.79	0.97	0.96
Gore	0.87	0.80	0.98	0.97
Browne	0.99	0.99	1.00	1.00
A.2. Non-absentee (regular) ballots				
Bush	0.90	0.85	0.99	0.98
Gore	0.88	0.82	0.99	0.98
Browne	0.99	0.98	1.00	1.00
B. Votomatic marked ballots				
B.1. Absentee ballots				
Bush	0.85	0.78	0.98	0.97
Gore	0.86	0.78	0.99	0.98
Browne	0.99	0.99	1.00	1.00
B.2. Non-absentee (regular) ballots				
Bush	0.89	0.84	0.99	0.98
Gore	0.87	0.81	0.99	0.98
Browne	0.99	0.98	1.00	1.00
C. Datavote marked ballots				
C.1. Absentee ballots				
Bush	0.77	0.66	0.86	0.79
Gore	0.85	0.77	0.92	0.88
Browne	0.99	0.98	1.00	1.00
C.2. Non-absentee (regular) ballots				
Bush	0.98	0.97	0.98	0.97
Gore	0.94	0.92	0.94	0.92
Browne	1.00	1.00	1.00	1.00
D. Optical scan marked ballots				
D.1. Absentee ballots				
Bush	0.97	0.95	0.98	0.97
Gore	0.97	0.95	0.97	0.96
Browne	0.99	0.99	1.00	0.99
D.2. Non-absentee (regular) ballots				
Bush	0.96	0.93	0.98	0.97
Gore	0.93	0.90	0.98	0.97
Browne	0.99	0.98	1.00	1.00

Table 7.4   Reliability statistics by candidate by outcome variable by design format: optical scan marked ballots

	Dimple or greater		Two corners or greater	
Candidate	$\overline{P}_2$	$\overline{P}_3$	$\overline{P}_2$	$\overline{P}_3$
	A. Arrow design			
Bush	0.95	0.92	0.98	0.97
Gore	0.92	0.88	0.99	0.98
Browne	0.99	0.98	0.99	0.99
	B. Oval design			
Bush	0.97	0.95	0.98	0.98
Gore	0.95	0.93	0.97	0.96
Browne	0.99	0.99	1.00	1.00
	C. Oval design, single column			
Bush	0.96	0.94	0.98	0.97
Gore	0.95	0.93	0.97	0.96
Browne	0.99	0.98	1.00	1.00
	D. Oval design, split column			
Bush	0.97	0.96	0.99	0.98
Gore	0.94	0.91	0.96	0.95
Browne	0.99	0.99	1.00	1.00

agreement for both Bush and Gore, and a smaller number of counties with differential agreement.

The analysis by county is limited by the fact that, in general, different teams of coders worked in different counties. Because there was no explicit random assignment of coder-teams to counties, county effects and coder-team effects are confounded with one another.

Reliability of overvote ballots. Finally, we turn to a brief examination of overvote ballots. Recall that we decided to use three coders for overvotes in only three of Florida's counties, and one coder per ballot in the remaining 64 counties.[145] They examined 2,114 Votomatic ballots in Pasco County, 1,294 Datavote ballots in Nassau County, and 668 optical scan ballots in Polk County, or 4,076 ballots overall. All reliability statistics are very high, regardless of candidate, ballot type, or outcome variable. From Table 7.5 we see that pairwise agreement is about 0.97, 1.00, and 0.99

[145][This decision was made because tests determined that overvotes were easier to code and led to less variability; hence, as a cost-saving measure, only one person was assigned to code overvotes in each of 64 counties. (See 2.2.1 in this volume.)]

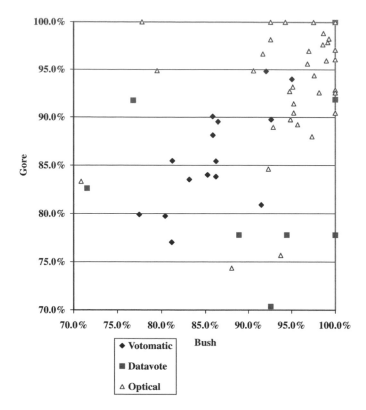

Figure 7.1 Scatter plot of Florida counties, Gore $\overline{P}_2$ versus Bush $\overline{P}_2$, dimple or greater standard: total marked ballots.

for Votomatic, Datavote, and optical scan ballots, respectively. Apparently, all ballot systems have similarly high reliability given ballots have been classified as overvotes.

Effects of Coder Characteristics on Coder Reliability. The coding of the data and the supervision of the coding were carried out with care and with the purpose of preventing any deliberate miscoding of ballots. However, there is always a concern that some characteristic of the coders may be associated systematically with some failure of accurate coding; for example, older coders might have more difficulty with their eyesight and might miss more non-null codes than their younger co-workers. Of greater concern would be the possibility of systematic favoritism toward a particular candidate, either deliberate or subconscious. Though coders did not have access to the relationship between chads and candidates in Votomatic counties, this information would not be difficult to obtain and could, in any case, become known to coders in an innocent manner.

Though NORC collected and recorded "extra-role" characteristics of coders—that is, characteristics of the coders that are not related to their role as coders [see, e.g.,

Table 7.5   Reliability statistics by candidate by outcome variable by type of ballot: overvote ballots

Candidate	Dimple or greater		Two corners or greater	
	$\overline{P}_2$	$\overline{P}_3$	$\overline{P}_2$	$\overline{P}_3$
	A. Three-county total			
Bush	0.98	0.97	0.98	0.97
Gore	0.98	0.98	0.98	0.98
Browne	0.98	0.97	0.98	0.97
	B. Votomatic ballots (Pasco County)			
Bush	0.97	0.96	0.97	0.96
Gore	0.97	0.96	0.98	0.96
Browne	0.97	0.96	0.97	0.96
	C. Datavote ballots (Nassau County)			
Bush	1.00	1.00	1.00	1.00
Gore	1.00	1.00	1.00	1.00
Browne	1.00	0.99	1.00	1.00
	D. Optical scan ballots (Polk County)			
Bush	0.99	0.98	0.99	0.98
Gore	0.99	0.98	0.99	0.98
Browne	0.98	0.97	0.98	0.97

Sudman and Bradburn (1974) for analogous characteristics of survey interviewers]— they were not used in any way to allocate work to coders. Thus, in some counties and precincts all three coders might have been similar in age and socio-economic characteristics. In other counties and precincts the three coders might have differed on all the characteristics. In still others, they might have differed on some characteristics and not on others.

This nonsystematic work allocation means that it is very difficult to establish the possible effect of extra-role characteristics of coders. It is not sufficient simply to compare the overall rate of codings for one candidate or another across all the ballots. Ballots differ from one another not only in level of difficulty (e.g., by voting system, quality of voting machines, maintenance of equipment, and haphazard variation in level of difficulty of coding), but also systematically by the voting propensities in the precinct to which they belong.

It is, therefore, necessary to construct a model to estimate the potential impact of coder characteristics. What is needed is a model that estimates the effect of each characteristic taking into account (controlling for) all the effects of the other characteristics. Thus, if we wish to estimate the effect of age of coder, for example, we would like to contrast the codings of younger and older coders while coding equivalent sets of ballots. The analysis would also need to control for all the other characteristics simultaneously; thus only young and old coders with otherwise identical characteristics (gender, socio-economics, party affiliation) could be used in the comparison. To carry out the contrast only in such equivalent sets would mean discarding most of the data.

Statistical analysis generally deals with this class of problem by using some form of regression analysis—fitting a model that accounts simultaneously for a set of characteristics. There are two particular issues that complicate the analysis for these data.

First, the outcome variable (the code) is an attribute, a binary variable that takes only two values: yes (there is a mark that has been coded) and no (there is no mark that has been coded). This outcome variable implies the use of logistic regression. Second, the data have a hierarchical structure that arises in two conceptually different ways:

1. Ballots are grouped within precinct; precincts are natural groupings of ballots that have their own (mostly unknown) characteristics. Precincts are grouped within counties; counties are also natural groupings with potentially distinctive features. Groups of counties share voting technology.

2. The three codings of the ballots are repeated measurements on the set of underlying ballots. Furthermore, teams of coders operate within precinct (coders are nested within precinct); generally, the same team of coders coded all the ballots in a precinct.

The hierarchical structure of the data implies violations of some of the usual assumptions of linear regression that are needed to obtain good estimates of effects and, in particular, good estimates of the precision of the estimators. The overall model is, therefore, a repeated measures multilevel model. These models were explored in depth by Goldstein, Healy, and Rasbash (1994) and Goldstein (1995).

The model we fit is described in the following. At each level of the hierarchy we allow for variation in the intercept of the regression equation. The slope is invariant across levels of the model. Thus, we estimate a single regression coefficient for the impact of each of the explanatory variables in the model. The outcome variable for a given candidate is defined by:

$$
\begin{aligned}
Y_{ijkp} \ &= \ 1, \quad \text{if the } i\text{th coder in the } j\text{th precinct in the } k\text{th county sees a} \\
&\qquad \text{mark for the candidate on the } p\text{th ballot;} \\
&= \ 0 \ \text{ otherwise.} \qquad\qquad\qquad\qquad\qquad\qquad\qquad\qquad\qquad (7.5)
\end{aligned}
$$

The outcome variable is modeled as a Bernoulli random variable with parameter $\phi_{ijkp} = \Pr\{\text{the } i\text{th coder, } j\text{th precinct, } k\text{th county sees a mark for the candidate on the } p\text{th ballot}\}$.

Define the log-odds by $\eta_{ijkp} = \log\{\phi_{ijkp}/(1 - \phi_{ijkp})\}$, and define a hierarchical, generalized linear model for the log-odds, as follows:

Level 1: Coder Level

$$\eta_{ijkp} = \beta_{0jkp} + \sum_q \beta_{qjkp} X_{qijkp},$$ (7.6)

where $q = 1, 2, \ldots, Q$ indexes the coder level characteristics denoted by X;

Level 2: Ballot Level

$$\beta_{0jkp} = \lambda_{0jk0} + w_{0jkp},$$ (7.7a)

$$\beta_{qjkp} = \lambda_{qjk0},$$ (7.7b)

where $w_{0jkp} \sim N(0, \kappa)$ and κ is the between-ballot variability within the same precinct;

Level 3: Precinct Level

$$\lambda_{0jk0} = \alpha_{00k0} + u_{0jk0}$$ (7.8a)

$$\lambda_{qjk0} = \alpha_{q0k0},$$ (7.8b)

where $u_{0jk0} \sim N(0, \tau)$ and τ is the between-precinct variability within the same county; and

Level 4: County Level

$$\alpha_{00k0} = \gamma_{0000} + \nu_{00k0}$$ (7.9a)

$$\alpha_{q0k0} = \gamma_{q000},$$ (7.9b)

where $\nu_{00k0} \sim N(0, \xi)$ and ξ is the between-county variability of coder-level effects.

In this notation, the subscript value "0" in a certain subscript position signifies that the parameter in question is fixed over entities represented by that position. For example, β_{0jkp} varies over precincts, counties and ballots, but not over coders, and α_{00k0} varies only over counties.

The following coder characteristics were included in our analyses as X variables: gender (binary: female as reference category); marital status (binary: not married as reference category); age (in years); education (binary: not a college graduate as reference category); income (binary: income $< \$50,000$ as reference category); race (binary: nonwhite as reference category); political affiliation (three categories: Democrat as reference category; Republican; neither Democrat nor Republican).

The basic purpose of the models is to partition the variability in the data so that we can estimate the impact of each coder characteristic, while simultaneously taking into account both the differences between precincts and counties and the clustering of ballots within precincts and counties. The full model is a three-level model with repeated measures at the first level, which is in effect a four-level model for estimation purposes. We used the current (2001) version of the MLWin software (Goldstein et al. 1998) to fit the models.

If we were to fit a model to all the Florida data simultaneously, we would need a five-level model:

1. Repeated measures,

2. Within ballot,

3. Within precinct,

4. Within county,

5. Within voting technology.

The five-level model presented technical problems in estimation, and there seemed insufficient advantage in combining different voting technologies in a single analysis. As it happens, the results and the implications are different for the different technologies and, thus, it is best in any case to look separately at the results. There were insufficient cases in the Datavote counties for the models to run successfully; consequently, we report separately for only Votomatic and optical scan ballot systems.

As outcome variables we considered "any mark" for Bush; "any mark" for Gore; and the difference between the coding for Bush and the coding for Gore. We should point out that the final dependent variable (the difference) is not a binary variable and we used a linear rather than a logistic regression in estimating this model.

In each case we ran

1. Single-level (nonhierarchical) models, ignoring the hierarchical structure;

2. Two-level models, taking into account only the repeated measures aspect of the data;

3. Three-level models, incorporating the precinct clustering into the analysis; and

4. Four-level models, taking into account the clustering of precincts into counties.

We carried out the analyses separately for optical scan counties as a group and Votomatic counties as a group.

For optical scan counties, the one-level (no structure) and two-level (repeated measurement structure) models showed significant effects of coder characteristics on the outcome of the coding operations. If we were to accept these results as valid, we would conclude that coder characteristics do have a significant impact on the results of the coding process. In particular, we would believe that the coding is sensitive to the extra-role characteristics of the coders. However, once we included the hierarchical structure in the model, all the significant effects disappeared. Thus, there is no evidence of systematic effects of coder characteristics on the outcome of the coding process in optical counties.

For Votomatic counties, the one-level (no structure) and two-level (repeated measurement structure) models again showed significant effects of coder characteristics on the outcome of the coding operations. For those counties, however, even when we introduced the hierarchical structure, the effects did not disappear. The coefficients and their standard errors are shown in Table 7.6. For example, the log-odds of seeing a mark for Bush is significantly higher for male coders than for female coders. The

Table 7.6 Multilevel repeated measures (four-level: coder-ballot-precinct-county) regression in Votomatic counties: coefficients (standard errors)

Outcome variable	Gender	Marital status	Age	Education	Income	Race	Political affiliation	
							Republican	Neither
Bush	0.335*	0.250*	0.002*	−0.026	−0.211*	−0.186*	0.220*	0.128*
	(0.019)	(0.019)	(0.001)	(0.018)	(0.021)	(0.028)	(0.019)	(0.021)
Gore	0.066*	−0.010	−0.002*	−0.046*	0.016	−0.041	−0.040*	0.014
	(0.018)	(0.0184)	(0.001)	(0.017)	(0.020)	(0.026)	(0.018)	(0.019)
Difference	0.031*	0.027*	0.000	0.005	−0.026*	−0.008	0.031*	0.012*
	(0.004)	(0.004)	(0.000)	(0.004)	(0.005)	(0.006)	(0.004)	(0.005)

*Coefficient is statistically significant at the 0.05 level.

Table 7.7   Odds ratios (with confidence intervals) and adjusted probabilities of coding by coders of each party affiliation for Votomatic undervote ballots

Outcome variable/ political affiliation of the coder	Odds ratio	95% confidence interval	Adjusted probability
Bush/			
Republican	1.247	(1.201, 1.295)	0.192 = Pr (Mark for Bush | Coder is Affiliated with Republican Party)
Neither	1.137	(1.092, 1.184)	0.178 = Pr (Mark for Bush | Coder is Affiliated with neither party)
			0.160 = Pr (Mark for Bush | Coder is Affiliated with Democratic Party)
Gore/			
Republican	0.961	(0.927, 0.996)	0.182 = Pr (Mark for Gore | Coder is Affiliated with Republican Party)
Neither	1.014	(0.976, 1.053)	0.190 = Pr (Mark for Gore | Coder is Affiliated with neither party)
			0.188 = Pr (Mark for Gore | Coder is Affiliated with Democratic Party)

log-odds of recording a mark for Gore is also higher for males than for females. But the sex effect is smaller in the Gore model than in the Bush model.

For the logistic regression models, the dependent variable is the log of the ratio of the proportion coding for a particular candidate to the proportion not coding for that candidate; this ratio is the odds of coding "yes" for the candidate. The coefficients in the logistic model express the additive impact of each factor on the log of the odds. The underlying model is a multiplicative model for odds; by exponentiating the (additive) coefficient in the logistic regression, we can find the multiplicative effect on the odds themselves. For each explanatory variable (factor) there is a reference category. The impact of the factor is expressed as the ratio of the odds when a coder belongs to a particular category relative to the odds when a coder is in the reference category. In particular, for gender the reference category is female; the impact of gender in the model is the ratio of the odds that a male coder will code "yes" to the odds that a female coder will code "yes." This is the odds ratio.

Table 7.7 considers a detailed breakdown of the impact of political affiliation on the probability of the outcome of the coding operation in the Votomatic counties. (Similar analysis could be done of the gender effect.) The reference category is Democrat—in other words the other two affiliations are contrasted with Democrats in the model. For the Bush model, the coefficient for Republican in the logistic regression model is 0.22. This translates into an odds ratio of 1.25 in Table 7.7; the 95% confidence interval for this odds ratio is (1.20, 1.29). A null value for the odds ratio would be 1.00 (no effect on the odds).

To understand its substantive significance, we need to translate this odds ratio into an effect on the probability of the coding outcome. This is done in the far right column in of the table. The adjusted probability is the estimated proportion of ballots that will be coded positively (i.e., marked) for the candidate. There are three adjusted probabilities for each outcome. The first is the proportion of ballots that would be coded for the candidate if the coder were Republican. The second is the corresponding proportion if the coder is neither Republican nor Democrat. The third is the proportion if the coder is Democratic. Thus, we can see that in interpreting Bush chads, Republican coders will, on average, code 19.2% of the ballots as marks for Bush; Democratic coders will code only 16% for Bush; and the others will code 17.8% for Bush. In interpreting Gore chads, the position is reversed, though the size of the effect is much less. Here, 18.2% of Republicans will code a mark for Gore; 18.8% of Democrats will do so; and 19% of the others will code for Gore.

The linear regression model is one way of interpreting the impact of these findings on the difference between for Bush and for Gore. The estimated impact of being Republican, rather than Democrat, can be seen in the coefficient of 0.03 for Republican in the equation. This implies that, on average, Republican coders found 3% more marks for Bush relative to Gore than did Democrat coders.

The results of the multilevel analysis are important in two ways. First, the contrast between the hierarchical and the nonhierarchical analyses show the importance of taking the data structure into account. Failing to take the structure into account could lead to making the wrong inference from the data. Second, there is a clear difference between the outcome in the optical scan counties and the Votomatic counties. In the Votomatic counties there is clear evidence of sensitivity of the coding outcome to the extra-role characteristics of the coders.

References

Bailar, B. A., and Tepping, B. J. (1972). *Effects of Coders*, Series ER, 60 (9), Washington, DC: U.S. Bureau of the Census.

Fleiss, J. L. (1965). Estimating the Accuracy of Dichotomous Judgments, *Psychometrika*, 30 (469).

————— (1971). Measuring Nominal Scale Agreement Among Many Raters, *Psychological Bulletin*, 76 (378).

Goldstein, H. (1995). *Multilevel Statistical Models*, London, England: Edward Arnold.

Goldstein, H. M.; Healy, J. R.; and Rasbash, J. (1994). Multilevel Time Series Models with Applications to Repeated Measures Data, *Statistics in Medicine*, 13 (1643).

Goldstein, H.; Rasbash, J.; Plewis, I.; Draper, D.; et al. (1998). *A User's Guide to MlwiN*, London, England: Institute of Education.

Kalton, G. and Stowell, R. (1979). A Study of Coder Variability, *Applied Statistics*, 28 (276).

Orlando Sentinel (2001). The Final Report of the Florida Ballot Project, November 12, 2001, B5.

Wall Street Journal (2001). In Election Review, Bush Wins with No Supreme Court Help; Majority of Florida Voters Would Have Picked Gore but for Poor Ballot Design; Both Backed Wrong Strategy,' November 12, 2001, A14.

Sudman, S. and Bradburn, N. M. (1974). *Response Effects in Surveys*, Chicago, IL: Aldine.

The third excerpt we present here provides the technical appendix for the paper that forecasts outcomes in Congress during mid-term elections. That paper appears in Chapter 4 of this volume. (See 4.2.3.)

7.2.3 Appendix to: Forecasting House Seats from Generic Congressional Polls

Joseph Bafumi, Robert S. Erikson, and Christopher Wlezien

Appendix: Simulating Votes and Seats. For each integer value of the vote from 50% Democratic to 60% Democratic in generic ballot polls, we obtain 1,000 computer simulations of the seat distribution. Each batch of 1,000 simulations is based on a value of the national vote, N_{gj}.

$$N_{gj} = P_g + e_j, \tag{7.10}$$

where $P_g =$ the projected national Democratic percent of the vote, given the prediction from the generic ballot result g, from the 15-election regression equation displayed below as Equation 7.11 and $e_j =$ a random draw ($j = 1$ to 1,000) from the distribution of error signified by the root mean squared error (RMSE) in Equation 7.11.

Nat'l % Dem Vote $= 24.38 + 0.51 *$ Generic Ballot% $- 1.09 *$ Presparty $+ e_j$ (7.11)

Adjusted R-squared $= 0.75$; RMSE $= 1.90$, $N = 15$ mid-term elections, 1946–2002 where Presparty $= 1$ if a Democratic President and -1 if a Republican President.

For each simulated value of the national vote, we need to simulate the outcome in 435 congressional districts. The district vote (D_{gjk}) is:

$$D_{gjk} = N_{gj} + L_k + u_k = P_g + L_k + e_j + u_k, \tag{7.12}$$

where $L_k =$ the district (local) component of the expected district vote and $u_k =$ the simulation of the district k error.[146] The formula for L_k differs for open seats and incumbent races. For open seats (no incumbent running), the district vote simulation is from Equation 7.13. For seats where the incumbent seeks reelection, the district vote simulation is from Equation 7.14.

Open Seats: $D_{gjkm} = -42.61 + 0.89 * \%\text{Kerry}_k + P_g + e_j + u_{k\_\text{open}}$, (7.13)

where $\%\text{Kerry}_k =$ the Kerry percent of the two-party vote in district k in 2004.

Incumbent races: $D_{gjkm} = -45.91 + 0.94 * \%\text{Dem}(2004)_k$
$$+6.58 * \text{Frosh}_k + P_g + e_j + u_{k\_\text{incumbent}}, \tag{7.14}$$

where $\%\text{Dem}(2004)_k =$ the Democratic vote for the House in district k in 2004; $\text{Frosh}_k = 1$ if a Freshman Democrat in 2006 and -1 if a 2006 Freshman Republican; otherwise 0.

---

[146] Actually, all simulated errors e_j and u_k are drawn fresh with each iteration.

The numerical parameters of Equations 7.13 and 7.14 are derived from regression equations predicting the 2004 district vote for 2004 open seats and 2004 incumbent races.[147] The constant terms of Equations of 7.13 and 7.14 (-42.61 and -45.91) offset P_g, the projected national vote from the generic polls. The constants represent the 2004 regression intercepts minus the 2004 national vote.[148]

Importantly, the root mean squared errors from the two 2004 regression equations provide the standard deviations for u_{k_open} and $u_{k\_incumbent}$. For open seats, this value is 7.56. For incumbent races, it is 5.20. These values represent the uncertainty around the district vote after taking into account the projected district vote from the 2004 presidential vote (open seats) or 2004 congressional vote (incumbent seats).

The final step is to tally the 1,000 simulated outcomes for each of eleven values of the generic vote, 50% Democratic to 60% Democratic. For generic ballot result g, we obtain 1,000 estimates of the national vote N_{jg}, each with 435 estimates of the district vote D_{gjk}. For each value of the generic vote j, we record the mean seat outcome, the distribution of 1,000 seat outcomes, and the frequency with which the seat outcome for the Democrats equals or surpasses 218, the number for a numerical majority. This provides the basis for the graph of Figure 4.3 [See Bafumi, Erikson, and Wlezien (2006) in Chapter 4 of this volume (4.2.3).]

7.3 APPENDIX: SURVEY QUESTIONS

The following provides the lifestyle questions that are used to create weights for nonprobability Internet samples in a survey in Sweden. The balance of the paper is presented in Chapter 5 (Reading 5.6.1).

---

[147]For incumbent races, the incumbent's past vote is an obvious benchmark. For open seats, we substitute the district presidential vote as a measure of underlying district partisanship.

[148]For the adjustment, we assume that the mean district vote swing (for districts contested in both 2004 and 2006) equals the net 2004–2006 national vote swing in terms of votes counted. The 2004 open seat equation, based on the 2004 presidential vote, provides a baseline equation. With no net swing 2004–2006, this equation would apply unmodified for 2006. For specific hypothetical values of the vote (P_g), the constant is the original intercept minus 48.6 (the national 2004 congressional vote). Note that the equation adds the hypothetical 2006 national vote (P_g). For incumbent races, the procedure for determining the constant term is the same except that some algebra must be applied so that the net swing of the mean district vote is consistent with the national vote. For incumbent races, the constant equals the intercept for the 2004 incumbent race equation minus 48.6 (2004 national vote) minus 1.72. The 1.72-point adjustment is necessary so that the projected mean district swing across all districts equals the difference between the projected national vote (P_g) and 48.6. Given our methodology, all seats that were uncontested in 2004 must be assigned to the 2004 winner. Bernie Sanders's former seat as an Independent is assigned to the Democratic candidate.

7.3.1   A Comparison Between Using the Web and Using Telephone to Survey Political Opinions
Annica Isaksson and Gösta Forsman

Appendix: The Lifestyle Questions Used for Propensity Score Weighting

Q: Does the risk that someone can abuse or spread personal information about you make you abstain from... *(Check one alternative per row)*

	Yes	No Know	Don't to answer	Don't want
... buying a product or service over the phone?	☐	☐	☐	☐
... paying a restaurant bill with credit card?	☐	☐	☐	☐
... shopping on the Internet with credit card?	☐	☐	☐	☐
using a cash machine?	☐	☐	☐	☐
... leaving information about yourself, that can be used for offering you tailored services, products, or information on Web pages?	☐	☐	☐	☐

Q: Today, many companies collect information about their customers' hobbies and lifestyle in order to tailor information, services and products accordingly.

Do you consider this adaptation to the individual positive or negative, or do you not have an opinion? *(Check one)*
☐ Yes, individual adaptation is a positive thing
☐ No, individual adaptation is not a positive thing
☐ Don't know
☐ Don't want to answer

Q: Some people feel that they fail to notice things that happen around them. Do you feel this way? *(Check one)*
☐ Yes
☐ No
☐ Don't know
☐ Don't want to answer

Q: Have you, during the last month... *(Check all that apply)*

☐ ... watched a documentary on TV?
☐ ... gone away?
☐ ... read a book?
☐ ... none of the above?

Q: Some people feel that there is too much information available today through TV, radio, magazines, newspapers and the computer. Others want to have access to as much information as possible. How do you feel? *(Check one)*
☐ There is too much information
☐ I like to have access to as much information as possible
☐ Don't know
☐ Don't want to answer

7.4   ADDITIONAL METHODOLOGICAL PAPERS

The following three additional papers complement other contributed readings that appear earlier in the book. They are more methodological in nature and relate to the statistical basis for the articles that are provided in prior chapters.

First is a reading by Lindeman, Liddle, and Brady (2005), which further examines the issues that contributed to misleading projections after the 2004 election in Ohio. The methodological discussion here complements the work by Lindeman (2006) presented in Chapter 3 of this volume (3.2.3).

7.4.1   Investigating Causes of Within Precinct Error in Exit Polls: Confounds and Controversies
Mark Lindeman, Elizabeth Liddle, and Richard Brady

Introduction.   No one could have anticipated that, in the aftermath of the 2004 U.S. presidential election, the acronym "WPE" (Within-Precinct Error) would enter into public discourse, albeit less conspicuously than 2000's "pregnant chads." But then, no one expected a widely publicized discrepancy between the exit poll results and the official returns. The National Election Pool's exit poll interviews, conducted by Edison Media Research and Mitofsky International (Edison/Mitofsky or E/M), indicated that John F. Kerry led incumbent George W. Bush by over 6 percentage points in the crucial swing state of Ohio, which Bush ultimately won (E/M, 2005:22). The weighted national exit poll results[149] showed Kerry leading Bush by 3 percentage points, whereas the final official popular vote showed Bush ahead by 2.5 points (E/M,

---

[149] Ordinary-English statements about "exit poll results" must be read with caution, due to the proliferation of meaningful but incommensurable "results." In this case, the figure is derived from the national exit poll (which includes a subsample of precincts from the various state exit polls). This estimate incorporates geographic and nonresponse demographic weights, but does not incorporate prior survey estimates or official vote returns. In contrast, Steve Freeman (2005a, 2005b) argues that the exit polls indicate that Kerry won the popular vote by about 4.6 points.

2005: 20). Although in no case was a state called incorrectly for Kerry, some observers believe to this day that the exit polls demonstrate massive electoral fraud favoring Bush. While this paper makes no attempt to address the fraud debate directly, it is indirectly motivated by analytical efforts and debates growing out of the fraud controversy. We find that analysis of exit poll error faces complications that have generally been overlooked by practitioners and political scientists. In particular, we conclude that the intuitively appealing percentage-point measure of WPE suffers from a confound that renders it unsuitable for inferential analysis—and we commence the search for viable alternatives.

'Error' Analysis in a Fraud Debate. By convention, total error in exit polls (as well as pre-election polls) is measured under the assumption that the official returns are accurate, or at least approximately so.[150] Obviously, some observers are not willing to accept this assumption regarding the 2004 U.S. presidential election. Skeptics variously cite evidence of security holes in electronic voting machines and central tabulators, reports of touch screens that misrecorded Kerry votes as Bush votes, and many other facts and anecdotes—including, of course, the exit poll results—to argue that John Kerry may, indeed, have won a majority of votes cast (or intended to be cast). The substantive merits of these arguments are generally beyond the scope of this paper but, as a matter of logic, in analyzing exit poll "error" one must acknowledge the possibility of error in the vote count. Unfortunately, it is hard to find a "fraud-neutral" language that is universally acceptable, succinct, and clear. We, therefore, use "error" in a manner that encompasses the possibility of vote count error. In brief, the signed error, or bias, is the discrepancy between the exit poll and the official vote count that observers on both sides of the debate are trying to understand.

Most of the debate about the 2004 exit polls has hinged on three basic categories of error: sampling error, nonresponse error, and vote count error. Despite debate on specifics, it is widely agreed that sampling error alone cannot account for the discrepancy between the exit polls and the official returns (Freeman, 2004; Brady, 2005; E/M, 2005). Whereas the Edison/Mitofsky evaluation report proposed nonresponse error as the "most likely" cause (E/M, 2005: 3), critics argued the case for vote count error (U.S. Count Votes, 2005). To our knowledge, all critics accept Edison/Mitofsky's conclusion that the precinct sample does not account for the error (E/M, 2005: 3, 28-30; U.S. Count Votes, 2005: 7) and, therefore, controversy has centered on the interpretation of within-precinct error or "Precinct Level Discrepancy" (Freeman, 2005a). Coverage error and measurement error are generally not regarded as *major* explanations of the exit poll error.[151]

---

[150] For instance, the National Council on Public Polls offers a summary of presidential pre-election poll accuracy from 1936 through 2000, reporting mean "margin errors" and "candidate errors" based on the official returns (*http://www.ncpp.org/1936-2000.htm*). Similarly, Edelman and Merkle (1995:11) calculate exit poll error by subtracting "the actual Dem-Rep difference (i.e., the Democratic vote percentage minus the Republican vote percentage)" from the exit poll margin. We are not aware of past complaints about using official returns to assess poll accuracy.

[151] Here we allude to Groves' (1989) influential categorization of Total Survey Error into sampling error, nonresponse error, coverage error, and measurement error. Coverage error [occurs] when a segment of the

Limitations of the WPE Measure. E/M's evaluation report, and most criticisms thereof, rely on a specific measure of within-precinct error (WPE) as a difference of margin proportions. This measure can be traced to Frederick Mosteller et al.'s 1949 assessment of 1948 pre-election polls. Mosteller et al. described several measures, of which WPE was the fifth. Mosteller et al. (1949: 57) wrote, "There is nothing intrinsically wrong with the fifth method, except complexity of explanation"—but in hindsight, simplicity of explanation seems to be WPE's greatest virtue.

The WPE is the difference between the actual and the predicted margins for the two leading candidates. In E/M's analyses, the actual margin is calculated as the proportion of votes counted for the Democratic candidate minus the vote proportion for the Republican candidate. The predicted margin similarly equals the predicted proportion of votes for the Democrat minus the predicted proportion for the Republican. The predicted margin is then subtracted from the counted margin. WPE can be expressed algebraically as:

$$WPE = (V_D - V_R) - (P_D - P_R) \tag{7.15}$$

where V_D and V_R are the actual vote proportions, in a given precinct, of the Democratic and Republican candidates, respectively, and P_D and P_R are the predicted proportions, based on the sampled respondents. In E/M's analysis and, therefore, in this paper, the predicted proportions simply equal the observed exit poll proportions. We will assume that only two candidates receive votes, so that $V_D + V_R = 100\%$. (In actual elections these numbers may total less than 100% and, perhaps, much less—although generally not in the 2004 presidential election, wherein the two major candidates received about 99% of all presidential votes cast.) Thus, WPE will equal, subject to the sign convention, twice the error in either candidate's vote proportion. For instance, consider a precinct in which the Democrat receives 48% of the vote and the Republican 52%—a −4% margin. Suppose that in the exit poll, 53% of respondents support the Democrat and 47% the Republican—a 6% margin. The calculated WPE will be −10.0% or −0.10.

Measuring WPE as a difference in proportions makes it readily interpretable as a descriptive statistic, but not conceptually persuasive for purposes of causal inferential analysis. For instance, consider E/M's finding that in 2004, the mean WPE was −8.0 in precincts where interviewers reported that the precinct officials were not cooperative (E/M, 2005: 38). The descriptive interpretation is straightforward—the exit poll overstated Kerry's vote share margin in these precincts by 8 percentage points (or his vote share by about 4 points) on average. But WPE fails because it is confounded with

population is not included in the sampling frame. (For instance, coverage error was surely extensive in Vermont, where 19% of voters cast absentee ballots and were never surveyed. In twelve other states, E/M conducted telephone surveys to measure early and absentee voters.) Measurement error arises from inaccurate responses to survey instruments. Inaccurate responses can have many causes, including questionnaire design, cognitive effects, and social-psychological. A well-known exit poll example comes from the 1989 Virginia gubernatorial election, where the Mason-Dixon exit poll overstated the Democratic candidate's victory margin by almost 10 points. This error has been attributed to social desirability; in particular, white Democrats may have been reluctant to admit to interviewers that they had voted for the black Republican (Traugott and Price, 1992). Note that, in practice, nonresponse error may be hard to distinguish from measurement error due to false reporting.

the prior distribution of voters. A given magnitude of WPE where the vote share margin is small represents a much smaller relative bias than the same WPE in a heavily partisan district. Indeed, a WPE of -8.0 is arithmetically impossible for some values of vote share margin.

An Alternative Error Measure. We propose a different starting place for error measurement, rooted in a specific interpretation of the discrepancy between the exit poll and the official vote count. The root of E/M's (2005: 31) conjecture that "hypothetical completion rates of 56% among Kerry voters and 50% among Bush voters overall would account for the entire Within Precinct Error that we observed in 2004" is non-response bias. Nonresponse bias implies a difference in the propensities of voters for various candidates to *complete* the exit poll when asked. It is also possible that interviewers had differential propensities to *select* (sample) Kerry voters and Bush voters. Although the interview protocol calls for interviewers to adhere rigorously to a prescribed interviewing rate—to attempt to interview "every kth voter"—empirical and anecdotal evidence indicates that some interviewers did not (Blumenthal, 2005; E/M, 2005: 36).[152] Selection bias and nonresponse bias may both contribute to what we will refer to as "representation bias"—the difference between the rates at which poll responses for each candidate are represented by counted votes.

Representation bias will, thus, also arise from vote count error, and can be conceptualized as the ratio[153] of propensities for each candidate's voters to be represented by votes. This propensity ratio can be estimated by the ratio between the *achieved sampling* rates, i.e., observed representation rates. The representation (achieved sampling) rate for each candidate equals the proportions of exit poll responses divided by votes for that candidate. If there is no selection bias, nonresponse bias, or vote count error, representation propensities will be the same for both groups of voters, yielding a 1:1 ratio, and the ratio of achieved sampling rates will have a median close to 1 in large samples. We will state this observed ratio of achieved sampling rates as α, defined (by arbitrary convention) so that ratios greater than 1 imply voters for the Democratic candidate either participate in the poll at higher rate than voters for the Republican candidate or that their votes are counted at a lower rate. This is expressed algebraically in:

$$\alpha = \frac{P_D/V_D}{P_R/V_R} = \frac{S_D}{S_R} \qquad (7.16)$$

---

[152] Interviewers are instructed: "If the targeted voter declines to participate or if you miss the voter and do not get a chance to ask him or her to participate, you should mark them as a 'Refusal' or 'Miss' on your Refusals and Misses Tally Sheet"—and not to interview another voter instead (Blumenthal, 2005). However, Blumenthal (2005) reports that at least some interviewers did deviate from this instruction. Such deviations are more likely as the interviewing rate increases. The observed increase in (negative) magnitude of WPE as interviewing rate increases tends to support the hypothesis that selection bias accounts for some portion of representation bias; the generally lower miss rates associated with high interview rates may also support the hypothesis (E/M, 2005: 36).

[153] An arithmetic difference measure is conceivable, but inter alia would complicate the partitioning of representation bias into selection bias and completion bias.

where S_D is the representation (achieved sampling) rate of the voters who voted for the Democrat and S_R is the representation rate of the voters who voted for the Republican. For instance, in 2004, a α value equal to 2 would indicate that the representation rate for Kerry voters was double that of Bush voters (2:1); an α of .5 would indicate a representation rate for Kerry voters was only half that of Bush voters (1:2).

In principle, α is independent of precinct vote share margin, whereas WPE is not. For instance, let us posit two precincts in which $\alpha = 1.25$, indicating that Kerry votes are represented by Kerry respondents at 1.25 the rate at which Bush votes are represented by Bush respondents. The first precinct has an equal number of Kerry votes and Bush votes; its WPE will be -11.1.[154] The second precinct is heavily Republican, with only 10% votes; its WPE will be just -4.4.[155] Note that sampling error will influence the observed representation ratios α—by chance, the Democratic (and Republican) interviewing rates will vary. We will later explore the effects of sampling error on observed values of α.

In empirical analysis, it is often convenient to calculate α as an odds ratio of survey proportions and vote proportions:

$$\alpha = \frac{P_D/P_R}{V_D/V_R} \qquad (7.17)$$

Note that α is undefined in the case where $P_R = 0$ (no Republican voters are interviewed).[156]

The ratio α is asymmetrical: the inverse of the representation ratio 0.8:1 is 1:0.8 or 1.25, and 0.8 and 1.25 obviously are not equidistant from 1 (= no net bias). To generate a symmetrical "bias index," Liddle (2005) earlier proposed to take the natural logarithm (ln) of α. (For instance, for alpha values of 0.8, 1.0, and 1.25, the corresponding ln α are -0.223, 0.000, and +0.223, respectively.) Martin et al. (2005) independently proposed using the natural log of a similar odds ratio to assess the accuracy of pre-election polls (where the error variance is typically much smaller than in exit poll data). However, there is no evident theoretical warrant for a logarithmic transformation; in particular, it is not obvious why a given percentage increase in α (for instance, from 1.0 to 1.1 to 1.21) should yield a linear change in the bias index. It is also awkward that ln is numerically unbounded and becomes undefined in the case where $\alpha = P_D = 0$. We now propose an alternative bias index based on arctangent:

$$\tau = \mathrm{atan}(\alpha) - \mathrm{atan}(1) \qquad (7.18a)$$

---

[154]For instance, suppose that the precinct contains 400 Kerry voters, of whom 12.5% = 50 complete the survey, and 400 Bush voters, of whom 10% = 40 complete the survey. The net vote margin is, of course, 0. The Bush predicted (or poll) share is 44.44...%, and the Kerry predicted share is 55.55...%, yielding a predicted margin of 11.1%. The WPE equals $0 - 11.1\% = -11.1\%$.

[155]The Kerry vote margin is -80%. The Kerry proportion in the poll can be computed as $(0.1)(1.25)/(0.1)(1.25) + 0.9 \approx 12.2\%$, so the Kerry predicted margin is $12.2 - 87.8 = -75.6\%$. The WPE $= -80 - (-75.6) = -4.4$.

[156]It is also undefined, or indeterminate, in the less likely case where $V_D = 0$. In this case, P_D should also equal 0—in the absence of measurement error or vote count error—and, therefore, α can be set to 1.

$$\alpha(\tau) = \tan(\tau + \text{atan}(1)) \qquad (7.18b)$$

Arctangent is commonly used to stabilize the variance of ratio measures. Unlike ln, it is bounded (by 0 and $\pi/2$, so that τ is bounded by $-\pi/4$ and $+\pi/4$); when $\alpha = 0$, $\tau = -\pi/4$. Moreover, in the degenerate case where α is undefined because its denominator equals 0, τ can be set at its upper bound, $+\pi/4$. Note the shorthand $\alpha(\tau)$ to indicate the value of α associated with a given value of τ.

Because α is conceived to measure bias in the rates at which poll responses are represented by counted votes, both differential completion rates and vote count error will result in values of α that depart from unity. For instance, given E/M's hypothetical completion rates of 56% for Kerry voters and 50% for Bush voters—and assuming no selection bias or other error source—α will equal $0.56/0.50 = 1.12$. However, vote count error will also induce bias inasmuch as people's reported votes differ from their recorded votes.[157] For instance, in a precinct where the *intended* votes were evenly divided, but 3 percentage points of Kerry votes were somehow changed to Bush votes (giving Bush a 53%–47% margin), the result – again assuming no other source of bias or error – would be $\alpha = 1.128$.[158]

Interestingly, simple electoral fraud mechanisms often generate very different values of α as a function of each candidate's vote share. *Shifting* votes from one candidate to another, as in the preceding example, could be an unobtrusive means of fraud, because the count of ballots cast will match the number of voters recorded at the precinct. A shift of 3 percentage points from Kerry to Bush in every U.S. precinct—corresponding to an expected WPE of -6.0 in every precinct—could, in the abstract, roughly account for the discrepancy between exit poll and vote results. (To our knowledge no one has proposed this specific scenario, which is contradicted by other evidence.) Such a shift produces strikingly different values of α at different levels of vote share—as noted above, $\alpha = 1.128$ if the intended vote is 50% Kerry, but $\alpha = 1.345$ if the intended vote is 90% Kerry (or if it is 13% Kerry).[159] Of course this is not to imply that fraud entails any particular relationship between α and vote share—nor vice versa, since α might covary with vote share for reasons other than fraud.

---

[157] While "electoral fraud" has no fixed meaning, some activities widely regarded as fraudulent would not be manifested as "bias" in exit poll analysis, because they would equally affect the exit polls and the official returns. For instance, if legitimate voters are prevented from voting because they were illegally stricken from the registration lists, they may not be represented in either the exit polls or the official returns. (However, some such voters may cast provisional ballots—which may or may not subsequently be counted. Also, some may participate in the exit poll even if they did not actually cast a ballot.)

[158] In this scenario, the representation ratio α is $(0.5 / 0.47) / (0.5 / 0.53) = 1.128$.

[159] The representation ratio in the 90% case is $(0.9 / 0.87) / (0.1 / 0.13) = 1.345$; in the 13% case it is $(0.13 / 0.10) / (0.87 / 0.90) = 1.345$, again. At least one simple mechanism does yield α uncorrelated with vote share: to increase one candidate's vote total by a given proportion (not number of percentage points), without altering the other candidate's total. For instance, if the Republican candidate's vote tally is arbitrarily increased by 10%, then α should equal 1.10, regardless of precinct vote share (except in the case where the Republican received no votes). As noted in the text, it should be possible to detect such fraud by comparing the number of ballots cast to the number of voters recorded.

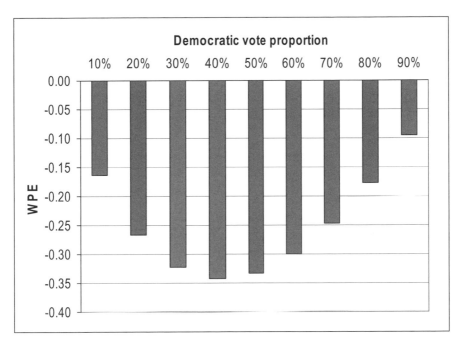

Figure 7.2   Expected WPE values (without sampling error) for a uniform extreme representation rate ratio α of 2:1 at various degrees of vote share.

Exploring the WPE/Vote Share Confound. Let us suppose a bias mechanism independent of precinct vote share—for instance, a uniform propensity for Democratic voters to participate at a higher rate than Republican voters. In the absence of sampling error, α will (by assumption) take on some constant value, but WPE will vary as a function of vote share. Liddle (2005) derives the relationship between WPE and vote share (as measured by Democratic vote share V_D) for a given value of α:

$$WPE = 2 \times \frac{V_D^2(1 - \alpha) - V_D(1 - \alpha)}{V_D(1 - \alpha) - 1} \tag{7.19}$$

Note that when $\alpha = 1$ (and the bias is, therefore, zero), WPE reduces to zero. As α departs from 1, the behavior of this function is increasingly non-obvious to most observers and, therefore, warrants investigation.

Figure 7.2 illustrates the WPEs that would theoretically result from a uniform and extreme representation ratio of $\alpha = 2$. By construction, the bias is identical in all precincts. Nonetheless, WPE varies drastically; its magnitude is over three times as large in the 50%-Democratic districts as in the 90%-Democratic districts. Notice also the asymmetry of the distribution; relative under-sampling of the minority voter group (as in the right side of Figure 7.2 in this example) produces a smaller WPE than relative under-sampling of the majority group.

The example depicted in Figure 7.2 is, of course, extreme (although observed values of α in individual precincts do sometimes exceed 2). It is also overly simplistic, because it is unlikely that any real-life nonresponse bias will be uniform. It is more likely that the magnitude and even the direction of the bias will vary across precincts. Liddle (2005) simulated the effects of normally distributed variation in nonresponse bias. We here state the modeling assumptions somewhat more formally: Assume that the propensity of Bush and Kerry voters to participate in an exit poll varies from precinct to precinct. This propensity may be influenced by characteristics of voters, of interviewers, of voter/interviewer interactions, or of precincts—but we initially assume that the variance is entirely unexplained. For simplicity, we assume that this variance does not affect selection rates, but only completion rates. For convenience, we assume that these perturbations are independent,[160] drawn from a truncated normal distribution, and equal in magnitude, so that the completion propensities in each precinct are:

$$c_{Bi} \sim TN(c_B, s^*, \min = c_B - 2.5s^*, \max = c_B + 2.5s^*) \qquad (7.20)$$

$$c_{Ki} \sim TN(a^* c_B, s^*, \min = a^* c_B - 2.5s^*, \max = a^* c_B + 2.5s^*) \qquad (7.21)$$

where c_{Bi} and c_{Ki} are completion propensities among Bush and Kerry voters, respectively, in precinct i; c_B is the expected mean completion propensity for Bush voters; a^* is a bias multiplier such that the expected mean completion propensity for Kerry voters $c_K = a^* c_B$; and s^* determines the magnitude of the random variance. We arbitrarily truncate the distributions at ± 2.5 standard deviations. In many of our simulations, $s^* = 0.1$, implying that completion propensities vary in a range of ± 0.25 or $\pm 25\%$. (In simulation studies, the value $s^* = 0.1$ yields variation in precinct-level bias, as measured by mean absolute WPE, that is similar to—but somewhat smaller than—the value observed in the 2004 E/M data.)

Using the method just described, Liddle (2005) simulated WPE distributions for nine values of condition (Figures 7.3a and b), the completion propensities for both Democratic and Republican voters in each Democratic vote proportion under two bias conditions, on the assumption of zero sampling error. Under that assumption (and the others inherent in the simulation), the poll proportions and—hence, the WPE—for each precinct can be calculated directly from vote proportion and (simulated) completion propensities. In the first, "no net bias" condition, $a^* = 1.00$, $c_B = c_K = 0.50$, $a^* = 1.00$, and $s^* = 0.1$; thus, the completion propensities for both Kerry and Bush voters had mean 50% and range [25%, 75%]. (Thus, the difference between Democratic and Republican completion rates was normally distributed with mean 0 and standard deviation approximately 14%.) In the second, "net Kerry bias" condition, $a = 1.12$, yielding Kerry completion propensity $c_K = 56\%$, corresponding with the E/M conjecture.[161] Summary results for 1 million simulations of each combination

---

[160] In practice, Bush and Kerry completion rates probably covary substantially. However, in our simulations, introducing covariance had minimal effect on error.

[161] Alternative specifications—for instance, running the model with mean completion rates of 53% for Kerry voters and 47% for Bush voters—gave similar results.

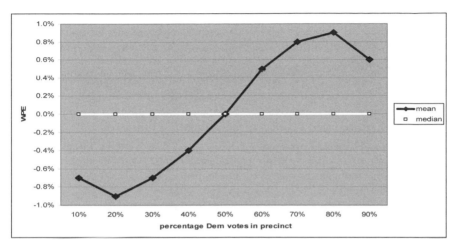

Figure 7.3a   Simulated distributions of WPE allowing variance in completion propensities: no net bias condition (mean Bush completion rate = 50%, mean Kerry completion rate = 50%). In the "no net bias" condition, the completion propensities for both Democratic and Republican voters in each precinct are $\sim N(50\%, 10\%)$, truncated at $z = \pm 2.5$ (i.e., 25% to 75%).

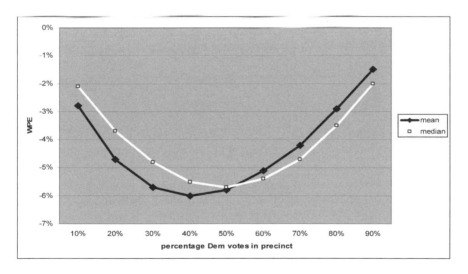

Figure 7.3b   Simulated distributions of WPE allowing variance in completion propensities: net Kerry bias condition (mean Bush completion rate = 50%, mean Kerry completion rate = 56%). In the "net Kerry bias" condition, the Democratic completion rates are $\sim N(56, 10)$, similarly truncated to the range 31% to 81%. (Note the difference in vertical scales.)

of vote proportion and bias are displayed in Figure 7.3a and b. Notice that in each case, the mean diverges from the median across much of the vote share spectrum, and that WPE is positively correlated with Democratic vote proportion. Note also that the covariation of mean WPE with vote share increases markedly when net representation bias is introduced.

In the same simulation, if α is calculated for each simulated precinct (see Equation 7.17), the mean values of both $\ln \alpha$ and τ are essentially invariant with regard to vote share, and τ in particular is approximately normally distributed. In ten million trials with $\alpha* = 1.12$, τ had mean 1.0566, skewness 0.0013, and kurtosis -0.1237. From Equation 7.18b, $\alpha(1.0566) = 1.1202$ (the truncation induces a very slight bias). Thus, if the completion propensities were directly observable (i.e., without sampling error), τ would be a symmetrical measure of bias unconfounded with vote share.

The τ/Vote Share Confound in the Presence of Sampling Error.   In practice, unfortunately, completion propensities are not directly observable—with parlous consequences for τ. To illustrate the problem, let us consider a hypothetical precinct with a total of 200 voters, of whom 30 (or 15%) voted for Bush and the remaining 170 for Kerry. Assume that interviewers select exactly 100 of these voters without bias. Thus, the number of selected Bush voters (attempted interviews) a_B is sampled from a hypergeometric distribution $H(100, 30, 200)$, and the number of selected Kerry voters $a_K = 100 - a_B$. Suppose further that $c_B = 0.56, c_K = 0.50$ (thus, $\alpha* = 1.12$), and $s* = 0.1$. c_{Bi} and c_{Ki} are drawn from truncated normal distributions as usual, and counts of Bush respondents and Kerry respondents are drawn from binomial distributions:

$$r_{Bi} \sim B(a_{Bi}, c_{Bi})$$
$$r_{Ki} \sim B(a_{Ki}, c_{Ki}) \tag{7.22}$$

The expected means are $E(a_{Bi}) = 15$, $E(a_{Ki}) = 85$, $E(c_{Bi}) = 0.50$, $E(c_{Ki}) = 0.56$, $E(r_{Bi}) = 7.5$, and $E(r_{Ki}) = 47.6$. These expected values of the response counts r_{Bi} and r_{Ki} imply a Bush proportion of $7.5/(7.5+47.6) \approx 0.1361$. However, in actual simulation results (one million runs), the mean Bush proportion is larger (0.1388), and the median Bush proportion is slightly smaller (0.1346), than the implied proportion.

More to the immediate point, both the mean and the median τ are distinctly larger than the value of 0.0565 that would be associated with $\alpha = 1.12$. The mean τ of 0.0733 implies $\alpha \approx 1.158$. Simulation results consistently show that the mean τ is greater than the value implied by the propensity ratio (in this case, $\alpha = 1.12$ implying $\tau \approx 0.0565$) whenever the proportion of Bush voters in the precinct is less than 0.5, and less than the expected value whenever it is greater than 0.5. Thus, τ itself is confounded with vote share in the presence of individual sampling error.

The confound of τ with vote share can be more simply demonstrated from the binomial distribution alone, producing even larger estimates of the distortion than in the complicated mixture model of the preceding paragraph. Whereas the preceding scenario assumed 200 voters of whom 100 were selected (and some smaller number completed the survey), now assume a precinct with arbitrarily many voters, 15% of whom voted for Bush. Sample $r = 55$ respondents (note that $E(r_{Bi}+r_{Ki}) = 55.1$ in the

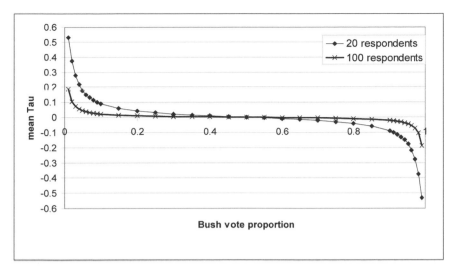

Figure 7.4 Expected value of τ as a function of Bush vote proportion, assuming no nonresponse bias (equal Bush and Kerry completion propensities), for two different numbers of completed interviews.

preceding scenario), where the probability of any one of them having voted for Bush, derived from $\alpha^* = 1.12$, is $0.15 / (0.15 + (1 - 0.15)(1.12)) \approx 0.1361$. The expected value of τ can be derived computationally as

$$E(\tau) = \sum_{b=0}^{r} P(b)\tau(b) \tag{7.23}$$

where $P(b)$, the probability of obtaining b Bush voters in r respondents conditional on α^* and the Bush vote proportion, here equals binom$(b, 55, 0.1361)$; and $\tau(b)$ is the value of τ associated with the Bush poll proportion $P_R = b / r$, via Equations 7.17 and 7.18a. The distribution $P(b)$ is, of course, skewed with a long right tail; $\tau(b)$ is more positive for small values of b. The expected value $E(\tau) \approx 0.0786$, implying $\alpha \approx 1.171$ (compare the value of 1.158 in the preceding simulation).

We can use this computational approach to illustrate the confounding effect of sampling error on τ across the vote share range. Figure 7.4 shows the expected value of τ for $r = 20$ completions (the smallest size included in most of E/M's analyses) and for $r = 100$ completions (a relatively large size) assuming no actual bias in completion propensities ($\alpha^* = 1.0$). While most of the variation in the expected value occurs at the extremes of vote share, values as large as ± 0.047 (indicating $\alpha = 0.9$ or 1.1, a very substantial degree of spurious bias) occur with vote share values as moderate as 0.18 or 0.82 for $r = 20$.

In correlational analysis, the confound of τ with vote share induced by sampling error is actually stronger than the confound of WPE induced by variation in completion propensities (as in Figure 7.3a), but (substantively) in the opposite direction.

To explore this behavior, we extended the simulation based on completion propensities in several respects. We simulated 1,250 precincts per run (equaling the number of precincts in the Edison/Mitofsky tabular analyses), with number of completed interviews in each precinct $r_i \sim TN(78, 22, \min = 20)$, approximating the observed distribution, and Bush vote proportion $V_{Ri} \sim TN(0.5, 0.2, \min = 0.05, \max = 0.95)$.[162] In 1,000 runs with $c_B = 0.5$ and $c_K = 0.5$ ($\alpha^* = 1.0$) and $s^* = 0.1$, the average observed Pearson $r(V_R$, WPE) was -0.030 (s.e. < 0.001), implying typical pro-Kerry survey error in precincts with high Bush vote proportion. The average observed $r(V_R, \tau)$ was -0.055 (s.e. 0.001), implying typical pro-*Bush* ($\alpha <$ 1) survey error in the high-Bush precincts. We do not conclude that WPE is a better measure than τ. (As inspection of Figure 7.3b indicates, WPE fares somewhat better in correlational analysis largely because its distortion tends to be nonmonotonic—not a characteristic that inspires confidence.) Nonetheless, the bias in τ must be corrected if it is to be suitable for inferential analysis across the vote share spectrum.

Adjusting for the Vote Share Confound τ'. A straightforward and reasonably effective correction is to calculate τ' (tau prime) $= \tau - E(\tau)$, where $E(\tau)$ is calculated as in Equation 7.23 with $\alpha^* = 1.0$. Thus, τ' measures how far τ differs from its expected value in the absence of bias. This measure is theoretically flawed in several respects. First, the binomial estimate of $E(\tau)$ is larger than the result one would obtain from a hypergeometric distribution based on the counts of Bush and Kerry voters, so this measure overcompensates for bias (and therefore is expected to introduce a positive correlation). Second, while both τ and $E(\tau)$ are bounded by $\pm\pi/4$, τ' obviously is not, and therefore individual values of τ' cannot always be interpreted. Third, while it can be shown computationally that $E(\tau')$ is relatively uniform and that $\alpha(E(\tau'))$ approximates α^* fairly closely over much of the vote share spectrum, τ' has some tendency to understate bias especially at both vote share extremes. (For instance, when $\alpha^* = 1.1$, $\alpha(E(\tau')) \approx 1.098\pm0.0002$ for V_R between 0.3 and 0.9; $\alpha(E(\tau')) \approx$ 1.0936 at $V_R = 0.05$.)

It should also be noted that the variance of τ', like the variance of τ, increases markedly as V_R approaches 0 or 1, roughly proportional to $1/rV_RV_D$ over most of the vote share range (before necessarily falling to 0 for very extreme V_R). This heteroskedasticity is, in the main, not a measurement artifact, but rather reflects the fact that larger samples are needed in strongly partisan precincts to measure bias with a given degree of reliability. Weighted least squares can mitigate the effects of unduly influential outliers at partisan extremes.

Despite these caveats, τ' performs relatively well in correlational analysis in simple simulations. Using the same simulation conventions as in the preceding example (crucially, normally distributed completions) and $c_B = c_K = 0.5$, in a set of 1,000

---

[162]The actual distribution of precinct vote share is probably asymmetrical, with a mean Bush vote proportion somewhat greater than 0.5 but with more low values than high values. (For 11,369 Ohio precincts in 2004, the mean Bush proportion of the presidential vote was 49.5%—less than his vote share, reflecting the fact that Bush generally did somewhat better in larger precincts—but the 5th and 95th percentiles were 8.9% and 74.6% respectively.) We use a symmetrical simulated distribution to simplify the interpretation of non-zero correlations between vote share and bias.

simulation runs, the observed mean Pearson $r(\tau', V_R)$ was 0.005 (s.e. $= 0.001$), appreciably smaller than the correlation of either WPE or τ with V_R. In another set of simulations where $c_B = 0.5$ and $c_K = 0.56$ ($a^* = 1.12$), $r(\tau', V_R) = 0.006$. Also, the average mean τ' (i.e., the average across runs of the mean in each run) was 0.05564, yielding $a(\tau') = 1.1180$—a slightly biased but reasonably accurate estimate of a^*. The spurious correlation can be further reduced by basing the estimate of $E(\tau)$ on the hypergeometric distribution, although we have not fully explored this computationally intensive approach.[163]

Detecting Actual Correlation between Vote Share and Survey Error. An adequate error measure should not only retrieve reasonable estimates of mean error, and covary minimally (if at all) with vote share in the absence of an underlying causal correlation, but should yield substantial covariance when a causal relationship exists. This has potential importance, as one possible cause of representation bias—vote count error— will affect vote share itself. A measure that arbitrarily "flattened" error across the vote share spectrum would, therefore, be inappropriate. How does τ' fare, in comparison with other measures, in yielding significant correlations when it "should"?

To investigate this question, we consider the measures WPE, ln a, τ and τ' in a set of simulations similar to the ones above, except that c_K is made to co vary with vote share. Specifically, $c_B \sim N(0.5, 0.1)$ as before (thus $s^* = 0.1$), and $c_K \sim N(0.5a^*, 0.1) + \delta V_B$, where a^* is either 1.0 or 1.12, and δ is a covariance parameter, set to 0.0 or 0.1 in our simulations. (This deterministically linear bias mechanism is presumably unrealistic, but we have no particular theoretical basis for an alternative.) Mean correlations and standard deviations are reported for 1,000 runs of each condition (the nominal standard errors are approximately 0.001 in each case).

Each of these measures performs modestly well in identifying this degree of correlated bias. (Note that varying a^* has little effect; the most useful contrast is between the second and third columns of results [of Table 7.8].) For instance, for with WLS,[164] the observed means and standard errors imply that the correlated-bias condition in the last column will yield a correlation larger than the no-correlated-bias mean of 0.002

---

[163] First it is necessary to revise the simulation to incorporate vote counts, selection counts, and completion counts. The total votes $v \sim \text{int}(722e^{N(0,1)})$, constrained to the range [70, 6200], a distribution that approximates the observed distribution in 2004. The proportion of Bush votes V_B is again $\sim \text{TN}(0.5, 0.2, \min = 0.05, \max = 0.95)$, adjusted to yield integer counts of Bush and Kerry votes (v_B and v_K, respectively). The target interview rate I is round ($v/100$), constrained to the range [1, 10], and the total number of voters selected a is int(V/I). Selections are assumed unbiased, so that Bush voter selections $a_B \sim H(a, v_B, v)$. Completion propensities c_B and c_K are again normally distributed. $E(\tau)$ can then be estimated from v_B, v_K and number of respondents r—as opposed to just V_B and r in the binomial approach—by computing the probability $P(b)$ in Equation 7.23 (where b is a given number of Bush respondents) as hypergeo(b, r, v_B, v). (An even more elaborate computation using the number of selected voters can be imagined, but we eschew it in part because we do not believe interviewers' refusal and miss counts are entirely reliable.) The results of this simulation are not directly comparable to the results from the simpler simulation reported below. In one run of 500 simulated polls, we derived a mean r between vote share and τ' of 0.0126 (nominal s.e. $= 0.0014$) using the original binomial calculation for $E(\tau)$, or 0.0051 (s.e. $= 0.0014$) using the hypergeometric calculation.

[164] WLS equalizes the variance on the assumption that variance in τ' is proportional to $1/r V_R V_D$, as described in the text.

Table 7.8   Mean Pearson r (nominal standard errors in parentheses) for various bias measures, selected combinations of nonresponse bias α^* and vote share/bias covariation δ (1,000 runs each)

	Bias parameters		
Bias measure	$\alpha^* = 1.0$ $\delta = 0.0$	$\alpha^* = 1.12$ $\delta = 0.0$	$\alpha^* = 1.12$ $\delta = 0.1$
WPE	−0.030	−0.031	−0.068
	(0.023)	(0.023)	(0.023)
Ln α	−0.061	−0.066	−0.014
	(0.032)	(0.033)	(0.034)
τ	−0.054	−0.056	−0.007
	(0.032)	(0.033)	(0.033)
τ'	0.005	0.006	0.055
	(0.034)	(0.033)	(0.033)
τ'(WLS)[†]	0.003	0.002	0.048
	(0.028)	(0.028)	(0.028)

[†] WLS equalizes the variance on the assumption that variance in τ' is proportional to $1/r \, V_R V_D$, as described in the text.

about 95% of the time. However, by the same token, a two-tailed 95% confidence interval for τ' will include 0.002 more than half the time! Recall that the value $s^* = 0.1$ was chosen to approximate the observed absolute mean WPE in the exit poll data. Thus, *if* we assume that most of this variance is unexplained, then it is reasonable to conjecture that a correlated bias as strong as $\delta = 0.1$ may not be reliably detectable.

Conclusion. We do not expect this article to end the use of the percentage-difference WPE measure in summary description of exit poll error. Surely, it is genuinely useful, for instance, to compare mean signed WPE in the 2004 NEP [National Election Pool] exit poll (−6.0) to the overall error of estimate (about −5.5 points in the national exit poll), in order to illustrate that precinct-level error accounts for the apparent bias in results. Nevertheless, WPE is analytically inappropriate for investigating possible covariance between exit poll error and precinct vote share, or for multivariate analysis in the presence of such covariance. In our simulations to date, the proposed measure τ' substantially reduces the confound between bias and vote share inherent in WPE, while being similarly sensitive to actual induced covariance.

We cannot definitely assess implications for the 2004 election—no measure is much use without data and, at this time, no multivariate analysis of the NEP data— confounded or otherwise—has been released.[165] Certainly, our work offers another reason why debates about how to interpret large WPEs in "Republican strongholds"

[165][Note that this paper is reprinted here as it appeared in 2005.]

were doomed to futility.[166] It also invites caution about whether exit poll data are likely to have sufficient statistical power to identify actual relationships between bias—or vote count error—and vote share. At first blush it may seem perverse that a sample of over 100,000 respondents could yield insufficient statistical power, but the exit poll is better construed as a sample of fewer than 1,500 precincts with idiosyncratic political characteristics and interviewer effects. Some of the limitations in the data set can likely be mitigated by more rigorously training interviewers about the importance of adhering to the interviewing rate (E/M, 2005: 47), and by deploying more than one interviewer at least to large polling places. In particular, if we are concerned with the possible role of vote count error in determining the counted margin, as well as in the observed participation bias, we may require larger samples where the expected margin between the candidates is large.

Work remains to be done on developing solutions to the analytical problem. No new measure will provide a panacea for eliminating vote share confounds when measuring representation bias. The wide variety of possible vote count error mechanisms and other error sources further complicates the analytical challenges. While we here present a measure that is reasonably well-behaved in OLS [Ordinary Least Squares] regression, we believe that alternative estimation methods (including WLS, as briefly discussed above) may be required to delineate without confound the causes of precinct level discrepancy in exit polls.

References

Blumenthal, Mark (2005). Professor M, January 28, 2005, available on the Internet at: *http://www. mysterypollster.com/main/2005/01/professor_m.html* (last accessed 10/21/05).

Brady, Rick (2005). A Critical Review of "The Unexplained Exit Poll Discrepancy," March 28, 2005 (initially published March 20, 2005), available on the Internet at: *http://stonescryout.org/files/unexplained. pdf* (last accessed 10/21/05).

Edelman, Murray and Merkle, Daniel M. (1995). The Impact of Interviewer Characteristics and Election Day Factors on Exit Poll Data Quality, a paper presented at the annual meeting of the American Association for Public Opinion Research, Fort Lauderdale, FL, May 18-21, 1995.

Edison Media Research and Mitofsky International (E/M) (2005). Evaluation of Edison/Mitofsky Election System 2004, January 19, 2005, available on the Internet at: *http://www.vote.caltech.edu/media/ documents/EvaluationJan192005.pdf* (last accessed 02/21/08).

Freeman, Steven F. (2004). The Unexplained Exit Poll Discrepancy, December 29, 2004 (original edition published November 12, 2004), available on the Internet at: *http://center.grad.upenn.edu/center/get. cgi?item=exitpollp* (last accessed 10/21/05).

Freeman, Steve (2005a). "Polling Bias or Corrupted Count? Accepted Improbabilities and Neglected Correlations in 2004 U.S. Presidential Exit Poll Data, presentation to the Fall Meeting of the American Statistical Association, Philadelphia Chapter, October 14, 2005, available on the Internet at: *http://www.*

[166]The phrase "Republican strongholds" was sometimes used to describe precincts with a Democratic (Kerry) vote proportion less than 20% (the "High Rep" precincts in E/M, 2005: 36-37). One problem with this attribution is that, as noted in footnote 162, the "Republican strongholds" are by no means as "strong" as their Democratic counterparts. In Ohio, only 1.1% of precincts had a Kerry vote proportion less than 20%, and among these precincts the median proportion was 17.3%. As Figure 7.3b indicates, the expected mean WPE, and expected gap between mean and median WPE, are much larger—in the vicinity of 15% to 20% Kerry vote proportion –than they would be near 0. (In contrast, 9.3% of Ohio precincts had a Kerry vote proportion greater than 80% and the median among these precincts was 91.2%.)

appliedresearch.us/sf/Documents/ASAPImprobabilities%20&%20Neglected%20Correlations% 20051014%20text.pdf(last accessed 11/9/05).

Freeman, Steve (2005b). The Election Outcome Based on Exit Poll Reported Voting, handout accompanying presentation at the Fall Meeting of the American Statistical Association, Philadelphia Chapter, October 14, 2005, available on the Internet at: *http://www.appliedresearch.us/sf/Documents/ Election%20Outcome%20based%20on%20EP%20Respondent%20Reported%20Voting.pdf* (last accessed 11/7/2005).

Groves, Robert (1989). *Survey Errors and Survey Costs,* New York, NY: Wiley-Interscience.

Liddle, Elizabeth (2005). Edison/Mitofsky Exit Polls 2004: Differential Nonresponse or Vote Count Corruption? April 27, 2005, available on the Internet at *http://www.geocities.com/lizzielid/WPEpaper.pdf* (last accessed 10/21/05).

Martin, Elizabeth A.; Traugott, Michael W.; and Kennedy, Courtney (2005). A Review and Proposal for a New Measure of Poll Accuracy, *Public Opinion Quarterly,* 69 (3), Fall, 342–369.

Mosteller, Frederick; Hyman, Herbert; McCarthy, Philip J.; Marks, Eli S.; and Truman, David B. (1949). The Pre-Election Polls of 1948: Report to the Committee on Analysis of Pre-election Polls and Forecasts, New York, NY: Social Science Research Council.

Traugott, Michael W. and Price, Vincent (1992). A Review: Exit Polls in the 1989 Virginia Gubernatorial Race: Where Did they Go Wrong? *Public Opinion Quarterly,* 56 (2), Summer, 245–253.

U.S. Count Votes (2005). Analysis of the 2004 Presidential Election Exit Poll Discrepancies, April 12, 2005 (originally published March 31, 2005), available on the Internet at: *http://electionarchive.org/ ucvAnalysis/US/Exit_Polls_2004_Edison-Mitofsky.pdf* (last accessed 10/21/05).

In the next article, Bloom and Pearson (2008) examine the likely voter issue from a different perspective. In order to assess reliability of pre-election polls, they tackle the challenge of drawing a probability sample when the size of the population is not knowable and, hence, standard corrections for sampling error do not apply.

7.4.2 Reliable Compared to What? A Probability-Theory Based Test of the Reliability of Election Polls
Joel David Bloom and Jennie Elizabeth Pearson

Introduction: Reliable Compared to What? Over the last several election cycles, pollsters have often been favorite media punching bags. Surprisingly strong Democratic showings in the 1998 and 2000 congressional elections triggered strident criticisms of the polls and those who conducted them, as did the surprisingly strong Republican showing in 2002. (See, e.g., Huffington, 2002.) Indeed, in 2002 in particular, there appeared to be some major misses by the pollsters in the Senate and gubernatorial races.

The appropriate question, however, is not whether some polls were inconsistent with election results; that is to be expected. The question is whether the gap between poll results and election results was larger than we would expect when compared to theoretically sensible and statistically valid points of comparison.

Fortunately, we need not look far to find appropriate comparison points—in fact, we find them in the very probability theory on which every poll's published margin of error is based. Using data sets consisting of publicly available Senate polls in 2002 and 2006 and statewide presidential polls in 2004, we test the polls against two major markers:

1. The percentage of polls outside of their own margin of error and

2. The actual median poll error compared to the expected median poll error given the margins of error.

The key figures are derived from a standard Z-table, representing area beneath a normal curve, based on the fact that if we could know the true population value for a particular parameter, and if we could take an infinite number of polls on that question, the results of all the polls taken together should be distributed normally. According to the Z-table, the number 1.96 represents that point at which 2.5% of cases will be found on each tail of the distribution, or 5% total (Hanushek and Jackson, 1977). Thus, 1.96 represents the distance from the true population value where we would place the standard 5% confidence interval by which margins of error are reported. The median expected error for a poll is represented by 0.675 on the Z distribution, or the distance from the true population value within which we would find 25% of cases on each side, or 50% of cases total. Because 0.675 is just slightly more than one third of 1.96, as a general rule of thumb we can say that the median expected absolute value error for any poll should be roughly one third of that poll's margin of error (after corrections for deleting undecideds and a correction for rounding error, which we will explain below).

This principle is explained graphically in Figure 7.5, below, which shows a hypothetical case in which the true population value is 55%, and an infinite number of polls are taken that have a 3% margin of error. In this case, only 5% of polls should be either greater than 58% or less than 52%–that much is standard poll reporting. But the less obvious fact that we learn from probability theory is that the typical poll—the poll with the median error—should be off by only around one point.

Looking at our meta-sample of election polls, we find mixed results. The presidential polls in 2004 actually out-performed expectations by both measures. On the other hand, the Senate polls missed the mark—by a modest amount in 2006, but by a much larger amount in 2002.

Why We Might Expect Election Polls to Be Problematic. A number of factors suggest that election polls should be less accurate than polls on other issues. The most important reason for this is that, unlike standard opinion polls, election polls attempt to sample from an unknown population those who will vote on Election Day. This is an especially large factor in the lower-turnout off-year elections, but is also the case in presidential years.

Second, and perhaps more importantly, when pre-election polls are in the field, voters are a population that technically *does not yet exist*. While some individuals are nearly certain to vote and others are nearly certain not to (with a great many in between), the population of those who will actually vote, prior to the deadline for voting, is not yet a population, but *in the process of becoming one*. Thus, the population of voters is not a true population, but a *latent* one. Needless to say, attempting to

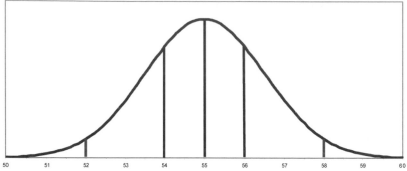

95% of polls should fall between 52% and 58% (within the error margin); 67% should fall between 53.5% and 56.5% (1/2 the margin); and 50% should fall between 54% and 56% (1/3). The median, or typical poll, should be off by 1%, or 1/3 the margin of error.

Figure 7.5   Area under a normal curve. If the true population value for a survey question is 55%, results of an infinite number of surveys on that item will resemble a normal curve centered around that figure (shown with 3% margin of error).

sample from a population that does not yet exist presents a unique set of challenges to election pollsters, with an impact on reliability that is both potentially very large and impossible to estimate in advance.

While this paper and its predecessors (Bloom, 2003; Bloom and Pearson, 2005; Bloom 2007) are the first to use the term "latent" to describe the target population in a pre-election survey, others have noted the problems associated with these features. As Traugott and Lavrakas put it:

> While there is strong scientific and statistical basis for drawing samples and constructing questionnaires, estimating who will vote on Election Day is an area where the practitioner's art comes into play. There is no standard, widely accepted way for estimating a person's likelihood of voting. Most polling organizations combine the answers to several questions to estimate the likely electorate, and some methods work better than others. (Traugott and Lavrakas, 2000:14)

Norman Ornstein put it more colorfully:

> We try to portray polling as a science, but it's a witchcraft kind of art. When it comes to the midterm elections, we're trying to predict how 35% of the electorate will vote, but we don't know which 35% will turn out. It's beyond embarrassing. (Neumann, 2002)

As a result, the largest differences observed among polls may not be due to differences in their polling methods (although these differences are considerable), but rather the methods for determining who is a likely voter and, sometimes, the weights applied to various demographic or political subgroups. (See, e.g., Traugott and Lavrakas, 2000; and Crespi, 1988.)

These post-survey manipulations are arguably necessary—after all, we know for a fact that not everyone with an opinion will vote. But even if they are not witchcraft, they are educated guesses and add potentially non-random bias to reported survey results. Unfortunately, since these techniques for determining likely voters are often closely-guarded trade secrets, analysis of differences resulting from these methods is impossible.

The fact that pollsters can't know in advance who will actually vote makes the laws of sampling error theoretically inapplicable. In other words, when we sample from a known population of one million, for example, we can use our standard sampling tables to say that a sample of 400 will be associated with a sampling error of 4.9%; 600 with 4.0%; 1,000 with 3.1%; and so on. And, indeed, election polls routinely report these figures as if they have drawn a sample from a known population.

The fact that they are actually *attempting* to sample from a *latent* population means that election polls might actually produce a range of results that appear to be less reliable than sampling error would predict, even if that observed unreliability might still be due only to sampling error (or sampling-*related* error). In such a framework, non-coverage bias and nonresponse bias can also play major roles.

At the same time that election polls face all these challenges not faced by standard attitude polling, they also face a much stricter test not faced by other surveys—barring large-scale problems with the vote count, the impending election serves as the presumed true population value of the parameter. As Martin, Traugott and Kennedy (2003) put it, "A peculiar position of pre-election polls is that they represent one of the few instances in which there is an external validation of the survey estimates—the actual outcome of the election."

If, despite these challenges, election polls, indeed, turn out to be as reliable as or more reliable than the standard response error formulae predict, this would be quite a tribute to the skills and talents of election pollsters. If, on the other hand, election polls are not as accurate as claimed, both public opinion professionals and consumers will need to consider that in their interpretation of election poll data.

The Data Sets. The means of data collection varied somewhat for the three years included in this [research]. Only publicly available polls that included sponsor or data collection firm, dates in the field, and either sample size or sampling error have been included. The data sets have been cross-checked, proof-read and copy-edited, but are certainly neither exhaustive nor error-free. In 2002 and 2006 only states that ended up being decided by a margin of 20% or less were included; in 2004 all states with poll data were included. In 2002, polls in the field at least one day on or after September 1st were included; in 2004 and 2006 the cutoff was October 1st, due to the larger numbers of polls in those years:[167]

- **2002.** We compiled a sample of 232 polls of U.S. Senate races from a number of publicly available Internet locations, including: D.C.'s Political Report, *National Journal's* Poll Track, Kiva Communication, and Our Campaigns, as well as the data set provided by the National Council on Public Polls (O'Neill, Mitofsky, and Taylor, 2002a and 2002b).

- **2004.** We compiled a meta-sample of 437 statewide presidential polls gathered from a variety of sources, including *PollingReport.com* and *OurCam-*

---

[167]We will be happy to share the data set with other researchers on request.

paigns.com, as well as a data set compiled by SurveyUSA.[168] In a number of instances of tracking polls in which rolling (i.e., overlapping) samples were reported daily; we included only polls without overlap, with one or two exceptions in which a one-day overlap was better than omitting several other days in the field.

- **2006.** Here we returned to Senate polls, using the data set of polls provided by Mark Blumenthal and Charles Franklin at *Pollster.com*, which they compiled from a variety of sources. [169]

How to Measure Error in Election Polls. Motivated by the pre-election poll debacle of 1948, Mosteller et al. (1949) discussed eight methods for evaluating pre-election poll accuracy, two of which have become the most commonly used methods for determining poll reliability and accuracy (Traugott, 2005). Revisiting Mosteller fifty years later, Mitofsky (1998) concludes that of the eight distinct measures, the best measures for determining the accuracy of pre-election polls are by either averaging the absolute difference between the prediction and the actual election outcome for each candidate (Mosteller Measure 3) or by examining the margin of difference between the two leading candidates in the pre-election polls and the election results (Mosteller Measure 5).

Mitofsky (1998) prefers Measure 5, as it is best able to account for more than two parties in a single election and uses the statistic most often reported by survey organizations and in the media—the margin between the two major parties (see also Martin, Traugott, and Kennedy, 2005). A limitation of all measures is that they require making a judgment regarding treatment of undecided voters; otherwise the average error may be inflated (Mitofsky, 1998).

Treatment of undecided voters. Because third-party votes are not important factors in many U.S. polls, and the individual candidate estimates, rather than the margin, is what is measured by probability theory behind the margins of error, we follow Traugott (2001) in using Mosteller 3—looking at absolute value difference between the survey estimates and election results for major party candidates. This decision leaves open the question of what to do with undecided voters.

In nearly all surveys conducted, the option for respondents to select "don't know," "undecided," or "no opinion" is generally available, although it is not always stated explicitly. However, for the elections themselves, "undecided" is not an option, although not voting at all certainly is. Sampling and margin of errors are influenced by the size of the sample, so if there is a substantial number of undecideds, they affect error calculations and the resulting comparisons (Martin, Traugott, and Kennedy, 2005; Mitofsky, 1998).

Mitofsky (1998) proposes four techniques for dealing with the undecided voters; two of the techniques involve allocating the undecideds evenly [among] all the

---

[168] Special thanks to Tom Silver of PollingReport.com and Jay Leve of SurveyUSA for making data sets available.

[169] We gratefully acknowledge their contribution to this research.

candidates or proportionally between the two major parties. Another technique is to allocate them all to the challenger if there is an incumbent in the race, and the fourth technique is to drop the undecideds altogether.

For our analyses, we chose to use Mitofsky's fourth option, also adopted by Crespi (1988) and Martin, Traugott, and Kennedy of dropping the undecideds and recalculating the sampling error, since dropping the undecideds has essentially the same affect on the resulting calculations as proportional allocation, because it "assumes a proportional distribution" (Martin, Traugott, and Kennedy, 2005: 349). This decision was based on the fact that keeping the undecideds in the analysis as such would unrealistically increase the apparent error as compared to the election results; allocating them other than proportionately has no statistically valid basis, while allocating them proportionately overstates the effective sample size. Omitting the "don't knows" and undecideds from our analyses resulted in an increase of the margin of error of around one tenth of one percentage point, although it had a greater impact on some polls than others.

Martin, Traugott, and Kennedy (2003, 2005) have developed an innovative method of determining poll accuracy, developed based on the logarithm of the odds ratio of the pre-election poll results and the final election outcome. This measure has distinct advantages over Mosteller's Measures—for example, it can assess individual polls for a single election as well as compare across different races with varying numbers of candidates. The results, however, are not as easily interpreted as poll accuracy measures that are based on percentage differences or probability theory, and currently the formula is only suitable for two-party elections.

Rounding error. There is one additional source of error that is almost never mentioned in discussions of survey error—simple rounding error. Most election polls correctly round up or down to the nearest full percentage point. That is as it should be; when a pollster gives even one decimal point, it implies that the poll is accurate up to that level of precision, which even the largest polls are simply not. However, rounding error adds 0.25% to the median poll error on candidate percentages and 0.50% on the margin. Because of the just about 3:1 ratio in the Z-table between 95% of cases (1.96, or the sampling error) and 50% of cases (0.675, or the median), this means that rounding error alone has the impact of adding 0.73% (0.25% * (1.96/0.675)) to any poll's effective sampling error. Thus, if we are to test election poll reliability against the predictions made by probability theory, it is necessary to compare it against a margin of error that is increased both slightly by the removal of undecideds and more dramatically by correcting for sampling error.

Analysis. Table 7.9 shows basic sample parameters for the three years of analysis. First of all, in none of the three years was there a substantial net partisan bias to the polls, although the Republicans in 2006 did around a point worse in the actual election results than they did in the polls. For all years, the average absolute value difference between the poll and the election outcome is larger than the median, as the average was raised by outliers. But probability theory makes predictions with regard to median, not mean.

Table 7.9 Comparisons of Republican candidate poll numbers vs. actual votes

Year	# of Polls	Average Rep. Poll %	Actual Rep. Vote %	Average Net Difference	Average MOE*	Average Abs. Value Difference	Median Abs. Value Difference	Expected Median Error	% Outside MOE
2002	232	52.1%	52.2%	-0.1%	5.1%	3.0%	2.1%	1.7%	18.5%
2004	437	51.4%	51.6%	-0.2%	4.8%	1.7%	1.3%	1.6%	2.5%
2006	267	47.8%	46.9%	0.9%	4.8%	2.2%	1.8%	1.7%	10.8%

*Average of reported margin of error adjusted to account for reduced sample size due to omission of undecideds and minor-party voters, plus the adjustment for rounding error.

For each survey, we calculate expected median error by multiplying the new adjusted margin of error (MOE) by the Z-statistic for the median divided by the Z-statistic for the tails (or MOE * (0.675/1.96)). This produces median expected errors of 1.6% to 1.7%, a figure which of course varies from poll to poll as well. As the table shows, for the two off-year elections, the median error was greater than the expected error, while for 2004, the median error was below the expected error. More telling, perhaps, are the percentages of polls outside their margins of error (adjusted for undecideds and rounding error). Here, once again, the 2004 presidential polls actually performed better than expectations, with only 3% of polls being off by more than their margins of error. Eleven percent of the 2006 Senate polls, however, and 19% of the 2002 Senate polls, were off by more than their margins of error.

Figure 7.6, below, shows the distribution of error (difference between Republican percentage of two-party intent in polls and Republican percentage of the two-party vote in the actual vote, rounded to the nearest percentage point) for the three years, and it is a mixed picture. The curve for 2002 shows graphically just how far removed the actual distribution was from the expected normal distribution that year, with the long and thick tails of the distribution particularly pronounced. The large groupings of polls more than 5% in error on each side show graphically the failure of election polls to meet claimed levels of reliability that year, even after we have adjusted reported margins of error upward to account for decreased effective sample sides and rounding error.

The curve for 2004, however, shows what the distribution should have looked like. The difference between the tightly grouped distribution of 2004 presidential polls in this chart and the haphazard and widely-spaced distribution for the 2002 Senate polls is truly striking. The curve for 2006 falls in between.

Calculating effective margins of error. The Z distribution, or area underneath a normal curve, represents the same concepts, but with greater precision. By comparing the actual error distribution to the expected error distribution, we can calculate what the effective margins of error are, on average, for the election polls. One can make this calculation based on any point in the distribution; for our purposes, we will calculate effective margins of error on the same two points in the error distribution that we have been discussing previously: a median-based effective error estimate based on median error and a tail-based effective error estimate based on the tails of a the distribution.

The median-based estimate is arrived at simply by dividing the actual median error by the expected median error and multiplying that figure by the original margin or error adjusted for sample size and rounding error. As shown in Table 7.10a, below, based on this measure, the effective margin of error for 2002 was 6.1%, well above the official or even adjusted margins of error. The same is true of the 2006 figure, although by a lesser degree, with the estimated effective margin of error at 5.3%. Strikingly, the figure for 2004 is 3.9%, which is not only lower than the margin of error adjusted for rounding error, but is even below the originally reported margin of error for the polls that year.

As shown in Table 7.10b, the tail-based estimation is just slightly more complicated. Here we take the actual percentage of polls that were outside the margin of

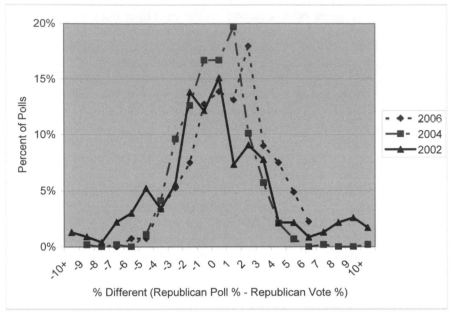

Figure 7.6   Difference between poll result and electoral outcome, 2002 and 2006 Senate polls, 2004 presidential polls.

error, divide it by 2 and subtract it from .50 to arrive at the percentage in the Z distribution to look up ("Z-Table %" in the table). Then, we find the Z-value for that figure; finally we divide 1.96 by that number and multiply that by the original margin of error adjusted for sample size and rounding error. Based on the tails of the distribution, all three years look worse than they did using the median-based estimate: 2002 is now at 7.5%, followed by 6.0% for 2006, and 4.2% for 2004. Again, we see both the 2002 and 2006 polls not passing the test of probability theory, while the 2004 polls do pass that test.

Additional Findings. In previous works (Bloom, 2003; Bloom and Pearson, 2005; Bloom, 2007) we tested a variety of additional hypotheses using the theory developed and tested above; space limitations, unfortunately prevent our including them here other than in summary form. Among our findings, which derive from both bivariate and multivariate analysis, are the following:

- Surveys do not become more reliable in a linear fashion as we approach Election Day; however, when we aggregate surveys by week, we do see trends representing the normal ups and downs of a campaign.

- Polls are more reliable in states with larger numbers of polls.

- Polls are more reliable in states with a race within 10 percentage points.

- Partisan polls are less reliable than non-partisan polls.

Table 7.10a Calculating median-based effective margins of error (MOE)

Year	Sample-Based 2-Party MOE	Adjusted 2-Party MOE	Actual Median Error	Expected Median Error	Ratio Actual/Expected	Estimated Effective MOE
2002	4.340	5.070	2.101	1.746	1.203	6.099%
2004	4.042	4.772	1.348	1.643	0.821	3.916%
2006	4.151	4.881	1.820	1.681	1.083	5.285%

Table 7.10b Calculating tail-based effective margins of error

Year	Original 2-Party MOE	Adjusted 2-Party MOE	Actual % Outside MOE	Expected % Outside MOE	Z-Table %	Z-Table Value	Ratio 1.96/ Z-Table Value	Estimated Effective MOE
2002	4.340	5.070	18.5%	5.0%	0.407	1.32	1.485	7.527%
2004	4.042	4.772	2.5%	5.0%	0.487	2.24	0.875	4.176%
2006	4.151	4.881	10.8%	5.0%	0.446	1.60	1.225	5.979%

- Polls conducted by prolific pollsters are more reliable.

- Polls conducted using automated voice-recognition (stocktickerAVR) technology are as reliable as more traditional telephone polls.

- On average, Internet polls are less reliable than telephone polls, but this included one practitioner who was well below average and another who was above average.

- Polls that use likely voter models are not more accurate than those that use registered voters.

Conclusion. This meta-analysis of survey data shows that the 2004 presidential election polls were far more reliable than the 2002 or 2006 Senate polls. Much more remains to be done here, including expanding the analysis to include Senate polls from 2004; this would allow us to determine whether there was something about polling in an off-year election that adversely affected reliability or whether Senate races might be just more difficult to achieve the levels of reliability found in presidential polling, due to the fact that Americans don't have as strong feelings about their Senate vote as they do their presidential vote.

While Martin et al. (2003, 2005) and Franklin (2003) contribute a great deal with their sophisticated multivariate methodologies, we believe that the analysis presented here is, in many ways, more appropriate and more broadly applicable in the public discourse about poll reliability. After all, pollsters make no claims about the logarithm of the odds ratio, but they do make claims about what percentage of cases should fall within particular distances of the election outcome.

At minimum, election polls should report a margin of error based on the number of "decideds" in their sample rather than the entire sample size; they should also be more forthcoming about the impact of rounding error. Even then, however, the actual error level is evidently sometimes substantially higher than claimed. On the other hand, in some ways the margin of error term, representing the tails of the distribution, presents a misleadingly bleak picture of the likely error. One possible solution would be to report the median expected divergence from the election results (in this case 1.75%) or the actual median divergence from previous elections (in these cases between 1.4% and 2.1%) in addition to the margin of error. This would give the poll consumer a better idea of how much polls *typically* diverge from election results. The question still remains, "reliable compared to what?" but we believe this chapter makes significant progress in answering that question.

Note: Some of the analyses contained in this article were originally presented at the 2003, 2005 and 2007 Annual Meetings of the American Association for Public Opinion Research. We will be happy to share the data set as well as more detailed analyses on request.

References

Bloom, Joel D. (2007). A Probability Theory-Based Test of the Reliability of 2006 Election Polls, a paper presented at the annual meeting of the American Association for Public Opinion Research, May 2007, Anaheim, CA.

Bloom, Joel D., and Pearson, Jennie E. (2005). A Probability-Theory Based Test of the Reliability of Election Polls, a paper presented at the annual meeting of the American Association for Public Opinion Research, May 2005, Miami Beach, FL.

Bloom, Joel D. (2003). Reliable Compared to What? Empirical Tests of the Accuracy of Election Polls, 2002, a paper presented at the annual meeting of the American Association for Public Opinion Research, May 2003, Nashville, TN.

Crespi, Irving (1988). *Pre-Election Polling: Sources of Accuracy and Error,* New York, NY: Russell Sage.

Franklin, Charles (2003). Polls, Election Outcomes and Sources of Error, a paper presented at the annual meeting of the American Association for Public Opinion Research, May 2003, Nashville, TN.

Hanushek, Eric A. and Jackson, John E. (1977). *Statistical Methods for Social Scientists,* San Diego, CA: Harcourt Brace.

Huffington, Ariana (2002). The Pollsters Can't Hear The Silent Majority, November 14, 2002, available on the Internet at: *http://www.ariannaonline.com/columns/files/111402.html.*

Martin, Elizabeth A.; Traugott, Michael W.; and Kennedy, Courtney (2003). A Review and Proposal for a New Measure of Poll Accuracy, a paper presented at the annual meeting of the American Association for Public Opinion Research, May 2003, Nashville, TN.

Martin, Elizabeth A.; Traugott, Michael W.; Kennedy, Courtney (2005). A Review and Proposal for a New Measure of Poll Accuracy, *Public Opinion Quarterly*, 69, 342-369.

Mitofsky, Warren J. (1998). The Polls-Review: Was 1996 a Worse Year for Polls than 1948? *Public Opinion Quarterly*, 62, 230-249.

Mosteller, F.; Hyman, H.; McCarthy, P.; Marks, E.; and Truman, D. (1949). *The Pre-Election Polls of 1948: Report to the Committee on Analysis of Pre-Election Polls and Forecasts,* New York, NY: Social Science Research Council.

Neumann, Johanna (2002). Looking to History, Pundits Never Saw This One Coming, *Los Angeles Times*, November 7, 2002.

O'Neill, Harry; Mitofsky, Warren; and Taylor, Humphrey (2002a). National Council on Public Polls Polling Review Board Analysis of the 2002 Election Polls, National Council on Public Polls press release, December 19, 2002.

O'Neill, Harry; Mitofsky, Warren; and Taylor, Humphrey (2002b). The Good and Bad of Weighting Data, a statement by the National Council on Public Polls' Polling Review Board.

Traugott, Michael W. (2005). The Accuracy of the National Pre-election Polls in the 2004 Presidential Election, *Public Opinion Quarterly*, 69, 642-654.

Traugott, Michael W. (2001). Assessing Poll Performance in the 2000 Campaign, *Public Opinion Quarterly*, 65, 389-419.

Traugott, Michael W. and Lavrakas, Paul J. (2000). *The Voter's Guide to Election Polls,* 2nd ed., New York, NY: Chatham House.

This last contribution relates to Rotz and Gracely (2007) and Mulrow and Scheuren (2008), both in Chapter 6 of this volume (6.5.1 and 6.5.2). It addresses the recommended need to audit votes, both to identify and help control intentional and unintentional errors in vote counts. In so doing, the authors point out the importance of using statistically-based procedures, in lieu of simply conducting audits on a predetermined percentage of the tallies in a voting precinct.

7.4.3   Percentage-based versus Statistical-power-based Vote Tabulation Audits[170]

John McCarthy, Howard Stanislevic, Mark Lindeman, Arlene S. Ash, Vittorio Addona, and Mary Batcher

Introduction. Electronic vote tally miscounts arise for many reasons, including hardware malfunctions, unintentional programming errors, malicious tampering, or stray ballot marks that interfere with correct counting. Thus, Congress and several states are considering requiring audits to compare machine tabulations with hand counts of paper ballots in randomly chosen precincts. Audits should be highly effective in detecting miscounts large enough to alter election outcomes; and they should be efficient—no larger than necessary—to confirm the winners. While financial and quality control audits set sample sizes that are very likely to detect errors large enough to cause harm, most proposed election auditing laws specify sampling fixed or tiered percentages of precincts. For example, Connecticut has just adopted a law (Public Act 07-194) requiring random audits of 10% of voting districts (precincts) in selected contests. We believe that the laws are written this way because most nonstatisticians have unrealistic fears about the inadequacy of small-percentage audits; because the authors have not measured the statistical effectiveness of percentage-based schemes in general; and because statisticians have—thus far—rarely been involved in drafting audit options for legislators. Statisticians, of course, know that we can measure the effectiveness of sampling strategies by their statistical power, which principally depends on the number of units sampled and the size of the effect to be detected. Thus, fixed-percentage audits are inefficient (too large) in the vast majority of contests, especially in statewide contests that involve many hundreds of precincts and that are not close; also, they are ineffective (too small) in the rare contests with small winning margins. However, most statisticians know little about election procedures and are not well-equipped to respond when asked "what percentage shall we put in the bill?" We hope that this article helps fill that gap.

Vote tabulation audits entail supervised hand-to-eye manual counts of all voter-verified paper ballots in a subset of precincts, randomly selected shortly after an election and before results are certified. We assume that a "hard copy" record of each voter's choice, one that was reviewable by the voter for accuracy before the vote was officially cast, is available for comparison with the electronic tally. In particular, this excludes Direct Recording Electronic (DRE) voting with no paper trail. This assumption is consistent with the draft standards of the Election Assistance Commission Technical Guidelines Development Committee (*http://www.eac.gov/vvsg*). For convenience we use the word "precinct" throughout, although the appropriate audit unit is the smallest cluster of votes that is separately tallied and reported. Thus, a batch of early votes cast at a central location is a "precinct." Or, if a precinct's votes combine tallies from two readily distinguished machines, then each machine could—and

[170]Percentage-based Versus Statistical-Power-Based Vote Tabulation Audits, by John McCarthy, Howard Stanislevic, Mark Lindeman, Arlene S. Ash, Vittorio Addona, and Mary Batcher, reprinted with permission of *The American Statistician*. Copyright ©2008: American Statistical Association. All rights reserved.

for efficiency should—be treated as a separate "precinct." From a legal perspective, a 100% audit may not equal a "recount" after a candidate disputes election results, since different procedures may apply. Whole-precinct audits are not designed to estimate the entire vote count directly, but rather to independently confirm (or challenge) the accuracy of precinct-level electronic tallies.

Manual audits should also be done in precincts with obvious problems, such as machine failure, and for routine fraud deterrence and quality improvement monitoring, even in the absence of doubt about who won. Importantly, election officials and candidates should be empowered to choose additional precincts with apparent anomalies for auditing, just as financial and quality audits examine "high-interest" units as well as random samples. Comprehensive auditing should also examine many other parts of the electoral process, as outlined by Marker, Gardenier, and Ash (2007) and Norden et al. (2007). Another important question not pursued here is the need for mandatory follow up when an initial audit casts doubt upon an electoral outcome.

Our simplified framework assumes that the outcome of an audit is dichotomous: if the audit sample includes one or more miscounted precincts, additional action is taken; otherwise, the election is confirmed. Here, the audit's power equals the probability of sampling at least one miscounted precinct *whenever there are enough precincts with miscounts to have altered the outcome*. Statistically based audit protocols should seek to fairly and efficiently use resources to achieve a pre-specified high power (99%, if feasible) in all elections—from statewide contests to those in a single Congressional district or county.

Model Assumptions. We assume that every vote is cast in one and only one precinct and that precincts to be audited will be chosen randomly (with equal chance of selection) *after* an election has taken place and after the unofficial vote counts for each auditable unit are publicly reported. We also assume that net miscounts in more than 20% of the votes cast in a precinct would trigger a suspicion-based "targeted audit." This implies that a *random* audit, to be successful, only needs to detect at least one of a set of precincts, with shifts of, at most, 20% each, that together would change the electoral outcome. The *Within-Precinct Miscount*, or WPM, of 20% is a parameter that could be reset as experience accumulates. Saltman (1975) described WPM as the "maximum level of undetectability by observation." Setting an upper bound for the WPM allows us to determine the *minimum* number of precincts that must contain miscounts if an election outcome with a reported margin of victory were to be reversed. The hypergeometric distribution can then be used to calculate the sample size needed to achieve a specified probability that at least one miscounted precinct appears in the sample. (While the same approach can be used without using WPM, it typically requires substantially larger samples.) Our general findings do not depend on any particular value of WPM; the 20% figure serves as a reasonable baseline for comparative purposes.

When a contest has outcome-altering miscounts, an initial audit is successful only if it finds at least one miscounted precinct, since, if it does not, the wrong outcome would be confirmed. This initial audit does not have to determine what the outcome should have been, so long as it triggers further actions to make that determination.

Procedures for deciding when an audit that uncovers small discrepancies should still confirm an election outcome is the subject of ongoing research. See, for example, Stark (2007).

Calculation of Statistical Power. Let:

- n = number of randomly selected precincts.

- N = total number of precincts.

- m = margin of victory in a particular election contest—that is, the difference between the number of votes for the winning candidate in a single seat contest (or the number of votes for the winning candidate with the smallest number of votes in a multi-seat contest) and the number of votes for the runner up candidate, divided by the total number of ballots cast.

- B_{min} = minimum number of miscounted precincts (out of the N precincts) needed to overturn the election result for a particular contest.

- X = number of miscounted precincts in the sample of n.

Then we define:

$$\text{Power} = P(X > 0| B_{min} \text{ miscounted precincts in the population}). \qquad (7.24)$$

The value of B_{min}, the minimum number of miscounted precincts that could alter the outcome, depends upon the possible extent of miscounts in each precinct. Assuming equally sized precincts with a maximum WPM of 20% (a 40-point shift in the percentage margin within that precinct), a proportional winning margin of m could be overcome by switching votes in

$$B_{min} = \left\lceil \left(N \cdot \frac{m}{2 \text{ WPM}} \right) \right\rceil, \qquad (7.25)$$

or

$$\lceil (N \cdot m/0.4) \rceil \qquad (7.26)$$

precincts. For instance, if all precincts contain the same number of votes, a 10-percentage-point margin ($m = 0.1$) could be overcome by switching 20% of votes in at least $0.25N$ or 25% of all precincts.

In practice, precincts contain varying number of votes. It is instructive to consider how B_{min} (and, therefore, audit power) is affected by the alternative assumption that all the miscounts reside in the largest precincts. For instance, under the "largest precinct assumption," for a margin of 10 points, B_{min} will be the smallest number of large precincts that together contain at least 25% of all *votes*. This figure can be calculated directly if the distribution of precinct-level vote counts is available (from a preliminary report of precinct-level election returns), or it can be estimated based on the distribution of precinct-level votes in the previous election; a reference distribution (such as the Ohio CD-5 distribution shown in Figure 7.9); or using an

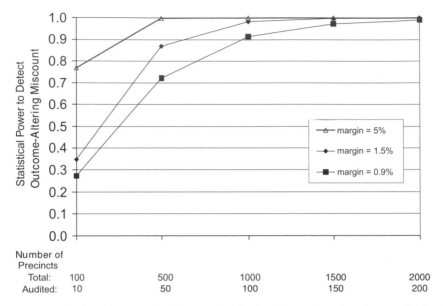

Figure 7.7   Statistical power of 10% audits for districts: by number of precincts audited and margin of victory. (Power = the probability of finding at least one miscounted precinct when the number of miscounted precincts equals the fewest average-sized precincts with 20% shifts needed to overturn the election.)

heuristic approximation. (For one such heuristic, see McCarthy et al. (2007).) As we demonstrate in the following discussion, this assumption can markedly reduce estimates of audit power.

Statistical Power of Fixed Percentage Audits. Figure 7.7 shows the power of a 10% audit for jurisdictions with varying numbers of precincts and winning margins. We assume here that all precincts contain the same number of votes or, equivalently, that the average number of votes for the precincts with miscounts is the same as for all precincts. Thus, for instance, to overcome a 5% margin, 20% (WPM) of votes in at least 12.5% of precincts must be switched from the reported winner to the reported loser. For context, Connecticut has 769 voting districts; New Hampshire's 1st Congressional District has only 114 reporting units (usually entire towns); West Virginia and Iowa have just under 2,000 precincts; and 28 states are off the scale in Figure 7.7 including five with over 10,000 precincts.

Clearly, 10% audits have limited power when electoral margins are close and/or when the total number of precincts is small. The fixed percentage approach also examines too many precincts when the election is not close and/or the number of precincts is large. For example, in California, a 10% state-wide audit would involve 2,200 precincts; this is an order of magnitude larger than needed, under an equal sized precinct assumption, even with a margin as small as 0.9%.

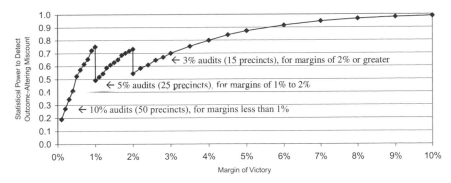

Figure 7.8 Statistical power of three-tiered 3–5–10% audits in a 500-precinct jurisdiction: by margin of victory. (Assumes the number of precincts with miscounts equals the minimum number needed to overturn the election if all miscounts are in average-sized precincts, each with a vote shift of 20%.)

Statistical Power of Tiered Percentage Audits. Tiered percentage audits specify a percentage of precincts to be audited, depending on the reported winning margin. For example:

1. Audit 3% of precincts if the winning margin is 2% or more of the total votes cast;

2. Audit 5% of precincts if the winning margin is at least 1% but less than 2%; and

3. Audit 10% of precincts if the winning margin is less than 1% of the total votes cast.

This approach addresses the need to audit more when margins are narrower, but does not solve the fundamental problem with fixed percentage audits. Most congressional districts contain fewer than 500 precincts. In these congressional districts, and in other modest-sized legislative districts, even sampling 10% of precincts does not achieve 75% power when the winning margin is at most 1%. A tiered audit also allows precipitous drops in power at the tier thresholds, which appear as "sawteeth" in Figure 7.8.

Variations in Precinct Size. The power of an audit also can be reduced when precincts differ markedly in numbers of voters (Saltman, 1975; Stanislevic, 2006; Dopp and Stenger, 2006; and Lobdill, 2006), because fewer large miscounted precincts can alter enough votes to change the outcome. For reference, Figure 7.9 shows the vote distribution in the 640 precincts of Ohio's Fifth Congressional District (CD-5) in the 2004 general election, where votes per precinct ranged from 132 to 1,637. Only about 8.6% (55) of Ohio CD-5's largest precincts are needed to encompass 15% of the vote. Applying Ohio CD-5's precinct size distribution to any number of precincts, we can

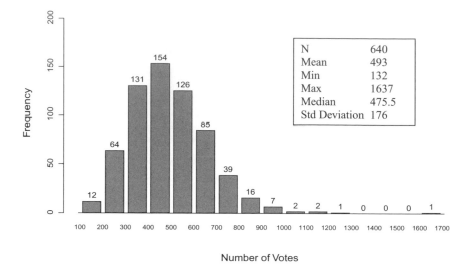

Figure 7.9   Distribution of votes counted in 2004 among the 640 precincts of Ohio's Fifth Congressional District.

estimate the power of audits in a "worst-case scenario" when miscounts occur in the smallest possible number of precincts (making it harder to find "at least one" in the audit sample). For instance, consider an election with 500 total precincts and a 6% margin, so that B_{min} equals the number of precincts containing at least 15% of the vote. A 3% audit sample is estimated to have 92% power if miscounts are equally common among small and large precincts, but just 75% power if they occur in the largest precincts. (In the first case, $B_{min} = 0.15$ (500) = 75; in the second, $B_{min} = 0.086$ (500)= 43.)

Statistical Power-based Audits. Statisticians and a growing number of election experts have urged replacing fixed percentage audits with audits that employ a statistically grounded criterion of efficacy. Here we present a power-based audit which determines the number of precincts that must be sampled to achieve a specified power level for each election contest. In addition to the desired power, the sample size will depend on the reported victory margin, the value of WPM, and both the number and size-distribution of the precincts. Given the total number of precincts in a contest and the minimum number of miscounted precincts that could alter the outcome, the sample size for a given power can be obtained using the hypergeometric distribution or an appropriate approximation to simplify calculations so they can be readily implemented by election officials (Aslam, Popa, and Rivest, 2007). In practice, slightly larger samples may be useful to contend with minimal miscount rates observed even in well-functioning systems (Stark, 2007).

Percentage-based Versus Power-based Audits. We have already seen that percentage-based audits can examine too few precincts in some contests, while sampling more

Table 7.11   Federal elections (2002–2006) achieving various levels of power by type of audit

	Type of audit					
	Tiered/fixed percentage				Power based	
	Tiered 3-5-10%	2%	3%	10%	99% power	95% power
Power of the Audit	Number of elections (percent)					
at least 99%	1152 (82.7%)	1089 (78.2%)	1152 (82.7%)	1292 (92.7%)	1393 (100%)	–
from 95% up to 99%	77 (5.5%)	98 (7.0%)	74 (5.3%)	31 (2.2%)	–	1393 (100%)
from 50% up to 95%	112 (8.0%)	137 (9.8%)	110 (7.9%)	51 (3.7%)	–	–
less than 50%	52 (3.7%)	69 (5.0%)	57 (4.1%)	19 (1.4%)	–	–
Total hand-counted votes (in millions)	20.5	15.3	19.4	57.6	23.0	19.0

NOTE: Results for 1,393 election contests for U.S. President, Senate, and House of Representatives. Vote margins were calculated from FEC [Federal Election Commission] data for 2002 and 2004; Dr. Adam Carr's Psephos archive for 2006. Numbers of precincts per election were estimated based on the 2004 Election Assistance Commission's Election Day Survey (*http://www.eac.gov/election_survey_2004/intro.htm*); for House elections, the number of precincts in each state was divided by the number of congressional districts to estimate the precincts per district. A minimum of one precinct per county is assumed to be audited under each rule. Power is calculated to protect against miscounts residing in the largest precincts, as described in "Variations in Precinct Size" above.

than necessary in others. To compare the cumulative costs of different types of audits, we studied all 1,393 federal election contests in 2002, 2004, and 2006, using the precinct size distribution of Ohio CD-5 in 2004 to estimate the extra auditing needed to account for variations in precinct size. Table 7.11 shows power and resource requirements using tiered, fixed, and power-based audits. Each audit was also made to fulfill a common legislative mandate: to include at least one precinct per county.

For example, since Iowa has 99 counties and 1,966 precincts, all rules assume audits of at least 99 precincts (5%) for every statewide election there. While 2% audits would have used the fewest resources (15 million ballots), they achieve less than 95% power in almost 15% of all contests. At the other extreme, 10% audits would have had at least 95% power in 95% of contests, but require auditing 57.6 million ballots, while still leaving 19 contests where election-altering miscounts are more likely to be missed than detected. Power-based audits could have achieved 99% power in all federal contests while examining only 23 million ballots.

Conclusions. Effective electoral oversight requires routine checks on the entire voting process, timely publication of the original tallies, and thorough reporting of the methods, data, and conclusions from all audits conducted. Since a key purpose of a post-election vote tabulation audit is to provide a check on the original tabulations, procedures should verify election results without trusting any part of the software used in voting. Reporting should include a post-hoc power calculation that fully describes the alternative hypothesis, so as to be explicit about how confident we are that the winner in the initial electronic tally is the same as would be identified in a 100% hand count of voter-verified paper ballots. Of course, all audits should be conducted within a larger framework of good practice, including following well-specified, publicly observable procedures to ensure public (as well as statistical) confidence in the process, examining both randomly selected precincts and targeted precincts with observed anomalies; and generating publicly accessible auditing data. Exactly one feature distinguishes power-based audits from percentage-based: they sample just enough precincts to make it very likely to detect a miscount if an election-changing miscount has occurred.

Election audits have been conducted in some state and local jurisdictions for years. The e*lectionline.org* briefing paper "Case Study: Auditing the Vote" discusses the auditing experiences of several states, including California and Minnesota, and Saltman (1975) cited the 1%"manual tally" still used in California. Although professional auditors, statisticians and computer scientists should advise on standards and procedures, competent election officials and staff can implement power-based audit sample size calculations, detailed in McCarthy et al. (2007), without special assistance. We encourage the use of statistical power-based audits as a preferred alternative to percentage-based audits. At the very least, we hope the laws being adopted now will allow states to use procedures that achieve high power to confirm all electoral outcomes.

References

Aslam, J. A.; Popa, R.A.; and Rivest, R.L. (2007). On Estimating the Size and Confidence of a Statistical Audit, April 22, 2007, available on the Internet at: *http://theory.csail.mit.edu/~rivest/AslamPopaRivest-OnEstimatingTheSizeAndConfidenceOfAStatisticalAudit.pdf*.

Dopp, K. and Stenger, F. (2006). The Election Integrity Audit, September 25, 2006, available on the Internet at: *http://electionarchive.org/ucvAnalysis/US/paper-audits/ElectionIntegrityAudit.pdf*.

Electionline.org (2007). Case Study: Auditing the Vote, March 2007, available on the Internet at: *http://www.electionline.org/Portals/1/Publications/EB17.pdf*.

Lobdill, J. (2006). Considering Vote Count Distribution in Designing Election Audits, Revision 2, November 26, 2006, available on the Internet at: *http://vote.nist.gov/Considering-Vote-Count-Distribution-in-Designing-Election-Audits-Rev-2-11-26-06.pdf*.

Marker, D.; Gardenier, J.; and Ash, A. (2007). Statistics Can Help Ensure Accurate Elections, *Amstat News*, June 2007.

McCarthy, J.; Stanislevic, H.; Lindeman, M.; Ash, A.; Addona, V.; and Batcher, M. (2007). Percentage-based vs. SAFE Tabulation Auditing: A Graphic Comparison, November 2, 2007, available on the Internet at: *http://www.verifiedvotingfoundation.org/auditcomparison*.

National Institute of Standards and Technology Technical Guidelines Development Committee (2007). Voluntary Voting System Guidelines Recommendations to the Election Assistance Commission, draft available for comment online at *http://www.eac.gov/vvsg*.

Norden, L.; Burstein, A.; Hall, J.; and Chen, M. (2007). Post-Election Audits: Restoring Trust in Elections, available on the Internet at: *http://brennancenter.org/dynamic/subpages/downloadfile50227.pdf*.

Saltman, R. G. (1975). Effective Use of Computing Technology in Vote-Tallying, National Bureau of Standards, Final Project Report, March 1975, prepared for the General Accounting Office. (See, particularly, Appendix A, Mathematical Considerations and Implications in Selection of Recount Quantities.) Available on the Internet at: *http://csrc.nist.gov/publications/nistpubs/NBSSP500-30.pdf*.

Stanislevic, H. (2006). Random Auditing of E-Voting Systems: How Much is Enough? VoteTrustUSA E-Voter Education Project, August 16, 2006, available on the Internet at: *http://www.votetrustusa.org/pdfs/VTTF/EVEPAuditing.pdf*.

Stark, P. (2007). Conservative Statistical Post-Election Audits, Technical Report 741, Department of Statistics, University of California, Berkeley, available on the Internet at: *http://statistics.berkeley.edu/~stark/Preprints/conservativeElectionAudits07.pdf*.

7.5 SUMMARY OBSERVATIONS

Chapter 7 concludes this volume. It provides a few examples of some of the more technical methodological work that is being conducted in efforts to improve and enhance pre-elections surveys, voter entry polls, and exit polling. Clearly, there is a lot more research ongoing than that presented here, but time and space constraints forced us to limit our selections. Nevertheless, the intent is to give readers a sense of the kinds of techniques that are being explored. With the many challenges already discussed—from discrepancies between polling results and official vote counts, to new technologies that are impacting polling—including cell phones, data mining, and the Internet—and increasing concerns about privacy and rising nonresponse issues, it is important to realize that many advances in polling research are only effective when based on sound theory and methodology. Like the more descriptive research provided in the earlier chapters, such studies are heartily encouraged.

Editors' Additional References

Bafumi, Joseph; Erikson, Robert S.; and Wlezien, Christopher (2008). Forecasting House Seats from Generic Congressional Polls, unpublished paper.

Bloom, Joel David and Pearson, Jennie Elizabeth (2008). Reliable Compared to What? A Probability-theory-based Test of the Reliability of Election Polls, unpublished paper.

Isaksson, Annica and Forsman, Gösta (2003). A Comparison Between Using the Web and Using Telephone to Survey Political Opinions, paper presented at the meeting of the American Association for Public Opinion Research, *Proceedings of the Section on Survey Research Methods*, American statistical Association, 100-106.

Lindeman, Mark (2006). Beyond Exit Poll Fundamentalism: Surveying the 2004 Election Debate. Paper presented at the annual meeting of the American Association for Public Opinion Research, Montreal, Canada, May 19, 2006.

Lindeman, Mark; Liddle, Elizabeth; and Brady, Richard (2005). Investigating Causes of Within Precinct Error in Exit Polls: Confounds and Controversies, *Proceedings of the American Statistical Association, Biometrics Section*, 282-292.

McCarthy, John; Stanislevic, Howard; Lindeman, Mark; Ash, Arlene S.; Addona, Vittorio; and Batcher, Mary (2008). Percentage-based Versus Statistical-power-based Vote Tabulation Audits, *The American Statistician*, February, 62(1), 11-16.

Mulrow, Edward and Scheuren, Fritz (2007). Producing a 2008 National Election Voter Scorecard, *Elections and Exit Polling*, Scheuren, Fritz and Alvey, Wendy (eds.), New York, NY: John Wiley & Sons, Inc.

Murray, Greg R; Riley, Chris; and Scime, Anthony (2007). Predicting Likely Voters: Using a New Age Methodology for an Age-Old Problem, paper presented at the annual meeting of the American Association for Public Opinion Research in Anaheim, CA.

Rotz, Wendy and Gracely, Edward (2007). General Principles for Statistical Election Auditing, unpublished paper

Wolter, Kirk; Jergovic, Diana; Moore, Whitney; Murphy, Joe; and O'Muircheartaigh, Colm (2003). Reliability of the Uncertified Ballots in the 2000 Presidential Election in Florida, *The American Statistician*, February, 57 (1), 1-14.

Speed Bump reprinted with permission. Copyright ©2000: Dave Coverly

Author Index

Index